STATISTICS

CONCEPTS AND CONTROVERSIES

FOURTH EDITION

David S. Moore

Purdue University

W. H. FREEMAN AND COMPANY

NEW YORK

Acquisitions editor:	Holly Hodder
Project editor:	Diane Cimino Maass
Cover and text designer:	Diana Blume
Cover illustration:	Mark Geo
Illustration coordinator:	Susan Wein
Production coordinator:	Paul Rohloff
Illustration and composition:	Publication Services
Manufacturing:	R. R. Donnelley & Sons Company

Library of Congress Cataloging-in-Publication Data

Moore, David S.
    Statistics: concepts and controversies / David S. Moore—4th
ed.
      p. cm.
    Includes index.
    ISBN 0-7167-2863-X (pbk.)
    1. Statistics. I. Title.
    QA276.12.M66 1996   96-44489
    519.5—dc21   CIP

Printed in the United States of America

First printing, 1996

STATISTICS

Does not He see my ways,
   and number all my steps?
      —Job

But even the hairs of your head
   are all numbered.
      —Jesus

Hell is inaccurate.
      —Charles Williams

CONTENTS

8 INFERENCE: CONCLUSIONS WITH CONFIDENCE 457

▶▶ To the Teacher:
Statistics as a Liberal Discipline

Statistics: Concepts and Controversies is a book on statistics as a liberal discipline, that is, as part of the general education of relatively non-mathematical students. I wrote and revised the book while teaching a class of freshmen and sophomores from Purdue University's School of Liberal Arts. These students come from many disciplines in the humanities and social sciences. They are fulfilling a quantitative requirement that not all of them welcome. Some elect the course as preparation for a later course in statistical methods, but most will never again encounter statistics as a discipline. For such students, statistics is not a technical tool but part of the general intellectual culture that educated people share.

Statistics among the Liberal Arts

Statistics has a widespread, and on the whole well-earned, reputation as the least liberal of subjects. When statistics is praised, it is most often for its usefulness. Health professionals need statistics to read accounts of medical research; managers need statistics because efficient crunching of numbers will find its way to the bottom line; citizens need statistics to understand opinion polls and the Consumer Price Index. Because data and chance are omnipresent, our propaganda line goes, everyone will find statistics useful, and perhaps even profitable. This is in fact true. But the utilitarian argument for studying statistics is at odds with the ideals of the liberal arts. The root idea of a liberal education is that it is general rather than vocational, that it aims to strengthen and broaden the mind rather than to prepare for a specific career.

A liberal education should expose students to fundamental intellectual skills, that is, to general methods of inquiry that apply in a wide variety of settings. The traditional liberal arts present such methods: literary and historical studies, the political and social analysis of human societies, the probing of nature by experimental science, the power of abstraction and deduction in mathematics. The case that statistics belongs among the liberal arts rests on the position that reasoning from uncertain empirical data is a similarly general intellectual method. *Data* and *chance*, the topics of this book, are pervasive aspects of our experience. Though we employ the tools of mathematics to work with data and chance, the mathematics implements ideas that are not strictly mathematical. In fact, psychologists argue convincingly that mastering formal mathematics does little to improve our ability to reason effectively about data and chance in everyday life.

Statistics: Concepts and Controversies is shaped, as far as the limitations of the author and the intended readers allow, by the view that statistics is an independent and fundamental intellectual method. The focus is on statistical thinking, on what others might call quantitative literacy or numeracy.

THE NATURE OF THIS BOOK

There are books on statistical theory and books on statistical methods. This is neither. It is a book on statistical ideas and statistical reasoning, and on their relevance to public policy and to the human sciences from medicine to sociology. I have included many elementary graphical and numerical techniques to give flesh to the ideas and muscle to the reasoning. Students learn to think about data by working with data. I hope, however, that I have not allowed technique to dominate concepts. My intention is to teach verbally rather than algebraically, to invite discussion and even argument rather than mere computation, though some computation remains essential. The coverage is considerably broader than that of traditional statistics texts, as the table of contents reveals. In the spirit of general education, I have preferred breadth to detail.

Despite its informal nature, *Statistics: Concepts and Controversies* is a textbook. It is organized for systematic study and has abundant exercises, many of which ask students to offer a discussion or make a judgment. Even those admirable individuals who seek pleasure in uncompelled reading should look at the exercises as well as the text. Teachers should be aware that the book is more difficult than its low mathematical level suggests. The emphasis on ideas and reasoning asks more of the reader than many recipe-laden methods texts.

Some mathematicians and statisticians who have taught from earlier editions noted the challenge of teaching nonmathematical material. I have tried to provide detailed help for teachers in the Instructor's Guide. Here are a few general suggestions:

▶ Try to establish a "humanities course" atmosphere, with much discussion in class. Many exercises require discussion; modify them to ask "Come to class prepared to discuss...."

▶ Don't spend all your time at the board working problems, though some of that is necessary.

▶ Supplement this book with other material that is current (examples from the news) or of special interest to your students. The Instructor's Guide suggests some resources.

THE FOURTH EDITION

I am pleased that the first three editions of *Statistics: Concepts and Controversies* have been widely read, and read by people from many disciplines and diverse backgrounds. I am grateful to many teachers and readers for helpful comments and suggestions. David Bernklau of Stuyvesant High School (New York), Jack Fraenkel of San Francisco State University, Gudmund Iversen of Swarthmore College and Steve Rigdon of Southern Illinois University deserve special thanks for their detailed reviews. The staff of W. H. Freeman and Company, especially Diane Cimino Maass and Diana Blume, have done their usual excellent job in designing and producing the book.

This fourth edition has been thoroughly rewritten with an eye to both contemporary issues and accessibility to readers. The logical units of the exposition are more clearly marked, and the topical examples carry more of the exposition. There are more examples and more exercises; brief titles help readers see at a glance what topics these concern. In Chapter 1, for example, readers will find examples titled "Telephone sampling," "Computer-assisted interviewing," and "The census under-count," as well as a new section on "Data ethics." Because so many exercises are suitable for either class discussion or short writing assignments, I have not marked these with icons. I have, however, added "Writing Projects" at the end of each chapter. These are intended for more substantial assignments that may include looking up background material.

I hope that this fourth edition continues to be true to my original purpose: to present statistics to nonmathematical readers as an aid to clear thinking in personal and professional life.

▶▶▶ INTRODUCTION:
WHAT IS STATISTICS?

Statistics is the science of gaining information from numerical data. We study statistics because the use of data has become ever more common in a growing number of professions, in public policy, and in everyday life. Here are some of the statistical questions this book discusses:

▶ The government reports that the unemployment rate last month was 6.5%. What exactly does that figure mean? How did the government obtain this information? (Neither you nor I were asked if we were employed last month.) How accurate is the official unemployment rate?

▶ A Gallup poll reports that 45% of American adults are afraid to go outside at night because of crime. Where did that information come from? How accurate is it?

▶ Medical experiments tell us that taking aspirin regularly reduces the risk of a heart attack. Why are these experiments convincing? How strong is the effect of taking aspirin?

▶ Does taking the drug Bendectin to relieve nausea during pregnancy cause birth defects? Does smoking cause lung cancer? In both cases, the evidence is said to be "statistical." What kind of evidence is this? Why is the statistical evidence against smoking strong and the evidence against Bendectin weak?

▶ As state lotteries and casinos multiply, more people are losing money by misunderstanding chance. What does "the law of averages" say to a gambler? Is there a system for beating the house or winning the lottery?

The goal of statistics is to gain understanding from data. Data are numbers, but they are not "just numbers." *Data are numbers with a context.* The number 10.5, for example, carries no information by itself. But if we hear that a friend's new baby weighed 10.5 pounds at birth, we congratulate her on the healthy size of the child. The context engages our background knowledge and allows us to make judgments. We know that a baby weighing 10.5 pounds is quite large, and that it is not possible for a human baby to weigh 10.5 ounces or 10.5 kilograms. The context makes the number informative.

Because data are numbers with a context, doing statistics means more than manipulating numbers. I hope that as you read this book you will do more thinking than calculating. That can be uncomfortable if you are expecting "another math course" full of recipes. Statistics isn't math (although we will use basic math), so relax and look for new ideas.

WHERE WE ARE GOING

In pursuit of our goal of gaining understanding from data, we will divide our study into three parts:

  I. Producing data

  II. Organizing data

  III. Drawing conclusions from data

In the first part of this book, we will look at statistical designs for producing good data. The ideas behind samples and experiments, though simple and nonmathematical, are among the most important concepts in statistics. The second part of the book presents graphical and numerical tools and strategies for exploring data. This topic is often called *data analysis*, the practical art of understanding what data say and communicating your understanding to others. The third part of the book is devoted to *statistical inference*. Inference uses the language of probability to draw conclusions from data, and to accompany those conclusions by a formal statement of how confident we are that they are correct.

This skeleton outline of our subject supports the flesh and blood of reasoning about data and chance. We will look at the influence of opinion polls in politics, the ethics of experiments on human subjects, the tricks of those who use numbers to mislead us, the subtle question of evidence for causation, and much more.

Your goals in reading this book should be threefold. First, reach an understanding of statistical ideas in themselves. Ideas for reasoning about data and chance are major intellectual accomplishments that are worthy of your attention. Second, acquire the ability to deal critically with numerical arguments. Many people are unduly credulous when numerical arguments are used. They are impressed by the solid appearance of a few numbers and do not attempt to penetrate the substance of the argument. Others are unduly cynical. They think numbers are liars by nature and never trust them. Numerical arguments are like any others. Some are good, some are bad, and some are irrelevant. A bit of quantitative sophistication will enable you to hold your own against the number-slinger. Third, gain an understanding of the impact of statistical ideas on public policy and in other areas of academic study.

DEALING WITH UNCERTAINTY

Data vary. Individual people, animals, and things are variable; repeated measurements on the same individual are variable. Conclusions based on data are therefore *uncertain.* Statistics faces the variability and uncertainty of the world directly. Statistical reasoning can produce data whose usefulness is not destroyed by variation and uncertainty. It can analyze data to separate systematic patterns from the ever-present variation. It can form conclusions that, although not certain—nothing in the real world is certain—have only a little uncertainty. More important, statistical reasoning allows us to say just how uncertain our conclusions are.

Statistical ideas and techniques emerged only slowly from the struggle to work with uncertain data. Almost two centuries ago astronomers and surveyors faced the problem of combining many observations which, despite the greatest care, did not exactly match. Their efforts to deal with variation in their data produced some of the first statistical techniques. As the social sciences emerged in the nineteenth century, old statistical ideas were transformed and new ones were invented to describe the variation in individuals and societies. The study of heredity and of variable populations in biology brought more advance. The first half of the twentieth century gave birth to statistical designs for producing data and to formal inference based on probability. By mid-century it was clear that a new discipline had been born. As all fields of study place more emphasis on data and increasingly recognize that variability in data is unavoidable, statistics has become a central intellectual method. Every educated person should be acquainted with statistical reasoning. Reading this book will enable you to make that acquaintance.

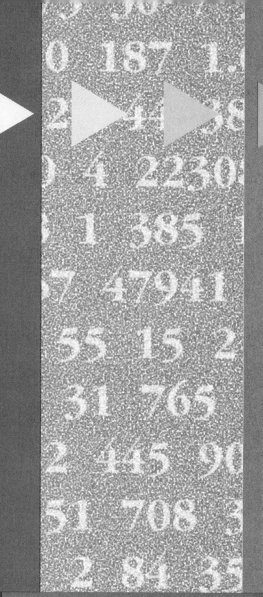

I'M FILLING OUT A READER SURVEY FOR *CHEWING* MAGAZINE.

SEE, THEY ASKED HOW MUCH MONEY I SPEND ON GUM EACH WEEK, SO I WROTE, "$500." FOR MY AGE, I PUT "43", AND WHEN THEY ASKED WHAT MY FAVORITE FLAVOR IS, I WROTE "GARLIC/CURRY."

THIS MAGAZINE SHOULD HAVE SOME AMUSING ADS SOON.

I LOVE MESSING WITH DATA.

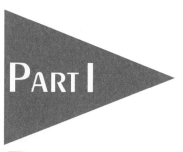

PART I

PRODUCING DATA

You want data on a question of interest to you. Perhaps you want to know the social and economic backgrounds of college undergraduates. Perhaps you want to know what causes of death are most common among young people. How will you get helpful data?

It is tempting to do without any systematic data. Instead, we base conclusions on our own experience. We think (without really thinking) that the students at our own college are typical. Or we recall an unusual incident that sticks in our memory exactly because it is unusual. We remember an airplane crash that killed several hundred people. We may ignore the fact that data on all flights show that flying is much safer than driving.

A better tactic is to head for the library or the Internet. There we find abundant data, not gathered specifically to answer our questions, but available for our use. The annual *Statistical Abstract of the United States*, for example, contains data that shed light on both the population of college students and causes of death broken down by age. You can learn, for example, that 30.8% of college students are also employed full time and that AIDS is the leading cause of death among men aged 25 to 44 but ranks only fifth in killing women in that age group. Sometimes you may even produce your own data, by handing out a questionnaire or by doing an experiment.

Whether you use data produced by others or produce your own, the quality of the information you get depends on the quality of the data. Good data are as much a human product as wool sweaters and compact disc players. Sloppily produced data will frustrate you as much as a sloppily made sweater. You examine a sweater before you buy, and you don't buy if it is not well made. Neither should you use data that are not well made. The first part of this book shows how to tell if data are well made. It describes statistical designs for producing good data, and alerts you to pitfalls that result in untrustworthy data.

CHAPTER 1

SAMPLES

"You don't have to eat the whole ox to know that the meat is tough." That is the idea of sampling: to gain information about the whole by examining only a part. How can we choose a sample that can be trusted to represent the whole?

▶ 1 SAMPLING BASICS

Here is the vocabulary that statisticians use to discuss sampling:

> **The vocabulary of sampling**
>
> **Population**–the entire group of people, animals, or things about which we want information.
>
> **Unit**–any individual member of the population. If the population consists of people, we often call its members **subjects**.
>
> **Sample**–a part of the population from which we actually collect information, used to draw conclusions about the whole.
>
> **Sampling frame**–the list of units from which we choose the sample.
>
> **Variable**–a characteristic of a unit, to be measured for those units in the sample.

Notice that we define the *population* in terms of our desire for information. If we want information about all U.S. college students, that is our population even if students at only one college are available for sampling. To make sense of any sample result, you must know what population the sample represents. Did that pre-election poll, for example, ask the opinions of all adults? Citizens only? Registered voters only? Democrats only? The *sample* consists of the people we actually have information about. If the poll can't contact some of the people it selected, those people aren't in the sample.

The distinction between population and sample is basic to statistics. The following examples illustrate this distinction and also introduce some major uses of sampling. These brief descriptions also indicate the variables measured for each unit in the sample. They do not state the sampling frame. Ideally, the sampling frame should be a list of all units in the population. As we shall see, obtaining such a list is one of the practical difficulties in sampling.

EXAMPLE 1. Public opinion polls, such as those conducted by Gallup and many news organizations, ask people's opinions on a variety of issues. The *variables* measured are responses to questions about public issues. Though most noticed at election time, these polls are conducted on a regular basis throughout the year. For a typical opinion poll:

Population: U.S. residents 18 years of age and over. Noncitizens and even illegal immigrants are included.

Sample: Between 1000 and 1500 people interviewed by telephone.

EXAMPLE 2. Government economic and social data are often produced by large *sample surveys* of a nation's individuals, households, or businesses. The monthly Current Population Survey (CPS) is the most important government sample survey in the United States. Many of the variables recorded by the CPS concern the employment or unemployment of everyone over 16 years old in a household. The government's monthly unemployment rate comes from the CPS. The CPS also records many other economic and social variables.[1] For the CPS:

Population: All 97 million U.S. households. Notice that the units are households rather than individuals or families. A household consists of all individuals who share the same living quarters, regardless of how they are related to each other.

Sample: About 60,000 households interviewed each month.

EXAMPLE 3. **Market research** is designed to discover consumer preferences and usage of products. One example of market research is the television-rating service of Nielsen Media Research. The Nielsen ratings determine how much advertisers will pay to sponsor a program and ultimately whether or not the program remains on the air. For the Nielsen national TV ratings:

Population: All 95 million U.S. households that have a television set.

Sample: About 5000 households that agree to use a "people meter" to record the TV viewing of all people in the household.

The *variables* recorded include the number of people in the household and their ages and sex, whether the TV set is in use at each time period and if so, what program is being watched and who is watching it.

EXAMPLE 4. **Social science research** makes heavy use of sampling. The General Social Survey (GSS), for example, has been carried out almost every year since 1972 by the National Opinion Research Center at the University of Chicago. The GSS asks more than 500 questions that reflect the interests of social scientists. The *variables* cover the subject's personal and family background, experiences and habits, and attitudes and opinions on subjects from abortion to war.[2]

Population: Adults (age 18 and over) living in households in the United States. The population does not include adults in institutions such as prisons and college dormitories. It also does not include persons who cannot be interviewed in English.

Sample: About 1400 adults interviewed in person in their homes.

There are many more uses of sampling, some bordering on the bizarre. For example, a radio station that plays a song owes the song's

"Opinion poll, market research, social survey, whatever. The answer is still NO! "

composer a royalty. The organization of composers (called ASCAP) collects these royalties for all its members by charging stations a license fee for the right to play members' songs. ASCAP has four million songs in its catalog and collects $420 million in fees each year. How should ASCAP distribute this income among its members? By sampling: ASCAP tapes about 60,000 hours from the 53 million hours of local radio programs across the country each year. The tapes are shipped to New York, where professional trivia experts who recognize all four million songs count how often each song was played in the sample. This sample count is used to split royalty income among composers, depending on how often their music was played. Sampling is a pervasive, though usually hidden, aspect of modern life.

WHY SAMPLE?

Why look at only a part of the population? Why not take a *census?*

> **Census**
>
> A **census** is a sample consisting of the entire population.

For example, the U.S. Constitution requires a census of the U.S. population every 10 years. If the population is large, a census is expensive and takes a long time. Even the federal government, which can afford a census, uses samples such as the CPS to produce timely data on employment and many other variables. If the government asked every adult in the country about his or her employment, this month's unemployment rate would be available next year rather than next month.

There are also less obvious reasons for preferring a sample to a census. If you are testing fireworks or fuses, the units in the sample are destroyed. Moreover, a relatively small sample often produces more accurate data than a census. A careful sample of an inventory of spare parts will almost certainly give more accurate results than asking the clerks to count all 500,000 parts in the warehouse. Bored people do not count accurately.

The experience of the Census Bureau reminds us that a more careful definition of a census is "an *attempt* to sample the entire population." The bureau estimates that the 1990 census missed 1.8% of the American population. These missing persons included an estimated 4.6% of the black population, largely in inner cities. A census is not foolproof, even with the resources of the government behind it. Why take a census at all? Only a census can give detailed information about every small area of the population. The government needs block-by-block population figures to create election districts with equal population. The main function of the U.S. census is to provide this local information.

HOW TO SAMPLE BADLY

So sample we must. It is, alas, easier to sample badly than to sample well. Suppose that I sell your company several crates of oranges each week. You examine a sample of oranges from each crate to determine the quality of my oranges. It is easy to inspect a few oranges from the top of each crate, but these oranges may not be representative of the entire crate. Those on the bottom are more often damaged in shipment. And once I learn your method of sampling, I will be sure that the rotten oranges are packed on the bottom with some good ones on top for you to inspect. If you sample from the top, your sample results are *biased*—the sample oranges are systematically better than the population they are supposed to represent.

Biased sampling methods

The design of a statistical study is **biased** if it systematically favors certain outcomes.

Selection of whichever units of the population are easiest to reach is called **convenience sampling**.

A **voluntary response sample** chooses itself by responding to a general appeal. Write-in or call-in opinion polls are examples of voluntary response samples.

Convenience samples and voluntary response samples are often biased.

EXAMPLE 5. **Interviewing at the mall.** Squeezing the oranges on the top of the crate is one example of convenience sampling. Mall interviews are another. Manufacturers and advertising agencies often use interviews at shopping malls to gather information about the habits of consumers and the effectiveness of ads. A sample of mall shoppers is fast and cheap. But people contacted at shopping malls are not representative of the entire U.S. population. They are richer, for example, and more likely to be teenagers or retired. Moreover, the interviewers tend to select neat, safe-looking individuals from the stream of customers. Mall samples are biased: they systematically overrepresent some parts of the population (prosperous people, teenagers, and retired people) and underrepresent others. The opinions of such a convenience sample may be very different from those of the population as a whole.

EXAMPLE 6. Write-in opinion polls. Ann Landers once asked the readers of her advice column, "If you had it to do over again, would you have children?" She received nearly 10,000 responses, almost 70% saying "NO!" Can it be true that 70% of parents regret having children? Not at all. This is a voluntary response sample. People who feel strongly about an issue, particularly people with strong negative feelings, are more likely to take the trouble to respond. Ann Landers' results are strongly biased—the percent of parents who would not have children again is much higher in her sample than in the population of all parents.

Write-in and call-in opinion polls are almost sure to lead to strong bias. In fact, only about 15% of the public has ever responded to a call-in poll, and these tend to be the same people who call radio talk shows. That's not a representative sample of the population as a whole.

SECTION 1 EXERCISES

1.1   A sociologist wants to know the opinions of employed adult women about government funding for day care. She obtains a list of the 520 members of a local business and professional women's club, and mails a questionnaire to 100 of these women selected at random. Only 68 questionnaires are returned. What is the population in this study? What is the sampling frame? What is the sample?

1.2   Different types of writing can sometimes be distinguished by the lengths of the words used. A student interested in this fact wants to study the lengths of words used by Tom Clancy in his novels. She opens a Clancy novel at random and records the lengths of the first 250 words on the page. What is the population in this study? What is the sample? What variable is being measured?

1.3   A newspaper article about an opinion poll says that "43% of Americans approve of the president's overall job performance." Toward the end of the article, you read "The poll is based on telephone interviews with 1210 adults from around the United States, excluding Alaska and Hawaii." What variable did this poll measure? What population do you think the newspaper wants information about? What was the sample? Are there any sources of bias in the sampling method used?

In each of Exercises 1.4 to 1.6, briefly identify the **population** (What is the basic unit? Which units fall in the population?), the **variables** measured (What information is desired?), and the **sample**. If the situation is not described in enough detail to completely identify the population, complete the description of the population in a reasonable way. Be sure that from your description it is possible to tell exactly when a unit is in the population and when it is not.

Moreover, each sampling situation described in Exercises 1.4 to 1.6 contains a source of probable bias. In each case, state the **reason** you suspect that bias will occur and also the **direction** of the likely bias. (That is, in what way will the sample conclusions probably differ from the truth about the population?)

1.4   **Mail to a member of Congress.** A congressman is interested in whether his constituents favor a proposed gun control bill. His staff reports that letters on the bill have been received from 361 constituents and that 323 of these oppose the bill.

1.5   A flour company wants to know what fraction of Minneapolis households bake some or all of their own bread. The company selects a sample of 500 residential addresses in Minneapolis and sends interviewers to these addresses. The interviewers work during regular working hours on weekdays and interview only during those hours.

1.6   The Miami Police Department wants to know how black residents of Miami feel about police service. A sociologist prepares several questions about the police. A sample of 300 mailing addresses in predominantly black neighborhoods is chosen, and a police officer goes to each address to ask the questions of an adult living there.

1.7   **Ann Landers takes a sample.** Advice columnist Ann Landers once asked her female readers whether they would be content with affectionate treatment by men with no sex ever. Over 90,000 women wrote in, with 72% answering "Yes." Many of the letters described unfeeling treatment at the hands of men. Explain why this sample is certainly biased. What is the likely direction of the bias? That is, is that 72% probably higher or lower than the truth about the population of all adult women?

1.8 **Women and love.** In 1987, Shere Hite published a best-selling book called *Women and Love.* Hite distributed 100,000 questionnaires through various women's groups, asking questions about love, sex, and relations between women and men. She based her book on the 4.5% of the questionnaires that were returned. The women who responded were fed up with men and eager to fight them. For example, 91% of those who were divorced said that they had initiated the divorce. The anger of women toward men became the theme of the book.

Explain why Hite's sampling method is biased. Is that 91% probably higher or lower than the true percent of all divorced women who initiated the divorce themselves? Why?

1.9 **Call-in opinion polls.** Want to sample public opinion quickly and cheaply? The telephone company will set up "900" telephone numbers that record how many calls are made to each number. Announce your question on TV, give a 900 number for "Yes" and another for "No," and wait. The respondents need not say a word; they just dial a number. A small charge appears on the phone bills of those who call.

The ABC television network was the first to make important use of call-in polling. At the end of the first Reagan-Carter presidential election debate in October of 1980, ABC asked its viewers which candidate won. The call-in poll proclaimed that Reagan had won the debate by a 2 to 1 margin. But a random survey by CBS News showed only a 44% to 36% margin for Reagan, with the rest undecided. Why are call-in polls likely to be biased? Can you suggest why this bias might favor the Republican Reagan over the Democrat Carter?

▶ 2 SIMPLE RANDOM SAMPLING

Sampling badly is easy. Good samples require more care. The bias in convenience samples and voluntary response samples is due to human choice. The statistician's remedy for this bias is to eliminate human choice by allowing impersonal chance to choose the sample. The result is a *simple random sample*. The essential idea is to give each unit in the sampling frame the same chance to be chosen for the sample as any other unit. Here is the precise definition:

Simple random sample

A **simple random sample** of size *n* is a sample of *n* units chosen in such a way that every collection of *n* units from the sampling frame has the same chance to be chosen. (We will always use *n* as shorthand for the number of units in a sample.)

We will abbreviate "simple random sample" as SRS. The definition of an SRS doesn't describe any one sample. It describes instead the method for choosing the sample. An SRS is obtained by a method that gives every possible sample of size *n* the same chance of being the sample actually chosen. Such a method has a clear advantage over a convenience sample: it is fair, or **unbiased**. Rich and poor, black and white, Democrat and Republican—all have the same chance of being part of an SRS. We might by bad luck get one SRS that has too many rich black Republicans, but the method of choosing the sample is unbiased. That's an important distinction: we can't guarantee that the sample represents the population, but we can at least use a fair method to choose the sample.

To see the idea of an SRS, think of choosing names from a hat. Identify each unit in the sampling frame on an identical tag, mix the tags thoroughly in a hat, then draw one blindly. If the mixing is truly complete, every tag in the hat has the same chance of being chosen. The unit identified on the tag you draw is the first unit in our SRS. Now draw another tag without replacing the first. Every remaining tag has the same chance of being drawn. So every pair of tags has the same chance of being the pair you have now drawn—you have an SRS of size 2. To obtain an SRS of size *n*, keep on drawing until you have *n* tags. The *n* units named on these tags are an SRS of size *n*.

Drawing names from a hat makes clear what it means to give each unit and each possible set of *n* units the same chance of being chosen. That's the idea of an SRS. Unfortunately, drawing tags from a hat would be a bit awkward for a sample of the country's 97 million households. We need a practical way to choose an SRS.

RANDOM DIGITS

Picture a wheel spinning on a smooth bearing so that it does not favor any particular orientation when coming to rest. Divide the edge of the wheel into 10 equal sectors and label the sectors 0, 1, 2, 3, 4, 5, 6, 7, 8, and 9. Fix a stationary pointer at the wheel's rim and spin the wheel.

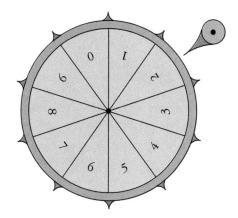

Slowly and smoothly it comes to rest. Sector number 1 is opposite the pointer. Spin the wheel again. It comes to rest with sector number 9 opposite the pointer. If you continue this process, you will produce a string of digits like these:

19223950340575628713...

On any one spin, the wheel has the same chance of producing each of the 10 digits. Because the wheel has no memory, the outcome of any one spin has no effect on the outcome of any other. You are producing a *table of random digits.*

Table A at the back of the book is a table of random digits. The table begins with the digits 19223950340575628713. To make the table easier to read, the digits appear in groups of five and in numbered rows. The groups and rows have no meaning—the table is just a long list of digits

Random digits

A **table of random digits** is a list of the digits 0, 1, 2, 3, 4, 5, 6, 7, 8, 9 that has the following properties:

1. The digit in any position in the list has the same chance of being any one of 0, 1, 2, 3, 4, 5, 6, 7, 8, 9.

2. The digits in different positions are independent in the sense that the value of one has no influence on the value of any other.

having the properties 1 and 2 described above. Some calculators and computer software will also produce random digits. Our goal is to use random digits, however we obtain them, to choose an SRS. We need the following facts about random digits, which are consequences of the basic properties 1 and 2:

3. Any *pair* of random digits has the same chance of being any of the 100 possible pairs, 00, 01, 02, . . ., 98, 99.

4. Any *triple* of random digits has the same chance of being any of the 1000 possible triples, 000, 001, 002, . . ., 998, 999.

5. . . . and so on for groups of four or more random digits.

CHOOSING AN SRS

These "equally likely" facts make it easy to use Table A to choose an SRS. Here is an example that shows how.

EXAMPLE 7. How to choose an SRS. Joan's small accounting firm serves 30 business clients. Joan wants to interview a sample of 5 clients in detail to find ways to improve client satisfaction. To avoid bias, she chooses an SRS of size 5.

Step 1: Label. Give each client a numerical label, using as few digits as possible. Two digits are needed to label 30 clients, so we use the labels

$$01, 02, 03, \ldots, 29, 30$$

It is also correct to use labels 00 to 29 or even another choice of 30 two-digit labels. Here is the list of clients, with labels attached.

01	A-1 Plumbing	16	Johnson Commodities
02	Accent Printing	17	JL Records
03	Action Sport Shop	18	Keiser Construction
04	Anderson Construction	19	Liu's Chinese Restaurant
05	Bailey Trucking	20	MagicTan
06	Balloons Inc.	21	Peerless Machine
07	Bennett Hardware	22	Photo Arts
08	Best's Camera Shop	23	River City Books
09	Blue Print Specialties	24	Riverside Tavern
10	Central Tree Service	25	Rustic Boutique
11	Classic Flowers	26	Satellite Services
12	Computer Answers	27	Scotch Wash
13	Darlene's Dolls	28	Sewer's Center
14	Fleisch Realty	29	Tire Specialties
15	Hernandez Electronics	30	Von's Video Store

Step 2: Table. Start anywhere in Table A and read two-digit groups. Suppose we begin at line 130, which is

69051 64817 87174 09517 84534 06489 87201 97245

The first 10 two-digit groups in this line are

69 05 16 48 17 87 17 40 95 17

Each successive two-digit group is a label. The labels 00 and 31 to 99 are not used in this example, so we ignore them. The first 5 labels between 01 and 30 that we encounter in the table choose our sample. Of the first 10 labels in line 130, we ignore 5 because they are too high (over 30). The others are 05, 16, 17, 17, and 17. The clients labeled 05, 16, and 17 go into the sample. Ignore the second and third 17s because that client is already in the sample. Now run your finger across line 130 (and continue to line 131 if needed) until 5 clients are chosen.

The sample consists of the clients labeled 05, 16, 17, 20, 19. These are Bailey Trucking, Johnson Commodities, JL Records, MagicTan, and Liu's Chinese Restaurant.

Using the table of random digits is much quicker than drawing names from a hat. What is more, computers can be programmed to choose an SRS from even a very large sampling frame almost instantly. That is what samplers do in practice. As Example 7 shows, choosing an SRS has two steps:

> **Choose an SRS in two steps**
>
> **Step 1: Label.** Assign a numerical label to every unit in the sampling frame.
>
> **Step 2: Table.** Use random digits to select labels at random.

Here are some hints for choosing an SRS: *Don't try to scramble the labels as you assign them.* Table A will do the required randomizing. You can assign labels in any convenient manner, such as alphabetical order for names of people. *Be certain that all labels have the same number of digits.* Only then will all individuals have the same chance to be chosen. *Use the shortest possible labels*: one digit for a population of up to 10 members, two digits for 11 to 100 members, three digits for 101 to 1000 members, and so on. You can start either with label 0 (or 00 or 000 as needed) or with label 1 (or 01 or 001) as you prefer. Because the table of random digits has no order, it is legal to read digits from it in any order—but we will always *read across rows from left to right* in the table, continuing to the following rows if needed. Don't forget to *ignore repeated labels and groups not used as labels*.

SECTION 2 EXERCISES

1.10   Use the table of random digits (Table A) to select an SRS of 3 of the following 25 volunteers for a drug test. Be sure to say where you entered the table and how you used it.

Agarwal	Garcia	Petrucelli
Andrews	Healy	Reda
Baer	Kim	Roberts
Berger	Lee	Shen
Brockman	Lynch	Smith
Casella	Menendez	Valdes
Frank	Moser	Wilson
Fuest	Musselman	
Fuhrmann	Navarro	

1.11 Your class in Ancient Ugaritic Religion is poorly taught, and has decided to complain to the dean. The class decides to choose four of its members at random to carry the complaint. The class list appears below. Choose an SRS of 4 using the table of random digits beginning at line 145.

Anderson	Gutierrez	Patnaik
Aspin	Green	Pirelli
Bennett	Harter	Rao
Bock	Henderson	Rider
Breiman	Hughes	Robertson
Castillo	Johnson	Rodriguez
Dixon	Kempthorne	Siegel
Edwards	Liang	Tompkins
Fernandez	Laskowsky	Vandegraff
Gupta	Olds	Wang

1.12 A food processor has 50 large lots of canned mushrooms ready for shipment, each labeled with one of the lot numbers below.

A1109	A2056	A2219	A2381	B0001
A1123	A2083	A2336	A2382	B0012
A1186	A2084	A2337	A2383	B0046
A1197	A2100	A2338	A2384	B1195
A1198	A2108	A2339	A2385	B1196
A2016	A2113	A2340	A2390	B1197
A2017	A2119	A2351	A2396	B1198
A2020	A2124	A2352	A2410	B1199
A2029	A2125	A2367	A2411	B1200
A2032	A2130	A2372	A2500	B1201

You want to choose an SRS of 5 lots for inspection. Use Table A to do this, beginning at line 139.

1.13 **An election day sample.** You want to choose an SRS of 25 of a city's 440 voting precincts for special voting-fraud surveillance on election day. Explain clearly how you would label the 440 precincts. Then use Table A to choose the SRS, and list the precincts you selected. Enter Table A at line 117.

1.14   **How do random digits behave?** Which of the following state-
       ments are true of a table of random digits, and which are false?

       (a)   There are exactly four 0's in each row of 40 digits.

       (b)   Each pair of digits has chance 1/100 of being 00.

       (c)   The digits 0000 can never appear as a group, because this
       pattern is not random.

1.15   **Sampling from a census tract.** Figure 1-1 is a map of a census
       tract in Cedar Rapids, Iowa. Census tracts are small geographical

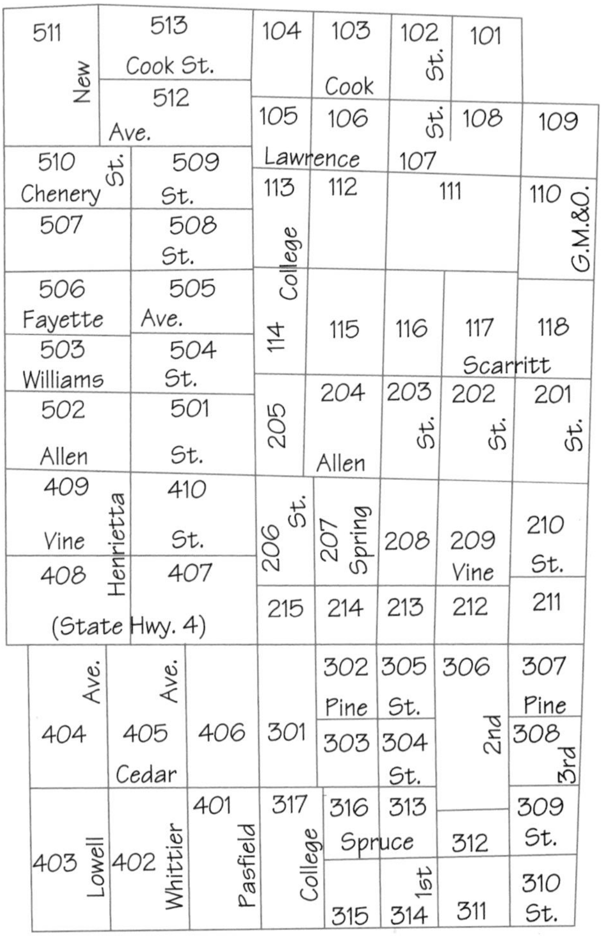

Figure 1-1 Map of a census tract in Cedar Rapids, Iowa, for Exercise 1.15.

areas averaging 5000 in population. On the map, each block is marked with a Census Bureau identification number. An SRS of blocks from a census tract is often the next-to-last stage in the multistage samples used by national sample surveys. Use Table A beginning at line 125 to choose an SRS of 5 blocks from this tract.

1.16   **A sampling experiment.** The following page contains a population of 80 circles. (They might represent fish in a pond or tumors removed in surgery.) Do a sampling experiment as follows:

(a)   Label the circles 00, 01, ..., 79 in any order, and use Table A to draw an SRS of size 4.

(b)   Measure the diameter of each circle in your sample. (All of the circles have diameters that are multiples of 1/8 inch. In decimal form, the possible diameters are $1/8 = 0.125$, $1/4 = 0.250$, $3/8 = 0.375$, $1/2 = 0.500$, $5/8 = 0.625$, $3/4 = 0.750$, and $7/8 = 0.875$. Record your results in decimal form.) Then use a calculator to compute the *mean* diameter of the four circles in your sample. The mean of the diameters d_1, d_2, d_3, d_4 is the ordinary average

$$\frac{d_1 + d_2 + d_3 + d_4}{4}$$

(c)   Repeat steps (a) and (b) three more times (four times in all), using a different part of Table A each time. Was any circle chosen more than once in your four SRSs? How different were the mean diameters for the four samples?

(d)   Draw an SRS of size 16 from this population using a part of Table A not yet used in this exercise. Measure the diameters of the 16 circles in your sample and find the mean (average) diameter.

As we will see in the next section, we expect to find less variation among means of samples with size 16 than for samples with size 4. Combining your results with those of other students should convince you that this is true for this sampling experiment.

▶ 3 POPULATION INFORMATION FROM A SAMPLE

The advice columnist Ann Landers once asked her readers, "If you had it to do over again, would you have children?" She received nearly 10,000 responses, almost 70% saying "NO!" Many of the letters included heart-rending tales of the torments inflicted by children on their parents. This is a blatant example of voluntary response. How blatant was suggested by a professional nationwide random sample commissioned by *Newsday*. That sample polled 1373 parents and found that 91% *would* have children again. A voluntary response sample can give 70% "No" when the truth about the population is close to 91% "Yes."

Does *Newsday*'s sample result really allow us to say that about 91% of all parents would have children again? As a newspaper reporter put it, "Far be it from us to question the validity of any statistic that we read in the papers, but we are talking somewhere in the neighborhood of a

"Hey, Pops, what was that letter you sent off to Ann Landers yesterday?"

1-in-50,000 sampling."[3] The reporter asks a perceptive question. Why can we trust a sample that interviews only one out of every 50,000 units in the population?

We know one advantage of an SRS—it has no bias. The *Newsday* poll gave all parents the same chance to respond, rather than favoring those who were mad enough at their children to write to Ann Landers. However, lack of favoritism is not enough when we are asked to draw conclusions about tens of millions of parents from information about only 1373 of them. We need to think more carefully about the process of using a sample to gain information about a population.

PARAMETERS AND STATISTICS

We take a sample to draw conclusions about the population, not about the sample itself. When Newsday asked a sample of parents if they would have children again, 91% said "Yes." That 91% describes the sample, the 1373 parents that Newsday actually interviewed. We want to know what percent of *all* parents would have children again. That unknown percent describes the population. Here is the vocabulary we use to keep straight whether a number describes a sample or a population.

Parameters and statistics

A **parameter** is a number that describes the **population**. A parameter is a fixed number, but in practice we do not know its value.

A **statistic** is a number that describes a **sample**. The value of a statistic is known when we have taken a sample, but it can change from sample to sample. We often use a statistic to estimate an unknown parameter.

So parameter is to population as statistic is to sample. Want to estimate an unknown parameter? Choose an SRS from the population and use a sample statistic as your estimate. That's what *Newsday* did.

EXAMPLE 8. **Would you have children again?** The proportion of all American parents who would have children again is a parameter describing the population of parents. Call it p, for "proportion." Alas, we do not know the numerical value of p. To estimate p, *Newsday* took a sample of 1373 parents. The proportion of the sample who would have children again is a statistic. Call it $\hat{p}$, read as "p-hat." It happens that 1249 of this sample of size 1373 would do it again, so for this sample

$$\hat{p} = \frac{1249}{1373} = 0.91$$

Because an SRS gives all parents the same chance to be asked, it is reasonable to use the statistic $\hat{p} = 0.91$ as an estimate of the unknown parameter p. Exactly 91% of the sample would have children again. We estimate that about 91% of all parents would have children again.

If we took a second random sample of 1373 parents, the new sample would have different people in it. It is almost certain that there would *not* be exactly 1249 positive responses. That is, the value of $\hat{p}$ will vary from sample to sample. This is called *sampling variability*. Aha! What is to prevent one random sample from finding that 91% of parents would have children again and a second random sample from finding that 70% would not? After all, we just admitted that the statistic $\hat{p}$ wanders about from sample to sample.

SAMPLING DISTRIBUTIONS

We are saved by a second property of random sampling, a property even more important than lack of bias. Imagine that we choose not just two SRSs of parents but thousands of SRSs. Each time we choose a sample, we ask those parents if they would have children again. The proportion who say "Yes" is the value of the statistic $\hat{p}$ for that sample. We end up with thousands of values of $\hat{p}$. These values have a regular pattern that enables us to see how trustworthy $\hat{p}$ is as an estimate of the unknown population proportion p.

> **Sampling variability and sampling distribution**
>
> **Sampling variability:** If we repeatedly choose samples from the same population, a sample statistic will take different values in different samples.
>
> **Sampling distribution:** A sample statistic from an SRS has a predictable pattern of sampling variability if we choose a large number of samples from the same population. The **sampling distribution** of the statistic describes this pattern.

EXAMPLE 9. **A sampling experiment.** To illustrate the sampling distribution of a statistic, let's do an experiment. Suppose that in fact (unknown to either Ann Landers or *Newsday*) exactly 80% of all parents would have children again. That is, the true population proportion is $p = 0.80$. Suppose I choose many SRSs of size 1373 from this population. In the first sample, 1100 of the 1373 parents say they would have children again. The sample proportion is

$$\hat{p} = \frac{1100}{1373} = 0.8012$$

The second sample has 1090 parents who would have children again. That is,

$$\hat{p} = \frac{1090}{1373} = 0.7939$$

The next few samples give proportions 0.8347, 0.7786, 0.7917, and so on. Already we are reassured: although the samples have different $\hat{p}$'s, the values are all close to the true population value, $p = 0.80$. None of these samples gives a really bad estimate of the population proportion.

I kept going until I had 1000 SRSs from this population, each of size 1373. That's 1000 values of the sample proportion $\hat{p}$, enough to show us how $\hat{p}$ behaves when we keep on choosing samples. Figure 1-2 displays the results in a picture.

The base of each bar in Figure 1-2 covers a range of possible values of $\hat{p}$. The height of the bar shows how many of my 1000 samples had a value in that range. For example, the tallest bar shows that 192 of

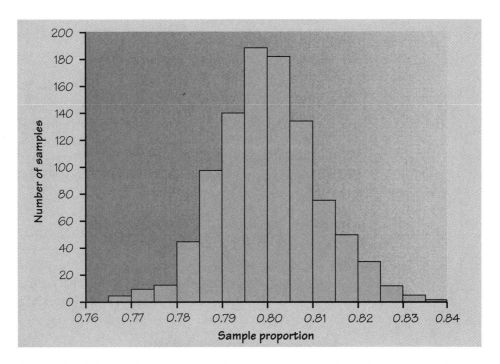

Figure 1-2 The results of 1000 simple random samples from the same population. This is the sampling distribution of the sample proportion $\hat{p}$ when the population proportion is $p = 0.80$.

the 1000 samples had a $\hat{p}$ between 0.795 and 0.80. Figure 1-2 shows what values the statistic $\hat{p}$ takes in many samples and how often it takes each value. That is, it displays the sampling distribution of $\hat{p}$. If we look carefully at Figure 1-2, here is what we find:

▶ **No bias.** The center of the values of $\hat{p}$ is very close to the population parameter $p = 0.80$. The statistic $\hat{p}$ as an estimate of p is too high in some samples and too low in others, but it has no tendency to be *usually* too high or *usually* too low when we take many samples. That reflects the absence of bias in simple random sampling.

▶ **Small variability.** All the samples have $\hat{p}$'s between 0.765 and 0.840. If 80% is the truth about the population, it appears that an SRS of size 1373 will almost never guess as far off as 75% or 85%. Moreover, 654 of the 1000 samples have $\hat{p}$'s between 0.79 and 0.81.

It appears that a clear majority of SRSs of size 1373 will give an estimate within 0.01 of the truth.

In short, the sampling distribution shows that *if we are satisfied with an answer that is approximately correct, we can trust an SRS of size 1373*. Almost all SRSs of this size give results that are close to the truth about the population. In this sampling experiment, we assumed we knew that p was 0.80, but the situation is much the same for any p. The sample statistic $\hat{p}$ is rarely exactly correct as an estimate of p, but its values are centered at the true p, and most samples give values quite close to p.

BIAS AND LACK OF PRECISION

In practice, we choose only one sample. We want to be confident that the statistic from our one sample is close to the true population parameter. It isn't enough to have no bias—we want almost all samples to give results close to the population truth. As you might guess, large samples are more trustworthy than small samples.

> **EXAMPLE 10. What if we use a smaller sample?** Could *Newsday* save money and effort by interviewing fewer than 1373 subjects? Not without paying a price. Figure 1-3 shows what that price is. The bottom graph is the sampling distribution of sample proportions from one thousand SRSs of size 1373 again. This is the same as Figure 1-2, but I have squeezed it a bit—each bar in Figure 1-3 contains the values from two bars in Figure 1-2. The top graph in Figure 1-3 shows the sampling distribution if we save money by interviewing only 300 people. The two graphs have the same scale, so they are easy to compare. *The results from smaller samples are still centered at the truth for the population, but they are much more spread out.* An SRS of size 300 is less trustworthy than an SRS of size 1373 because it is more likely to give an answer far from the truth about the population.

We need a final bit of vocabulary to describe the fact that lack of repeatability—the sample result wanders all over the barnyard—is as serious a flaw in a sampling method as is favoritism. Because we select a sample in order to gain information about a population, an error in

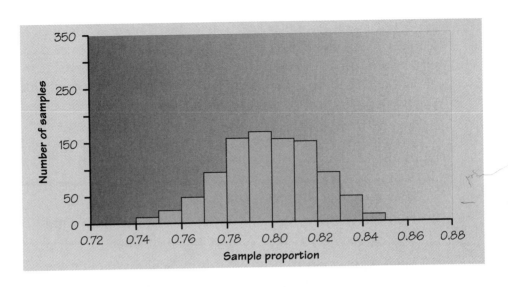

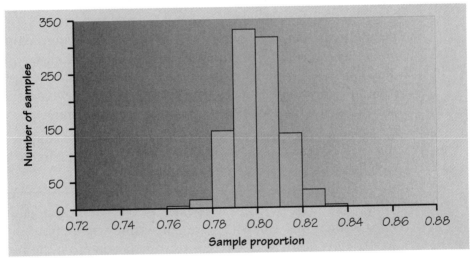

Figure 1-3 Larger samples have less sampling variability. These are the sampling distributions of the sample proportion $\hat{p}$ for samples of size 300 (top) and 1373 (bottom).

sampling means that the sample statistic misses the true value of the population parameter. Any method of sampling can have two types of error: *bias* and *lack of precision*.

We can think of the true value of the population parameter as the bull's-eye on a target, and of the sample statistic as a bullet fired at the bull's-eye. Both bias and lack of precision describe what happens when we take many shots at the target. *Bias* means that our sight is misaligned and we shoot consistently off the bull's-eye in the same direction. Our sample values do not center about the population value. *Lack of precision* means that repeated shots are widely scattered on the target. Repeated samples do not give similar results but differ widely among themselves. Figure 1-4 shows this target illustration of the two types of error.

Bias and lack of precision

Bias is consistent, repeated deviation of the sample statistic from the population parameter in the same direction.

Lack of precision means that in repeated sampling the values of the sample statistic are spread out or scattered. The result of sampling is not repeatable.

To reduce bias, use random sampling. When the sampling frame lists the entire population, simple random sampling produces *unbiased* estimates—the values of a statistic computed from an SRS neither consistently overestimate nor consistently underestimate the value of the population parameter.

To increase the precision of an SRS, use a larger sample. You can make the precision as high as you want by taking a large enough sample.

Notice that high precision (repeated shots are close together) can accompany high bias (the shots are consistently away from the bull's-eye in one direction). And low bias (the shots center on the bull's-eye) can accompany low precision (repeated shots are widely scattered). A good sampling scheme, like a good shooter, must have both low bias and high precision.

SAMPLING FROM LARGE POPULATIONS

Even *Newsday*'s sample of 1373 people is a very small fraction of the population of all parents. Does it matter whether 1373 is 1-in-100 or 1-in-50,000? No!

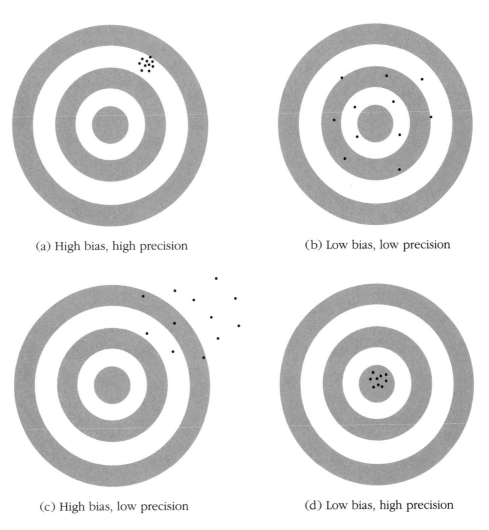

(a) High bias, high precision

(b) Low bias, low precision

(c) High bias, low precision

(d) Low bias, high precision

Figure 1-4 Bias and lack of precision in sampling. The bull's-eye represents the truth about the population, and the bullet holes represent the results of repeated samples.

The population size doesn't matter

The precision of a statistic from a random sample does not depend on the size of the population, as long as the population is much larger than the sample.

Why does the size of the population have little influence on the behavior of statistics from random samples? To see that this is plausible, imagine sampling harvested corn by thrusting a scoop into a lot of corn kernels. The scoop doesn't know whether it is surrounded by a bag of corn or by an entire truckload. As long as the corn is well mixed (so that the scoop selects a random sample), the variability of the result depends only on the size of the scoop.

This is good news for *Newsday*. Its sample of size 1373 has high precision because the sample size is large. It doesn't matter that the sample contained only 1-in-50,000 in the population. It is almost certain—Ann Landers to the contrary—that close to 91% of American parents would have children again.

However, the fact that the precision of a sample statistic depends on the size of the sample and not on the size of the population is bad news for anyone planning an opinion poll in a university or a small city. For example, it takes just as large an SRS to estimate the proportion of Ohio State University undergraduates who call themselves political conservatives as to estimate with the same precision the proportion of all adult U.S. residents who are conservatives. That there are about 31,000 Ohio State undergraduates and more than 194 million adults in the United States does *not* mean that a smaller SRS gives equally precise results at Ohio State.

SUMMING UP

This section has one big idea: to describe how trustworthy a sample is, ask "What would happen if we took a large number of samples from the same population?" If almost all samples give a result close to the truth, we can trust our one sample even though we don't know that it is close to the truth. The sampling distribution describes what would happen if we took many samples. Using a large SRS guarantees that almost all samples will give accurate results.

SECTION 3 EXERCISES

Each boldface number in Exercises 1.17 to 1.20 is the value of either a **parameter** or a **statistic**. In each case, state which it is.

1.17 The Bureau of Labor Statistics announces that last month it interviewed all members of the labor force in a sample of 60,000 households; **6.5%** of the people interviewed were unemployed.

1.18 A carload lot of ball bearings has an average diameter of **2.503** centimeters (cm). This is within the specifications for acceptance of the lot by the purchaser. The inspector happens to inspect 100 bearings from the lot with an average diameter of **2.515** cm. This is outside the specified limits, so the lot is mistakenly rejected.

1.19 A telephone sales outfit in Los Angeles uses a device that dials residential phone numbers in that city at random. Of the first 100 numbers dialed, **43** are unlisted numbers. This is not surprising, because **52%** of all Los Angeles residential phones are unlisted.

1.20 Voter registration records show that **68%** of all voters in Indianapolis are registered as Republicans. To test a random digit dialing device, you use the device to call 150 randomly chosen residential telephones in Indianapolis. Of the registered voters contacted, **73%** are registered Republicans.

1.21 **A call-in opinion poll.** Should the United Nations continue to have its headquarters in the United States? A television program asked its viewers to call in with their opinions on that question. There were 186,000 callers, 67% of whom said "No." A nationwide random sample of 500 adults found that 72% answered "Yes" to the same question. Explain to someone who knows no statistics why the opinions of only 500 randomly chosen respondents are a better guide to what all Americans think than the opinions of 186,000 callers.

1.22 Figure 1-5 shows the sampling distribution of a sample statistic in four situations. These figures are like Figures 1-2 and 1-3. That is, the heights of the bars show how often the sample statistic took various values in many samples from the same population. The true value of the population parameter is marked on each graph. Label each of the sampling distributions in Figure 1-5 as having high or low bias and as having high or low precision.

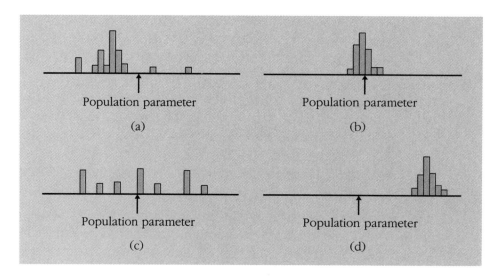

Figure 1-5 Four sampling distributions, for Exercise 1.22.

1.23 **Predict the election.** Just before a presidential election, a national opinion polling firm increases the size of its weekly sample from the usual 1500 people to 4000 people. Does the larger random sample reduce the bias of the poll result? Does it improve the precision of the result?

1.24 A management student is planning a project on student attitudes toward part-time work while attending college. She develops a questionnaire and plans to ask 25 randomly selected students to fill it out. Her faculty advisor approves the questionnaire, but suggests that the sample size be increased to at least 100 students. Why is the larger sample helpful?

1.25 **Sampling in the states.** An agency of the federal government plans to take an SRS of residents in each state to estimate the proportion of owners of real estate in each state's population. The population of the states ranges from about 485,000 people in Wyoming to more than 32 million in California.

(a) Will the precision of the sample proportion change from state to state if an SRS of size 2000 is taken in each state? Explain your answer.

(b)   Will the precision of the sample proportion change from state to state if an SRS of 1/10 of 1% (0.001) of the state's population is taken in each state? Explain your answer.

1.26   **A newspaper's opinion poll.** The *New York Times* conducted a national opinion poll on women's issues. The sample consisted of 1025 women and 472 men contacted by randomly selecting telephone numbers. One question was:

Many women have better jobs and more opportunities than they did 20 years ago. Do you think women have had to give up too much in the process, or not?

Forty-eight percent of the women and 33% of the men in the sample said that women have had to give up too much. The *Times* publishes complete descriptions of its polling methods. Here is part of the description for this poll:[4]

In theory, in 19 cases out of 20 the results based on the entire sample will differ by no more than three percentage points in either direction from what would have been obtained by seeking out all adult Americans.

The potential sampling error for smaller subgroups is larger. For example, for men it is plus or minus five percentage points.

Explain why the margin of error is larger for conclusions about men alone than for conclusions about all adults.

1.27   **The Census long form.** Not every household fills out the same form in the every-ten-years census of the United States. Basic questions about the number of people in the household and their age, sex, race, and so on appear on the basic form. Other questions appear only on a "long form" sent to a sample of 17% of households. The Census Bureau publishes summary statistics for various geographic areas. For the questions that appear on all forms, these statistics are published for areas with as few as 100 households. For questions appearing on the long form, the bureau does not publish statistics for areas with fewer than about 2000 households. Can you think of possible reasons for this policy?

1.28   **A sampling experiment.** Let us illustrate sampling variability in a small sample from a small population. Ten of the 25 club members listed below are female. Their names are marked with asterisks in the list. The club chooses five members at random to receive free trips to the national convention.

Agassiz	Darwin	Herrnstein	Myrdal	Vogt*
Binet*	Epstein	Jimenez*	Perez*	Went
Blumenbach	Ferri	Lombrosco	Spencer*	Wilson
Chase*	Gupta*	Moll*	Thomson	Yerkes
Chen*	Gutierrez	McKim*	Toulmin	Zimmer

(a)   Draw 20 SRSs of size 5, using a different part of Table A each time. Record the number of females in each of your samples. Make a graph like Figure 1-2 to display your results. What is the average number of females in your 20 samples?

(b)   Do you think the club members should suspect discrimination if none of the five tickets go to women?

1.29   **Simulation.** Random digits can be used to *simulate* the results of random sampling. Suppose that you are drawing simple random samples of size 25 from a large number of high school students and that 20% of the students are unemployed during the summer. To simulate this SRS, let 25 consecutive digits in Table A stand for the 25 students in your sample. The digits 0 and 1 stand for unemployed students, and other digits stand for employed students. This is an accurate imitation of the SRS because 0 and 1 make up 20% of the 10 equally likely digits.

Simulate the results of 50 samples by counting the number of 0's and 1's in the first 25 entries in each of the 50 rows of Table A. Make a graph like Figure 1-2 to display the results of your 50 samples. Is the truth about the population (20% unemployed, or 5 in a sample of 25) near the center of your graph? What are the smallest and largest counts of unemployed students you obtained in your 50 samples? What percent of your samples had either 4, 5, or 6 unemployed?

▶ 4 CONFIDENCE STATEMENTS

To learn how trustworthy a sample result is, we ask "What would happen if we took many such samples from the same population?" We saw that almost all SRSs of 1373 parents give results close to the truth about all parents, so we are quite confident that *Newsday*'s one sample of 1373 parents gave a result close to the truth. The final step in understanding how the results of sampling are reported is to make "quite confident" and "close to the truth" more exact. It will not surprise you to learn that we use numbers to replace the words when we want a more exact statement.

> **EXAMPLE 11. Using the sampling distribution.** Figure 1-6 is a copy of Figure 1-2. It shows how the sample proportion $\hat{p}$ behaves

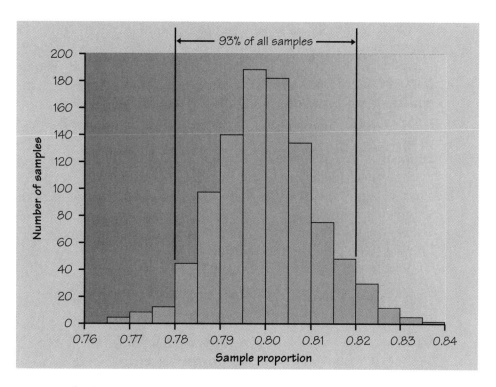

Figure 1-6 The sampling distribution of the sample proportion when the population proportion is $p = 0.80$. The central 93% of the sample results lie within ± 0.02 of the truth about the population.

in 1000 SRSs of size 1373 drawn from a population in which the true population proportion is $p = 0.80$. If we count the outcomes in the eight center bars, we find that 926 of the 1000 samples gave results between 0.78 and 0.82, the interval spanned by these bars. Here is a brief way to say that:

> *93% of all samples give a result within ±0.02 of the truth about the population.*

This is just the kind of statement that opinion polls, the Current Population Survey, and other sample surveys use to tell us how trustworthy their results are.

The claim of Example 11 is based on only 1000 samples, not all samples. Can we really say what would happen if we took samples forever? Yes—because mathematics comes to our rescue.

Drawing an SRS is a lot like dealing cards from a shuffled deck or rolling dice. The mathematics of probability describes what happens if we keep doing it forever. We can't predict the result of one SRS or one roll of the dice. We can, however, say exactly what pattern will emerge if we keep on sampling or keep on rolling dice forever. The real sampling distribution doesn't come from a thousand samples, but from a mathematical calculation that describes all possible samples. Of course, the result of a thousand samples is pretty close to the mathematical ideal. The math just gives us a shortcut. Once a sample survey's statisticians have done the math, we can understand their statements without going through the math ourselves. Look for an answer to the question "What would happen if we took very many samples from the same population?"

UNDERSTANDING THE NEWS

When you read or hear a news report about a sample survey, you are likely to encounter the phrase "margin of error." Here is an example from a newspaper account.[5]

> *As they look toward the year 2000, Americans expect to live better, according to a recent Gallup Poll. Asked if they expected their*

own life to be better by 2000, 77% of the 1234 adults questioned said yes. The poll had a margin of error of plus or minus four percentage points.

That "margin of error of plus or minus four percentage points" describes the accuracy of the poll, but it leaves much unsaid. As connoisseurs of data, we want to know the whole story. Here is what the opinion poll did.

> **EXAMPLE 12. Margin of error.** Gallup asked a random sample of 1234 adults, "Do you expect that the overall quality of your own life will be better by the year 2000?" The *parameter* of interest is the proportion p of all adult residents of the U.S. who expect life to be better in the future. In the sample, 950 answered "Yes." The *statistic* used to estimate p is the sample proportion
>
> $$\hat{p} = \frac{950}{1234} = 0.77$$
>
> Gallup's statisticians calculated the sampling distribution of $\hat{p}$. They then described the trustworthiness of the statistic as follows:
>
> *In 95% of all possible samples, the statistic $\hat{p}$ will take a value within ± 0.04 of the parameter p.*
>
> That is the exact statement that lies behind the news report's claim that "The poll had a margin of error of plus or minus four percentage points." Here is another form of the same statement:
>
> *With 95% confidence, the proportion of all adults who expect a better life by the year 2000 lies in the range*
>
> $$statistic \pm margin\ of\ error$$
> $$0.77 \pm 0.04$$
>
> "95% confidence" is a short way of saying "We used a method that gives correct results for 95% of all possible samples."

Gallup's statement in Example 12 is just like the statement we made in Example 11, and it comes from the same source, the sampling distribution of the statistic. It answers the question "What would happen if we took many samples?" The news report, on the other hand, gives only the *margin of error*, "plus or minus four percentage points." It leaves out the detail that this margin of error is achieved by only 95% of all samples.

CONFIDENCE STATEMENTS

Gallup made a *confidence statement*, and the news reported only one part of that statement.

Confidence statements

A **confidence statement** has two parts: a **margin of error** and a **level of confidence.** The margin of error says how close the sample statistic lies to the population parameter. The level of confidence says what percent of all possible samples satisfy the margin of error.

A confidence statement turns a fact about what would happen if we took all possible samples into a statement about how much we can trust the result of one sample. "95% confidence" means "We used a sampling method that gives a result this close to the truth 95% of the time." That leaves 5% of the samples that miss the truth by more than the margin of error. We don't know whether our sample is one of the 95% that hit or one of the 5% that miss. We can only say that we are 95% confident of a hit. Here are some hints for interpreting confidence statements:

▶ *The conclusion of a confidence statement always applies to the population, not to the sample.* We know exactly how the 1234 people in the sample feel, because Gallup interviewed them. The confidence statement uses the sample result to say something about the population of all adults.

▶ *Our conclusion about the population is never completely certain.* Gallup's sample *might* be one of the 5% that miss by more than four percentage points. If 95% is not good enough, we can demand higher confidence, say 99%. But we can't reach 100% confidence unless the margin of error covers all proportions from 0 to 1, which is useless.

▶ *If we want to be 99% confident, we must accept a larger margin of error than for 95% confidence.* There is a tradeoff between how closely we can pin down the population parameter (the margin of error) and how confident we are that a sample meets the margin of error.

▶ *It is usual to report the margin of error for 95% confidence.* If a news report gives a margin of error but leaves out the confidence level, it's pretty safe to assume 95% confidence.

▶ *Want a smaller margin of error with the same confidence? Take a larger sample.* Remember that the size of the sample controls the precision of the result. You can get as small a margin of error as you want and still have high confidence by paying for a large enough sample. A margin of error of four percentage points is not good enough for estimating the unemployment rate, for example. That's why the Current Population Survey takes a sample of 60,000 rather than 1234.

DETAILS FOR A NATIONAL POLL

Table 1-1 gives some details about the margin of error in samples taken by the Gallup poll. Gallup's sampling methods are typical of nationwide sample surveys, so we can use this table to understand how most national samples behave. All the margins of error in the table are for 95% confidence.

> EXAMPLE 13. **Gallup poll margin of error.** The Gallup poll interviews 1514 adults and finds that 53% of them oppose a longer school year. What confidence statement can we make?
>
> The sample size 1514 is close to 1500, so we use the table entries for a sample of size 1500. The table tells us that this margin of error is never greater than 3 percentage points ($\pm 3\%$). We are 95% confident that between 50% (that's $53 - 3$) and 56% (that's $53 + 3$) of adult Americans oppose lengthening the school year.

TABLE 1-1   Precision of the sampling procedure used by the Gallup Poll
as of 1972*

Population percentage	Sample size						
	100	200	400	600	750	1000	1500
Near 10	7	5	4	3	3	2	2
Near 20	9	7	5	4	4	3	2
Near 30	10	8	6	4	4	4	3
Near 40	11	8	6	5	4	4	3
Near 50	11	8	6	5	4	4	3
Near 60	11	8	6	5	4	4	3
Near 70	10	8	6	4	4	4	3
Near 80	9	7	5	4	4	3	2
Near 90	7	5	4	3	3	2	2

SOURCE: George Gallup, *The Sophisticated Poll Watcher's Guide* (Princeton Opinion Press, 1972), p. 228.

*The table shows the range, plus or minus, within which the sample percentage p falls in 95% of all samples. This margin of error depends on the size of the sample and on the population percentage p. For example, when p is near 60%, 95% of all samples of size 1000 will have p between 56% and 64%, because the margin of error is ±4%.

You can see from Table 1-1 that the margin of error does get smaller for larger samples. You can also see something else: the margin of error depends a bit on the parameter p, the "population percentage" in the table. Of course, we don't know the value of p. To be safe, just use the largest margin of error given in the column for your sample size. The margin of error is largest when the population is close to being evenly divided (population percentage "Near 50"). Moreover, it doesn't change much unless the true p is quite far from one-half. For example, a sample of 1000 persons gives a margin of error of ±4% when p is anywhere between 30% and 70%. Little is lost by always using ±4% as the margin of error.

Remember that Table 1-1 is for a complex national sample. The margins of error for simple random samples are a bit smaller than those in Table 1-1. The details are in Chapter 8, should you actually have to produce confidence statements. But you are already a sophisticated consumer.

SECTION 4 EXERCISES

1.30 **Who should get welfare?** A news article on a Gallup poll noted that "28 percent of the 1548 adults questioned felt that those who were able to work should be taken off welfare." The article also said, "The margin of error for a sample size of 1548 is plus or minus three percentage points." Explain to someone who knows no statistics what "margin of error plus or minus three percentage points" means.

1.31 **Is there a hell?** A news article reports that in a recent Gallup poll, 78% of the sample of 1108 adults said they believe there is a heaven. Only 60% said they believe there is a hell. The news article ends, "The poll's margin of sampling error was plus or minus four percentage points." Can we be certain that between 56% and 64% of all adults believe there is a hell? Explain your answer.

1.32 A national opinion poll asks a sample of 1500 adults whether they approve of the president's overall performance. Of these, 675 say "Yes." The news report says that the poll has a margin of error of plus or minus three percentage points.

(a) Make a complete confidence statement about the percent of all adults who approve the president's performance.

(b) Are you *certain* that the true percent for all adults lies inside the interval you found in (a)? Explain why or why not.

1.33 **Balancing the budget.** A poll of 1190 adults finds that 702 prefer balancing the budget over cutting taxes. The announced margin of error for this result is plus or minus four percentage points. The news report does not give the confidence level. You can be quite sure that the confidence level is 95%.

(a) What is the value of the sample proportion $\hat{p}$ who prefer balancing the budget? Explain in words what the population parameter p is in this setting.

(b) Make a confidence statement about the parameter p.

(c) A member of Congress thinks that 95% confidence is not enough. He wants to be 99% confident. How would the margin of error of a 99% confidence interval based on the same sample compare with the margin of error for 95% confidence?

(d)   Another member of Congress is satisfied with 95% confidence, but she wants a smaller margin of error than plus or minus four percentage points. How can we get a smaller margin of error, still with 95% confidence?

1.34   **Predicting an election.** A television commentator, talking about the 1988 presidential election, said:

> *The final opinion poll before the election showed that 53% would vote for Bush. The poll had a margin of error of only ±2%, so it was certain that Bush would win.*

Why was it *not* certain that Bush would win?

1.35   **Fear of crime.** The Gallup poll asked a random sample of 1493 adults, "Are you afraid to go out at night within a mile of your home because of crime?" Of the sample, 672 said "Yes." Use Table 1-1 to make a confidence statement about the percent of all adult Americans who fear to go out at night because of crime.

1.36   **Prayer in the schools?** A random sample of 1028 adults, chosen by the Gallup poll's sampling method, is asked "Do you favor an amendment to the Constitution that would permit organized prayer in public schools?" Of the sample, 678 say "Yes." Use Table 1-1 to make a confidence statement about the percent of all adults who favor a school prayer amendment.

1.37   **Will the future be better?** A national poll of 1433 adults finds that 46% feel that the country's future will be better than the present. No margin of error is given in the news account you read. If the poll used a method much like Gallup's (most do), what confidence statement can you make based on the information given?

1.38   **Safe on the streets?** A news report says, "The latest Harris survey on crime in America indicates that 26% of Americans feel less safe on the streets they they did a year ago." Can you make a confidence statement about this result? If so, make one. If not, explain why you cannot.

1.39   The final Gallup poll prior to a presidential election interviews 3509 people. This is a larger sample size than appears in Table 1-1. Is the margin of error of the result more or less than three percentage points? Why?

1.40   **The Current Population Survey.** Though opinion polls usually make 95% confidence statements, some sample surveys use other confidence levels. The monthly unemployment rate, for example, is based on the Current Population Survey of about 60,000 households. The margin of error in the unemployment rate is announced as about two-tenths of one percentage point with 90% confidence. Would you expect the margin of error for 95% confidence to be smaller or larger? Why?

▶ 5 THE PRACTICAL SIDE OF SAMPLING

The conclusions of the last two sections seem too good to be true. If random sampling eliminates bias and allows control of precision, why do we continue to read debates about the trustworthiness of pre-election polls? Alas, sampling in the real world is more complex and less reliable than drawing an SRS from a list of names in a textbook exercise. Confidence statements do not reflect all of the sources of error that are present in practical sampling.

Errors in sampling

Sampling errors are errors caused by the act of taking a sample. They cause sample results to be different from the results of a census.

Random sampling error is the deviation between the sample statistic and the population parameter caused by chance in selecting a random sample. The margin of error in a confidence statement includes *only* random sampling error.

Nonsampling errors are errors not related to the act of selecting a sample from the population. They can be present even in a census.

Most sample surveys are afflicted by errors other than random sampling errors. These errors can introduce bias that makes a confidence statement meaningless. Good sampling technique involves the art of reducing all sources of error. Part of this art is the science of statistics, with its random samples and confidence statements. In practice, however, good statistics isn't all there is to good sampling. Let's look at sources of errors in sample surveys and at how samplers combat them.

SAMPLING ERRORS

Random sampling error is one kind of sampling error. We can control random sampling error by choosing the size of our random sample. Another source of sampling error is the use of bad sampling methods, such as voluntary response. We can avoid bad methods. Other sampling errors are not so easy to handle. Any sample begins with a list of all the members of the population. This list is the *sampling frame*. It is often hard to obtain a frame that really does list the entire population. If the frame leaves out certain classes of people, even random samples from that frame will be biased.

> EXAMPLE 14. **Telephone sampling.** Most sample surveys are conducted by telephone rather than by face-to-face interviews. Using the telephone directory as a sampling frame would introduce strong bias. Unlisted telephones outnumber listed telephones in many large cities, so dialing only listed numbers would underrepresent city dwellers. Telephone polls therefore use *random digit dialing* (RDD). A survey organization chooses a sample from a list of all area codes and prefixes (the first three digits of a telephone number). RDD equipment then chooses the last four digits at random. The survey workers must face the fact that more than three-fourths of all possible telephone numbers are not assigned to households, as well as the fact that an increasing number of households have two telephone lines and so appear twice in the sampling frame. Nonetheless, RDD comes close to allowing random samples of all households with phones.
>
> However, about 6% of U.S. households have no telephone. These households do not appear in the frame for any telephone

survey. Does this omission bias survey results? Alas, yes. The percent without phones in the South is almost double that in other parts of the country. People living alone are less likely than other people to have a phone. So southerners and single-person households are underrepresented in telephone surveys. What is more, many telephone surveys omit Alaska and Hawaii to save expense. Because most people have phones, the bias in most telephone surveys is not large. Some bias is present, however, and the margin of error announced by the survey does not include this bias.

Good telephone surveys plan many callbacks to the chosen telephone numbers, because people who are easy to reach with one call differ from those who are hard to reach. It appears, for example, that Democrats are easier to reach than Republicans. One survey during the 1984 presidential campaign gave the Republican candidate Ronald Reagan a three-percentage-point lead over the Democrat Walter Mondale after one call. Reagan's lead grew to 13 percentage points after many callbacks contacted almost the entire original sample. If the survey wants individual opinions, it must also deal with the fact that most people who answer the phone are women. The New York Times/CBS News Poll reports that only 37% of the people reached by their telephone surveys on the first call to a household are men. To balance the responses, telephone polls often choose an adult from the household at random when the call is answered.[6]

NONSAMPLING ERRORS

Nonsampling errors are those that can plague even a census. They include **processing errors**, mistakes in mechanical tasks such as doing arithmetic or entering responses into a computer. The spread of computers has made processing errors less common than in the past.

EXAMPLE 15. Computer-assisted interviewing. The days of the interviewer with a clipboard are past. Contemporary interviewers carry a laptop computer for face-to-face interviews or watch a computer screen as they carry out a telephone interview. Computer software manages the interview. The interviewer reads questions from the computer screen and uses the keyboard to enter the responses. The computer skips irrelevant items—once a respondent says that she has no children, further questions about her children never appear. The computer can check that answers to related questions are consistent with each other. It can even ask questions in random order to avoid any bias due to always asking questions in the same order.

Computer software also manages the sampling process. It keeps records of who has responded and prepares a file of data from the responses. The tedious process of transferring responses from paper to computer, once a source of processing errors, has disappeared. The computer even schedules the calls in telephone surveys, taking account of the respondent's time zone and honoring appointments made by people who were willing to respond but did not have time when first called.

Processing errors can be nearly eliminated by sufficient care. This is not the case with *nonresponse,* the most serious source of nonsampling errors.

Nonresponse

Nonresponse is the failure to obtain data from a unit selected for a sample. The most common reasons for nonresponse in surveys of human populations are failure to contact a subject and the subject's refusal to cooperate.

Nonresponse is now the most serious problem facing sample surveys. People are increasingly reluctant to answer questions, particularly over the phone. Answering machines reduce the response rate for telephone surveys. Buildings guarded by doormen hinder face-to-face interviews. Nonresponse can certainly bias sample survey results, because different groups have very different rates of nonresponse. Poor people are harder to contact than the middle class. Refusals are higher in large cities and among the elderly. Bias due to nonresponse can easily overwhelm the random sampling error described by a confidence statement. Letters sent in advance improve the response rate of telephone surveys, but nonresponse is a problem that no amount of expertise can fully overcome. Even the U.S. census, with the resources of the government behind it, is afflicted by nonresponse.

> **EXAMPLE 16. The census undercount.** Every 10 years, the U.S. census mails questionnaires to all housing units on its master list of addresses. In 1980, 75% returned the census form; in 1990, the mail-back response rate dropped to 65%. In New York City, only 53% of the forms were returned. If nonresponse is not to cause severe bias, any survey must plan elaborate and expensive *follow-up*. The Census Bureau made six personal visits to each address that did not mail back the census form. In the 1990 census, this follow-up still failed to reach almost 20% of the no-mailback households in central cities. As a last resort, the Census Bureau tries to get basic information from neighbors. Nonetheless, the 1990 census missed about 1.8% of all the country's people—including about 3.8% of the residents of Los Angeles and about 4.6% of all blacks. Because census results decide the distribution of much federal aid as well as seats in Congress, this undercount led to lawsuits and much deep thinking about how to do better in 2000. One suggestion is to aim more intense and quicker follow-up at a sample of the households that don't mail back the census form rather than trying to contact all such households. Recall that careful samples can in fact give more accurate results than a census.[7]

Another type of nonsampling error is **response error**, which occurs when a subject gives an incorrect response. A subject may lie about her age or income. Or he may remember incorrectly when asked how many packs of cigarettes he smoked last week. Or a subject who does not understand a question may guess at an answer rather than show ignorance. Questions that ask subjects about their behavior during a fixed time period are particularly prone to response errors due to faulty memory. Most people cannot remember how many times they have visited a doctor in the past year, for example. That is a question on the National Health Survey—yet checking health records found that people fail to remember 60 percent of their visits to a doctor. Questions that involve sensitive issues are also subject to response errors, as the next example illustrates.

EXAMPLE 17. The effect of race. In 1989, New York City elected its first black mayor and the state of Virginia elected its first black governor. In both cases, samples of voters interviewed as they left their polling places predicted larger margins of victory than the official vote counts. The polling organizations were certain that some voters lied when interviewed because they felt uncomfortable admitting that they had voted against the black candidate.

WORDING THE QUESTIONS

A final influence on the results of a sample survey is the exact wording of questions. It is surprisingly difficult to word questions that are completely clear. A survey that asked about "ownership of stock" found

"Do I own any stock, Ma'am? Why, I've got 10,000 head out there."

that most Texas ranchers owned stock, though probably not the kind traded on the New York Stock Exchange. What is more, small changes in the wording of questions can significantly change the responses. Only 13% of adults think we are spending too much on "assistance to the poor," but 44% think we are spending too much on "welfare."[8]

The wording of questions always influences the answers. If the questions are slanted to favor one response over others, we have another source of nonsampling error. A favorite trick is to ask if the subject favors some policy as a means to a desirable end: "Do you favor banning private ownership of handguns in order to reduce the rate of violent crime?" and "Do you favor imposing the death penalty in order to reduce the rate of violent crime?" are loaded questions that draw positive responses from people who are worried about crime. Here is an example of the influence of a slanted question.

> **EXAMPLE 18. Campaign finance.** The financing of political campaigns is a perennial issue. Here are two opinion poll questions on this issue.
>
> *Should laws be passed to eliminate all possibilities of special interests giving huge sums of money to candidates?*
>
> *Should laws be passed to prohibit interest groups from contributing to campaigns, or do groups have a right to contribute to the candidate they support?*
>
> The first question was posed by Ross Perot, the third-party candidate in the 1992 presidential election. It drew a 99% "Yes" response in a mail-in poll. We know that voluntary response polls are worthless, so the Yankelovitch survey firm asked the same question of a nationwide random sample. The result: 80% "Yes." Perot's question almost demands a "Yes" answer, so Yankelovitch wrote the second question as a more neutral way of presenting the issue. Only 40% of a nationwide random sample wanted to prohibit contributions when asked this question.[9]

The use of slanted questions and voluntary response samples by various groups is growing. You should just ignore the results. Pollsters who want honest information test their questions with small groups before taking the big sample. They also work with psychologists, whose understanding of human behavior can help write questions that are clear and neutral.

QUESTIONS TO ASK BEFORE YOU BELIEVE A POLL

Opinion polls and other sample surveys can produce accurate and useful information if the pollster uses good statistical techniques and also works hard at preparing a sampling frame, wording questions, and reducing nonresponse. Many surveys, however, especially those designed to influence public opinion rather than just record it, do not produce accurate or useful information. Here are some questions to ask before you pay much attention to poll results.

▶ *Who carried out the survey?* Even a political party should hire a professional sample survey firm whose reputation demands that they follow good survey practices.

▶ *What was the population?* That is, whose opinions were being sought?

▶ *How was the sample selected?* Look for mention of random sampling.

▶ *How large was the sample?* It is even better to give a measure of precision, such as the *margin of error* within which the results of 95% of all samples drawn as this one was would fall.

▶ *What was the response rate?* That is, what percent of the original subjects actually provided information?

▶ *How were the subjects contacted?* By telephone? Mail? Face-to-face interview?

▶ *When was the survey conducted?* Was it just after some event which might have influenced opinion?

▶ *What were the exact questions asked?*

Major opinion polls, academic survey centers, and government statistical offices answer these questions when they announce the results of a sample survey. Editors and newscasters have the bad habit of cutting out these dull facts and reporting only the sample results. Many sample surveys by interest groups, newspapers, and TV stations don't answer these questions because their polling methods are in fact unreliable. If a politician, an advertiser, or your local TV station announces the results of a poll without complete information, be skeptical.

SECTION 5 EXERCISES

1.41   Which of the following are sources of *sampling error* and which are sources of *nonsampling error?* Explain your answers.

(a)   The subject lies when questioned.

(b)   A typing error is made in recording the data.

(c)   Data are gathered by asking people to mail in a coupon printed in a newspaper.

1.42   Each of the following is a source of error in a sample survey. Label each as *sampling error* or *nonsampling error*, and explain your answers.

(a)   The telephone directory is used as a sampling frame.

(b)   The subject cannot be contacted in three calls.

(c)   Interviewers choose people on the street to interview.

1.43   **A newspaper's opinion poll.** The *New York Times* includes a box entitled "How the poll was conducted" in news articles about its own opinion polls. Here are quotations from one such box (March 26, 1995) and some questions about them.

(a)   "The latest New York Times/CBS News poll is based on telephone interviews conducted March 9 through 12 with 1,156 adults around the United States, excluding Alaska and Hawaii." The box then describes random digit dialing, the method used to select the sample. What sources of sampling error are present in this poll?

(b)   "In addition to sampling error, the practical difficulties of conducting any survey of public opinion may introduce other sources of error into the poll. Variations in question wording or in the order of questions, for instance, can lead to somewhat different results." Now we are talking about nonsampling errors. What possible sources of nonsampling error, other than those mentioned by the *Times*, may influence this poll?

(c)   The box also announced a margin of error of plus or minus three percent, with 95% confidence. Which of the sources of error you listed in (a) and (b) are included in this margin of error?

1.44   **Crime data.** How bad is crime? It depends on which data you look at. The annual FBI publication *Crime in the United States* says that there were 93,825 forcible rapes in 1992. The FBI data list all crimes that are reported to law enforcement agencies and then reported by local law enforcement agencies to the FBI. The other main source of crime data is the National Crime Victimization Survey, a nationwide random sample of about 50,000 households. The survey asks people if they have been victims of a crime. According to the survey, there were $141,000 \pm 6000$ rapes in 1992. These two sets of data appear almost side by side in the *Statistical Abstract of the United States*.

(a)   Why does the National Crime Victimization Survey report many more rapes than the FBI?

(b)   No margin of sampling error can be attached to the FBI data. Why not?

(c)   Each source of data may be subject to nonsampling errors. What are some important sources of nonsampling error for the FBI report? For the National Crime Victimization Survey?

1.45   **Are we turning inward?** On the next page is part of a newspaper report of a public opinion poll, reprinted from the *New York Times* of September 21, 1994. At the end of Section 5 are some questions that a complete account of a sample survey should answer. Which of these questions does this newspaper report answer, and which does it not? Give the answers whenever the article contains them.

1.46   Section 5 mentions a number of things that well-run sample surveys do in order to reduce various sources of error. Briefly list as many of these procedures as you can find in the text. For

U.S. Voters Focus on Selves, Poll Says

By RICHARD L. BERKE

WASHINGTON, Sept. 20—As they turn inward and worry about their own financial difficulties, American voters have become less compassionate about the problems of the poor and minorities, a study of public opinion shows.

The finding was part of a wide-ranging study, conducted by the Times Mirror Center for the People and the Press and made public today, that found an electorate that is angry, self-absorbed and politically unanchored.

For the first time in the seven years of Times Mirror surveys, a majority of whites, 51 percent, say they agree that equal rights have been pushed too far; in 1992, only 42 percent shared that view.

The poll also found a striking decline in public support for social welfare programs. Fifty-seven percent said it was the responsibility of government to take care of people who cannot take care of themselves, down from 69 percent in 1992 and 71 percent in 1987.

There were some demonstrations of tolerance for minorities, people with AIDS and homosexuals, so long as people's pocketbooks were not affected. In the biggest shift, 65 percent agreed that is is all right for blacks and whites to date, a 22-point increase over the figure for 1987.

The Times Mirror Center has conducted a series of major surveys since 1987 to analyze public views in more depth than many public opinion surveys. For this survey, 3,800 adults nationwide were interviewed by telephone from July 12 to July 25. The poll's margin of sampling error is plus or minus two percentage points.

example, "Follow-up—Try several times to contact each member of the sample, rather than giving up after only one attempt."

1.47   **Internet users.** A survey of users of the Internet found that males outnumbered females by nearly 2 to 1. This was a surprise, because earlier surveys had put the ratio of men to women closer to 9 to 1. Later in the article we find this information:[10]

Detailed surveys were sent to more than 13,000 organizations on the Internet; 1,468 usable responses were received. According to Mr. Quarterman, the margin of error is 2.8 percent, with a confidence level of 95 percent.

(a)   What was the *response rate* for this survey? (The response rate is the percent of the planned sample that responded.)

(b)   Do you think that the small margin of error is a good measure of the accuracy of the survey's results? Explain your answer.

1.48   **Assisted suicide?** In 1995, *USA WEEKEND* printed a box like that below. Fine print said that a call would cost 50 cents. Do you consider the results of this opinion poll trustworthy? Explain your answer.

> ## VOTE NOW
>
> If you were terminally ill, would you want the right to end your life with a doctor's help?
>
> YES: 1-900-255-2257   NO: 1-900-255-2258

1.49   **TV ratings.** The method of collecting the data can influence the accuracy of sample results. The following methods have been used to collect data on television viewing in a sample household:

(a)   *The diary method.* The household keeps a diary of all programs watched and who watched them for a week, then mails in the diary at the end of the week.

(b)   *The roster-recall method.* An interviewer shows the subject a list of programs for the preceding week and asks which programs were watched.

(c)   *The telephone-coincidental method.* The survey firm telephones the household at a specific time and asks if the television is on, which program is being watched, and who is watching it.

(d)   *The automatic recorder method.* A device attached to the set records what hours the set is on and to which channel it is tuned. At the end of the week, this record is removed from the recorder.

(e)   *The people meter.* Each member of the household is assigned a numbered button on a hand-held remote control. Everyone is asked to push their button whenever they start or stop watching TV. The remote control signals a device attached to the set that keeps track of what channel the set is tuned to and who is watching at all times.

Discuss the advantages and disadvantages of each of these methods, especially the possible sources of error associated with each method. The Nielsen national ratings use Method (e). Local ratings (there are more than 200 local television markets) use Method (a). Do you agree with these choices? (Do not discuss choosing the sample, just collecting data once the sample is chosen.)

1.50 **Design your own sample survey.** You wish to determine whether students at your school think that faculty are sufficiently available to students outside the classroom. You will select an SRS of 200 students.

(a) What is the exact population? (Will you include part-time students? Graduate students?)

(b) How will you obtain a sampling frame?

(c) How will you contact subjects? (Is door-to-door interviewing allowed in campus residence halls?) What follow-up will you do when you cannot contact a subject?

(d) What specific question or questions will you ask?

1.51 Comment on each of the following as a potential sample survey question. Are any unclear or slanted?

(a) Has the amount of time you spend watching television increased, decreased, or stayed about the same over the past few years?

(b) Which of these best represents your opinion on gun control?
  (1) The government should confiscate our guns.
  (2) We have the right to keep and bear arms.

(c) In view of escalating environmental degradation and predictions of serious resource depletion, would you favor economic incentives for recycling resource-intensive consumer goods?

1.52 **Doubting the Holocaust?** An opinion poll conducted in 1992 for the American Jewish Committee asked:

Does it seem possible or does it seem impossible to you that the Nazi extermination of the Jews never happened?

When 22% of the sample said "possible," the news media wondered how so many Americans could be uncertain that the Holocaust happened. A second poll therefore restated the question:

Does it seem possible to you that the Nazi extermination of the Jews never happened, or do you feel certain that it happened?

Now only 1% of the sample said "possible."[11] Explain clearly why the two questions produced such different responses. Which response is more trustworthy?

1.53   **Closed versus open questions.** Two basic types of questions are closed questions and open questions. A closed question asks the subject for one or more of a fixed set of responses. An open question allows the subject to answer in his or her own words. The interviewer writes down the responses and sorts them later. For example, here are an open and closed question on the same issue, both asked by the Gallup poll within a few years of each other:[12]

Open: In recent years there has been a sharp increase in the nation's crime rate. What steps do you think should be taken to reduce crime?

Closed: Which two or three of the approaches listed on this card do you think would be the best ways to reduce crime?

   Cleaning up social and economic conditions in our slums and ghettos that tend to breed drug addicts and criminals.

   Putting more policemen on the job to prevent crimes and arrest more criminals.

   Getting parents to exert stricter discipline over their children.

   Improving conditions in our jails and prisons so that more people convicted of crimes will be rehabilitated and not go back to a life of crime.

Really cracking down on criminals by giving them longer prison terms to be served under the toughest possible conditions.

Reforming our courts so that persons charged with crimes can get fairer and speedier justice.

What are the advantages and disadvantages of open and closed questions? Use the example just given in your discussion.

1.54   In the 1992 presidential election, about 68% of the voting-age population registered to vote and about 61% actually voted. Just after the election, you ask a random sample of adults whether or not they voted. Do you expect bias? That is, do you expect about 61%, less than 61%, or more than 61% of the sample to claim that they voted? Why?

1.55   Many subjects are reluctant to give honest answers to questions about activities that are illegal or sensitive in some other way. One study divided a large group of white adults into thirds at random. All were asked if they had ever used cocaine. The first group was interviewed by telephone: 21% said yes. In the group visited at home by an interviewer, 25% said yes. The final group was interviewed at home, but answered the question on an anonymous form that they sealed in an envelope. Of this group, 28% said they had used cocaine.[13]

(a)   Which result do you think is closest to the truth? Why?

(b)   List three other activities that you think would be under-reported in a telephone survey.

▶ 6 MORE ON SAMPLING DESIGN

A reliable sample survey depends both on statistical ideas (random sampling) and on practical skills (following up, wording questions, skillful interviewing). When our goal is to sample a large human population, using an SRS is good statistics but too expensive to be good practice. First, a complete sampling frame is rarely available. Second, contacting a national SRS is too expensive. If the Bureau of Labor Statistics chose an SRS of households for the Current Population

Survey, it would be have to send interviewers off to Beetle, Kentucky and Searchlight, Nevada to find the addresses chosen. Even telephone interviewing is easier and cheaper when the numbers to be called are clustered in a few exchanges. The solution to these practical difficulties is to use a sampling design more complicated than an SRS and to start by sampling "not people but the map."[14] National sample surveys are almost always carried out in stages.

MULTISTAGE SAMPLES

The design of a sample survey that plans face-to-face interviews usually goes somewhat as follows:

Stage 1: Choose a sample from the 3141 counties in the United States. It is easy to get a list of counties to serve as the frame.

Stage 2: Select a sample of townships or wards in each of the counties chosen in Stage 1. A frame is again easy to find.

Stage 3: Using a map or an aerial photograph as the sampling frame, choose a sample of small areas (such as city blocks) within each ward from Stage 2.

Stage 4: Finally, choose a sample of households from each block in the Stage 3 sample. If the list of addresses in a block is not complete, the interviewers can fill in the gaps on the spot.

Such a **multistage sampling design** overcomes the practical drawbacks of an SRS. We don't need a list of all the residential addresses in the country, only address lists for the small areas arrived at by sampling counties, then wards, then blocks within wards. Moreover, all the addresses in the sample are clustered within these few small areas, making it much cheaper to collect the data. Telephone surveys also use multistage samples, with area codes and the prefixes (first three digits) of telephone numbers marking the first two stages. Random digit dialing then chooses the last four digits at random.

The sample selected at each stage of a multistage design may be an SRS, but may also be another type of random sample. For example, addresses within a block can be selected by a **systematic random sample.** Choose a starting point at random, then take every third or tenth unit in geographical or numerical order. See Exercise 1.62 for details. Systematic samples are fast, require no frame, and make geographical spread certain if the units are in geographical order. But there are pitfalls. A colleague of mine, working as an interviewer, was once told to visit every third address in a Chicago neighborhood. The block contained three-story walk-up apartment buildings, so all the addresses he visited were on the third floor. These households were poorer than those on the lower floors.

STRATIFIED SAMPLES

A multistage sample is a good idea, but an SRS of counties in the first stage is a terrible idea. Most counties are rural areas with few people. An SRS of counties could easily miss most of the nation's large cities. The first stage in a multistage sample usually selects a *stratified sample* of counties or telephone exchanges. Here is the idea.

Stratified sample

To choose a **stratified random sample:**

Step 1. Divide the sampling frame into distinct groups of units, called **strata.** Choose the strata because you have a special interest in these groups within the population or because the units in each stratum resemble each other.

Step 2. Take a separate SRS in each stratum and combine these to make up the complete sample.

We must of course choose the strata using facts about the population that are known before we take the sample. Counties may be stratified by number of residents and by whether they are primarily urban, suburban, or rural. The Current Population Survey stratifies geographical areas by population and includes *all* of the most populous areas in the first stage of its multistage sample. This is a stratified sample in which the sample in one stratum is a census.

Stratified samples have two advantages over an SRS:

▶ By taking a separate SRS in each stratum, a stratified sample allows us to gather separate information about each stratum.

▶ If the units in each stratum are more alike in the variable measured than is the population as a whole, estimates from a stratified sample are more precise than those from an SRS of the same size.

To understand the second advantage, think about the extreme case in which all the units in each stratum are exactly alike. Then a stratified sample of only one unit from each stratum would completely describe the population, but an SRS of the same size would have very low precision.

It may surprise you to notice that stratified samples can violate one of the most appealing properties of the SRS—stratified samples need not give all units in the population the same chance to be chosen. Some strata may be deliberately overrepresented in the sample.

EXAMPLE 19. Stratifying a sample by race. A large university has 30,000 students, of whom 3000 are black. An SRS of 500 students gives every student the same chance to be in the sample. That chance is

$$\frac{500}{30,000} = \frac{1}{60}$$

We expect an SRS to contain only about 50 blacks—because 10% of the students are black, we expect about 10% of an SRS to be black. A sample of size 50 isn't large enough to estimate black student opinion with reasonable precision. We might prefer a stratified random sample of 200 blacks and 300 other students.

You know how to select such a stratified sample. Label the black students 0001 to 3000 and use Table A to select an SRS of 200. Then label the other students 00001 to 27000 and use Table A a second time to select an SRS of 300 of them. These two SRSs together form the stratified sample.

In the stratified sample, each black student has chance

$$\frac{200}{3000} = \frac{1}{15}$$

to be chosen. Each of the other students has a smaller chance,

$$\frac{300}{27,000} = \frac{1}{90}$$

Because we have two SRSs, it is easy to estimate black and other student opinions separately. To use the entire stratified sample to estimate the opinion of all students together, more care is needed. To avoid bias, the samplers must correct for the deliberate overrepresentation of blacks in the sample. That isn't hard to do, because they know what chance each student had, even though the chances are not all the same.

PROBABILITY SAMPLES

Simple, systematic, and stratified random samples all use chance to select units from the population, as do multistage samples constructed from these building blocks. All fit the general statistical framework of *probability sampling.*

> **Probability sample**
>
> A **probability sample** is a sample chosen in such a way that we know what samples are possible (not all need be possible) and what chance, or probability, each possible sample has (not all need be equally probable).

We can leave the details of complex sampling designs to experts. (Strike a blow against poverty—hire a statistician.) What matters to users is that estimates from any probability sample share the nice properties of estimates from an SRS. The sampling distribution is known, and confidence statements can be made without bias and with increasing precision as the size of the sample increases. Nonprobability samples such as voluntary response samples do not share these advantages and cannot give trustworthy information about a population.

DESIGNING A SAMPLE SURVEY

It may be that you will never have the good fortune to participate in the design of a sample survey. But if you work in advertising, politics, or marketing, or if you use government economic and social data, you will surely have to use the results of surveys. We can summarize our study of sampling in an outline of the steps in designing a sample survey.

Step 1. Determine the population, both its extent and the basic unit. If you are interested in buyers of new cars, your unit could be new-car registrations, new-car owners (individuals), or households that have purchased new cars. You must also be specific about the geographic area and date of purchase needed to qualify a unit for this population.

Step 2. Specify the variables you will measure, and prepare the questionnaire or other instrument you will use to measure them. You will want to test your questionnaire on a pilot group of subjects to be certain it is clear and complete.

Step 3. Set up the sampling frame. This is related to Step 1. If you use a list of new-car registrations as a sampling frame because this list is easy to obtain, households that bought several new cars will appear several times on the list. The president of General Motors, to express his confidence in the industry in a period of poor sales, once bought *five* new GM cars in one year. His household appears five times in that year's list of new-car registrations. You see that an SRS of registrations is not an SRS of households.

Step 4. Do the statistical design of the sample. The design says how large the sample is and how to choose it. Multiple stages and stratification, for example, may be part of the design. You will often want to consult a statistician for expert advice.

Step 5. Attend to details, such as training interviewers and arranging the timing of the survey.

Much more might be said about each of these steps. A good deal is known about how to word questions, how to train interviewers, how to increase response in a mail survey, and so on. Much of this is interesting, and some is slightly amusing. For example, colorful commemorative stamps on the outer and return envelopes increase the response rate in a mail survey. In any case, a sample survey would show that you already know more about sampling than 99.9% of U.S. residents age 18 or over. Enough is enough.

SECTION 6 EXERCISES

1.56   A student at a large university wants to ask the opinion of students about whether the university should advertise itself primarily as a technical school. Students in the School of Engineering and the School of Liberal Arts will probably differ on this issue. How should this fact influence the design of the sample?

1.57   About 20% of the engineering students at a large university are women. The school plans to poll a sample of 200 engineering students about the quality of student life.

(a)   If an SRS of size 200 is selected, about how many women do you expect to find in the sample?

(b)   If the poll wants to be able to report separately the opinions of male and female students, what type of sampling design would you suggest? Why?

1.58   A university employs 2000 male and 500 female faculty members. The equal employment opportunity officer polls a stratified random sample of 200 male and 200 female faculty members.

(a)   What is the chance that a particular female faculty member will be polled?

(b)   What is the chance that a particular male faculty member will be polled?

(c)   Explain why this is a probability sample.

(d)   Each member of the sample is asked, "In your opinion, are female faculty members in general paid less than males with similar positions and qualifications?"

180 of the 200 females (90%) say "Yes."

60 of the 200 males (30%) say "Yes."

In all, 240 of the sample of 400 (60%) answered "Yes." The officer therefore reports that "Based on a sample, we can conclude that 60% of the total faculty feel that female members are underpaid relative to males." Explain why this conclusion is wrong.

(e)   If we took a stratified random sample of 200 male and 50 female faculty members at this university, each member of the faculty would have the same chance of being chosen. What is that chance? Explain why this sample is *not* an SRS.

1.59 A club has as members 25 students, named:

Abel	Fisher	Huber	Moran	Reinmann
Carson	Golomb	Jimenez	Moskowitz	Silvers
Cryer	Griswold	Jones	Neyman	Sobar
David	Hein	Kiefer	Ong	Thompson
Elashoff	Holland	Lamb	Perez	Valenzuela

and 10 faculty members, named:

Andrews	Fischang	Hernandez	Moore	Rabinowitz
Besicovitch	Gupta	Lightman	Phillips	Wu

The club can send four students and two faculty members to a convention. It decides to choose those who will go by random selection. Use Table A to choose a stratified random sample of four students and two faculty members.

1.60 You have alphabetized lists of the 2000 male faculty and of the 500 female faculty at the university described in Exercise 1.58. Explain how you would assign labels and use Table A to choose a stratified random sample of 200 female and 200 male faculty members. What are the labels of the first 5 males and the first 5 females in your sample?

1.61 You want to interview 10 male scholarship athletes at length about their attitude toward school and their future plans. Because you believe there may be large differences among sports, you decide to interview a stratified random sample of 7 basketball players and 3 golfers. Use the table of random digits, beginning at line 101, to select your sample from the team rosters below.

BASKETBALL

Abdul-Jabbar	Ewing	Robertson
Aguirre	Frazier	Robinson
Baylor	Johnson	Russell
Bird	Jordan	Stockton
Chamberlain	Malone	West
Cousy	Miller	Worthy
Erving	O'Neal	

GOLF

Ballesteros	Kite	Palmer
Faldo	Nicklaus	Snead
Floyd	Norman	Watson

1.62 Systematic random sample. The final stage in a multistage sample must choose five of the 500 addresses in a neighborhood. You have a list of the 500 addresses in geographical order. To choose a systematic random sample, proceed as follows:

Step 1. Choose one of the first 100 addresses on the list at random. (Label them 00, 01, ..., 99 and use a pair of digits from Table A to make the choice.)

Step 2. The sample consists of the address from Step 1 and the addresses 100, 200, 300, and 400 positions down the list from it.

If 71 is chosen at random in Step 1, for example, the systematic random sample consists of the addresses numbered 71, 171, 271, 371, and 471.

(a) Use Table A to choose a systematic random sample of 5 from a list of 500 addresses. Enter the table at line 130.

(b) What is the chance that any specific address will be chosen? Explain your answer.

(c) Explain why this sample is *not* an SRS.

1.63 Sampling library books. A group of librarians wants to estimate what fraction of books in large libraries falls in each of four height categories. This information will help them plan shelving. To obtain it, they plan to measure all of the several hundred thousand books in one library. Describe a sampling design that will save the librarians time and money. That is, outline steps 1 to 4 at the end of Section 6 for a sampling design (simple or complicated) you would suggest to the librarians.

1.64 Sampling college faculty. A labor organization wants to study the attitudes of college faculty members toward collective bargaining. These attitudes appear to be different at different types of colleges. The American Association of University Professors classifies colleges as follows:

Class I   Offer the doctorate, and award at least 15 per year.

Class IIA   Award degrees above the bachelor's, but are not in Class I.

Class IIB   Award only bachelor's degrees.

Class III   Two-year colleges.

Describe a sampling design that gathers information for each class separately as well as overall information about faculty attitudes.

1.65   **Agricultural sample surveys.** Many nationwide surveys "sample the map." That is, the sampling frame consists of identifiable geographic units rather than of a list of people or places. The Statistical Reporting Service of the U.S. Department of Agriculture makes extensive use of such "area frames" in its surveys of crops, farm economics, and so forth. The service prefers an area frame based on maps of rural areas for surveys of the acreage planted in each crop, but it uses a frame that lists farms or farmers for surveys of farm income. Can you explain why?

1.66   **The Current Population Survey.** An important government sample survey is the monthly Current Population Survey (CPS), from which employment and unemployment data are produced. The sampling design of the CPS is described in the *BLS Handbook of Methods,* published by the Bureau of Labor Statistics. Obtain the handbook from the library, and write a brief description of the multistage sample design used in the CPS.

▶ 7 ISSUES: OPINION POLLS AND THE POLITICAL PROCESS

Public opinion polls, especially pre-election "For-whom-would-you-vote?" polls, are the most visible example of survey sampling. They are also one of the most controversial. Most people are happy that sampling methods make employment and unemployment information rapidly available, and few people are upset when marketers survey the

"Seventy-three percent are in favor of one through five, forty-one percent find six unfair, thirteen percent are opposed to seven, sixty-two percent applauded eight, thirty-seven percent. . ."

buying intentions of consumers. The sampling of opinion on candidates or issues, however, is strongly attacked as well as strongly praised. We will briefly explore three aspects of polls and politics: first, polls of opinion on public issues; second, polls as a tool used by candidates seeking nomination or election; and third, election predictions for public consumption, designed to satisfy our curiosity about who's going to win. In each setting, I will summarize the arguments *for* and *against* polls. You draw your own conclusions.

POLLS ON PUBLIC ISSUES

Polls of public opinion tell us how the general public responds to questions on anything from gun control to tax cuts. Opinion polls do give accurate information, if we take care to ignore call-in polls and other abuses. The argument *for* opinion polls is that they offer a way for the public to be heard. Legislators are under constant pressure from interest groups who back their interests with lobbyists and campaign contributions. Opinion polls give the general public a chance to offset this pressure. Polls express public opinion clearly. The alternative is vague impressions and the loud voices of special interests.

Intelligent arguments *against* polling do not dispute that modern sampling methods guarantee that polls will give results close to the results we would get if we put the poll questions to the entire population. Critics ask instead what these opinions are worth. We know, for example, that small changes in the wording of questions can produce large changes in the response. Worse, the opinions reported by polls may be uninformed or uncaring. A question put by a telephone interviewer who calls as you are planning supper, a question with no responsibility attached to answering it, on an issue that you have not thought about, will get a low-quality opinion. If you cared about the issue, thought about it, and even did some study, you might have a different opinion. Your thoughtful opinion would probably be too complex to be properly described by an "agree" or "disagree" response to a poll question.

It is no surprise, then, that polls often produce confusing answers on major issues. Questions about government spending, for example, find broad support for cutting spending in general, but strong opposition to almost any specific cut that is large enough to have an impact on overall spending. The public likes reducing government spending in principle, but worries about the bad effects of any particular reduction. What *is* public opinion, anyway? That's a question worth pondering.

POLLS BY POLITICAL CANDIDATES

Running for major political office now requires a full team of experts, including a professional polling operation. Opinion polls conducted privately by candidates are a universal tool of campaign strategy. The purpose of these polls is to gather information for more effective campaigning. In what areas and with what groups of voters is the candidate weak? Which of the opponent's views are liabilities to be exploited? What arguments are most effective in advocating the candidate's views? Campaigners have always sought such information. Sample surveys replace vague impressions and intuition by reliable information about the voters.

The argument *against* polling by candidates is not aimed at polling specifically. Many people feel that political campaigns have become exercises in marketing—selling the candidate to the consumers. By market research (polls of voters), the campaign manager discovers what the consumers want. Then, using all the devices of advertising, the campaign cleverly presents the candidate as satisfying those wants. The polls also point to the opponent's weaknesses, which negative advertising can attack. The candidate with the better marketing strategy wins.

The argument *for* polling disagrees with this picture of campaigns as marketing. Candidates really do stand for different things, and it is fair both to point out your candidate's stands and to attack the opponent's views. Attempting to present the candidate in a false light to fit voter preferences will fail, and using poll results as guides to present the candidate's views most effectively is fair. Aren't we all entitled to use the best information available? The most serious ethical problems in election campaigns—such as negative advertising that slings mud at the opponent—are largely unrelated to the use of statistical sampling.

POLLS AS ELECTION PREDICTORS

Pre-election polls informing us that Senator So-and-so is the choice of 58% of Ohio voters annoy statisticians because the election often doesn't go the way the polls said it would. When the senator loses, the news reports say "the polls were wrong" and the public becomes yet more suspicious of statistics.

The key question asked in pre-election polls takes the form "If the election were held today, would you vote for Smith or Jones?" Modern sampling methods give us great confidence that a sample result from a professional poll is close to the truth about the population on the date of the poll. But the election is not being held today, and minds may change between the poll and the election. Some voters say that they are undecided, so the polling organization must decide what to do with the undecided vote. After all, these voters can't vote for "undecided" on election day. What is more, some people who announce their choice today will not take the trouble to vote at all on election day. The polling organizations try hard to determine how strongly their respondents hold their preferences and how likely they are to vote. They worry about how to allocate the "undecideds" in their samples. But the problems of changing opinions and low voter turnout cannot be entirely avoided, especially in primary elections. Election forecasting is one of the less satisfactory uses of sampling.

Exit polls based on interviews with voters as they leave the voting place are another matter. The sample consists of people who just voted—no ambiguity there. A good exit poll, based on a national stratified sample of election precincts, can often call a presidential election correctly long before the polls close. That fact sharpens the debate over our main issue, the political effects of election forecasts.

Arguments *against* public pre-election polls charge that they in-fluence voter behavior. Voters may decide to stay home if the polls predict a landslide—why bother to vote if the result is a foregone conclusion? Exit polls are particularly worrisome, because they in effect report actual election results before the election is complete. The U.S. television networks agree not to release the results of their exit surveys in any state until the polls close in that state. If a presidential election is not close, the networks may know the winner by mid-afternoon, but they will only forecast the vote one state at a time as the polls close across the country. Even so, a presidential election result may be known before the voting closes on the west coast. Some countries have laws restricting election forecasts. In France, no poll results can be published in the week before a presidential election. Belgium, Italy, and Portugal have similar laws.

The argument *for* pre-election polls is simple: democracies should not forbid publication of information. Voters can decide for themselves how to use the information. After all, supporters of a candidate who is far behind know that even without polls. Restricting publication of polls just invites abuses. In France, candidates continue to take private polls (less reliable than the public polls) in the week before the election. They then leak the results to reporters in the hope of influencing press reports.

SECTION 7 EXERCISES

1.67   **Homosexuals in the military?** Here are three opinion poll questions on the same issue, with the poll results:

"Do you approve or disapprove of allowing openly homosexual men and women to serve in the armed forces of the United States?" Result: 47% strongly or somewhat disapproved, 45% strongly or somewhat approved.

"Do you think that gays and lesbians should be banned from the military or not?" Result: 37% said they should be banned, 57% said not.

"Should President Clinton change military policy to allow gays in the military?" Result: 53% said no, 35% said yes.

Using this example, discuss the difficulty of using responses to opinion polls to understand public opinion.

1.68   **Gun control.** Opinion polls have long showed strong public support for stricter controls on firearms. Specific gun control proposals have often been favored by more than 80% of adults. Yet, in the face of strong opposition from a minority, little national gun control legislation has passed.

Why do you think this has occurred? Does this mean that opinion polls on issues do not really offer the public a means of offsetting special-interest groups? Explain your answer.

1.69   To see whether people often give responses on subjects about which they are entirely ignorant, ask several people (we won't require an SRS) the following questions:

(a)   Have you heard of the Bradley-Nunn bill on veterans' housing?

(b)   Have you heard of *Midwestern Life* magazine?

(A few years ago, 53% of a sample said yes to a question like (a) and 25% to (b) even though neither the bill nor the magazine ever existed.[15])

1.70   **Order of questions.** The order in which questions are asked can influence the results of polls on public issues. At the height of the cold war, a sample survey asked whether Soviet representatives should be allowed to appear on American TV to give the Soviet point of view. This question was sometimes asked after a question on whether American representatives should be allowed time on Soviet TV, and sometimes before that question. Which order do you think produced a result much more favorable to allowing Soviet representatives on American TV? Why?

1.71   **Should we restrict election polls?** Do you agree with the French law that forbids publication of election polls during the week before a presidential election? Defend your opinion.

1.72   **Exit polling.** Election-night television coverage includes exit polls that often allow quite precise predictions of the outcome before the polls have closed in the western part of the country. It is sometimes charged that late voters may stay home if the networks say that the national election is decided. Therefore, predictions based on samples of actual votes should not be allowed until the polls have closed everywhere in the country. Do you agree with this proposal? Explain your opinion.

1.73   An alternative solution to the problem of predicting the winner while the polls are still open is to require all states (except possibly Hawaii) to end voting at the same moment. The polls might close at 10 p.m. in the east and at 7 p.m. in the west, for example. Do you support this proposal? Explain your opinion.

1.74 In the final days before an election, the Gallup poll asks questions like these of all its respondents: "Do you know where to go to vote?" "Have you voted in recent elections?" "How much have you thought about the election?" Gallup throws out responses from people who do not answer these questions satisfactorily. Why?

1.75 **The woes of election polls.** Most of the major pre-election polls in the 1992 presidential contest got the winner's percent of the vote right to within their margins of error. (Bill Clinton won the election with 43% of the popular vote.) All of the polls, however, underestimated the vote for the third-party candidate, Ross Perot, who got 19% of the vote. What are some of the reasons why a pre-election poll can fail to forecast the outcome of an election?

1.76 During the 1968 presidential campaign, the segregationist George Wallace accused the polls of favoring moneyed interests and neglecting the common people. He would ask crowds at his rallies, "Have any of you-all ever been asked about this here election by Mr. Harris or Mr. Gallup?" The crowds shouted "No" and "Never."

There were about 150 million U.S. residents age 18 and over in 1968. If a poll selected 1500 at random, what was the chance that any one person would be interviewed? Suppose that the major polling organizations conducted 20 surveys during the campaign. What was the chance that a particular person would be interviewed at least once? (Being interviewed by an opinion poll is an unlikely event, and no accusations of bias are needed to explain why you have not been chosen.)

▶ 8 ISSUES: RANDOM SELECTION AS PUBLIC POLICY

Would you like to be the proud owner of the right to offer cellular mobile telephone service in your area? So would many other people. The Federal Communications Commission (FCC) faced the task of choosing among the many applicants in each area. During 1988 and 1989, the FCC conducted lotteries to award cellular telephone rights in 216 small cities and 428 rural areas. Anyone with enough financial backing could enter the lottery and hope to emerge with an exclusive right to offer cellular service.

What should a nation do if there are not enough volunteers for military service but only a small fraction of eligible youth are needed by the military? That question last arose during the Vietnam era. Beginning in 1970, a draft lottery was used to choose draftees by random selection. Although the draft was ended in 1976, young men must still register at age 18. The draft lottery may return if the army cannot attract enough volunteers.

WHY RANDOM SELECTION?

Both the cellular telephone lottery and the draft lottery are examples of using random selection as public policy. The idea of random selection is that of drawing lots—or taking an SRS. The goal is to treat everyone identically by giving all the same chance to be selected. To give another example, random selection has been used to allot space in public housing to eligible applicants when there are more applicants than available housing units.

When should random selection decide public issues? I claim that this is a policy question, not a statistical question. Random selection treats everyone identically. It is fair or unbiased in that sense. If we want a policy of identical treatment, random selection will carry out that policy. Debate over random selection should concentrate on whether or not we want to make distinctions among people in a certain situation.

For example, the FCC awarded cellular telephone licenses in the 90 largest cities after elaborate hearings designed to find the best candidate. The FCC used random selection in smaller areas because hearings would have been too slow and expensive. In the case of the military draft, Congress wanted to eliminate distinctions among young men, and so required random selection. For public housing, a federal court ruled that random selection can be used only when applicants are equally needy. Distinctions *should* be made among different degrees of need, the court said, so random selection can't be used for the entire pool of applicants for housing.

MAKING A RANDOM SELECTION

Actually making a random selection in the public arena isn't always simple. In principle, it is just like choosing an SRS. In practice, random digits can't be used. Because few people understand random digits, a lottery looks fairer if a dignitary chooses capsules from a glass bowl in front of the TV cameras. This also prevents cheating; no one can check the table of random digits in advance to see how cousin Joe will make out in the selection.

Physical mixing and drawing *looks* random, but it can be hard to achieve a mixing that really *is* random. There is no better illustration of this than the first Vietnam-era draft lottery, held in 1970.

> **EXAMPLE 20. The draft lottery.** Because an SRS of all eligible men would be hopelessly awkward, the draft lottery planned to select birth dates in a random order. Men born on the date chosen first would be drafted first, then those born on the second date chosen, and so on. Because all men ages 19 to 25 were included in the first lottery, there were 366 birth dates. The 366 dates were placed into identical small capsules, put into a bowl, and publicly drawn one by one.
>
> News reporters noticed that men born later in the year seemed to receive lower draft numbers. Statisticians showed that this trend was so strong that it would occur less than once in a thousand truly random lotteries. An inquiry found that the capsules had been filled one month at a time and not very thoroughly mixed. Birth dates in January tended to be on the bottom, and birth dates in December were filled last and tended to be on top.[16]

What's done is done, and off to Vietnam went too many men born in December. The next year, statisticians from the National Bureau of Standards were asked to design the lottery. Their design was worthy of Rube Goldberg. The numbers 1 to 365 (no leap year this time) were

"So you were born in December too, eh?"

placed in capsules in a random order determined by a table of random
digits. The dates of the year were placed in another set of capsules in
a random order determined again by the random digit table. The date
capsules were put into a drum in random order determined by a third
use of the table of random digits. The number capsules went into a
second drum, again in random order. The drums were rotated for an
hour. The TV cameras were turned on, and the dignitary reached into
the date drum: out came September 16. He reached into the number
drum: out came 139. So men born on September 16 received draft
number 139. Back to both drums: out came April 27 and draft number
235. And so on. It's awful, but it's random.[17] You can rejoice that in
choosing samples we have Table A to do the randomization for us.

SECTION 8 EXERCISES

1.77   A lottery for AIDS drugs? Only limited supplies of some experimental drugs to treat AIDS are available because the drugs

are very difficult to make. It is now usual to use a lottery to decide at random which AIDS patients can receive the drug. In 1995, for example, Hoffman-La Roche had enough doses of the drug Invirase for 2880 patients. Shortly after announcing this, the company had already received 10,000 calls from patients who wanted to enter the lottery.

Discuss this practice. Do you favor random selection? If not, how should recipients be chosen? By ability to pay? By value to society (as assessed by whom)? By age and family responsibilities? By some other method?

1.78   **A lottery for tickets?** A basketball arena has 8000 student seats, but 18,000 students would like to watch basketball games. Design a system of allotting tickets at random that seems fair to you. (All students can see some of the 12 home games if you use a rotation system. Will you give upperclassmen some preference? How many tickets may an individual buy? If your school actually does use a random drawing to allot seats, describe the details of the official system and discuss any changes you favor.)

1.79   **A lottery for medical school admission?** In 1975, the Dutch government adopted random selection for admissions to university programs in medicine, dentistry, and veterinary medicine. Applicants for these programs are much more numerous than available places. The random selection is stratified so that students with higher grades have a greater chance of being chosen. Do you favor such a system? Why?

1.80   **A lottery for military service?** Prior to 1970, young men were selected for military service by local draft boards. There was a complex system of exemptions and quotas that enabled, for example, farmer's sons and married young men with children to avoid the draft. Do you think that random selection among all men of the same age is preferable to making distinctions based on occupation and marital status? Give your reasons.

1.81   Give an example of a situation in which you definitely would *approve* random selection. Give an example of a situation in which you would definitely *disapprove* random selection.

▶ 9 ISSUES: DATA ETHICS

The production and use of data, like all human endeavors, raise ethical
questions. We won't discuss the salesperson who begins a telephone
sales pitch with "I'm conducting a survey." Such deception is clearly
unethical. It also enrages legitimate survey organizations, who find the
public less willing to talk with them. Neither will we discuss those few
researchers who, in the pursuit of professional advancement, publish
fake data. There is no ethical question here—faking data to advance
your career is just wrong. It will end your career when uncovered. But
just how honest must researchers be about real, unfaked data? Here is
an example that suggests the answer is "More honest than they often
are."

> **EXAMPLE 21. Missing details.** Papers reporting scientific research
> are supposed to be short, with no extra baggage. Brevity can al-
> low the researchers to avoid complete honesty about their data.
> Did they choose their subjects in a biased way? Did they report
> data on only some of their subjects? Did they try several sta-
> tistical analyses and only report the ones that looked best? The
> statistician John Bailar screened more than 4000 medical papers
> in more than a decade as consultant to *The New England Jour-
> nal of Medicine*. He says, "When it came to the statistical review,
> it was often clear that critical information was lacking, and the
> gaps nearly always had the practical effect of making the authors'
> conclusions look stronger than they should have."[18] The situa-
> tion is no doubt worse in fields that screen published work less
> carefully.

The most complex issues of data ethics arise when we collect data
from people. The ethical difficulties are more severe for experiments that
impose some treatment on people than for sample surveys that simply
gather information. Trials of new medical treatments, for example, can
do harm as well as good to their subjects. We will look at the ethics of
experiments with human subjects in the next chapter. Our topic here is
the ethical problems that accompany any data about people. Here are
the basic standards of data ethics, which must be obeyed by any study
that gathers data from human subjects.

> **Basic data ethics**
>
> The organization that carries out the study must have an **institutional review board** that reviews all planned studies in advance in order to protect the subjects from possible harm.
>
> All individuals who are subjects in a study must give their **informed consent** before data are collected.
>
> All individual data must be kept **confidential.** Only statistical summaries for groups of subjects may be made public.

The law requires that studies funded by the federal government obey these principles. But neither the law nor the consensus of experts is completely clear about the details of their application.

INSTITUTIONAL REVIEW BOARDS

The purpose of an institutional review board is not to decide whether a proposed study will produce valuable information, or whether it is statistically sound. The board's only purpose is to protect subjects. It reviews the plan of the study, including the questions to be asked in a sample survey. It also reviews the consent form to be sure that subjects are informed about the nature of the study and about any potential risks.

Who should serve on review boards? If a university's board, for example, consists entirely of faculty, their natural respect for new knowledge may reduce their willingness to criticize a study. Federal policy requires that the board contain at least one non-scientist and at least one representative of the local community. Some people want members of the "general public," or in medical research even "patient advocates." Others say that broader boards tend to criticize studies with quite small risks that would produce valuable knowledge.

INFORMED CONSENT

Both words in the phrase "informed consent" are important, and both can be controversial. Subjects must be *informed* in advance about the nature of a study and any risk of harm it may bring. In the case of a sample survey, physical harm is not possible. The subjects should be told what kinds of questions the survey will ask and about how much of their time it will take. They must then *consent* in writing before being asked the survey questions.

Are there some subjects who can't really give informed consent? It was once common, for example, to test new vaccines on prison inmates who gave their consent in return for good behavior credit. Now we worry that prisoners are not really free to refuse, and the law forbids medical tests in prisons. Children can't give fully informed consent, so the usual procedure is to ask their parents. Exercise 1.93 points out some of the problems this practice causes.

In the setting of experiments to test new medical treatments, almost every detail of informed consent is controversial. What about studies of new ways to help emergency-room patients who may be unconscious or have suffered a stroke? In most cases, there is not time even to get the consent of the family. Does the principle of informed consent in effect bar realistic trials of new treatments for unconscious patients? The difficulties of informed consent do not vanish even for patients who are fully conscious. Mentioning every possible risk leads to very long consent forms. "They are like rental car contracts," one lawyer said. Some subjects don't read forms that run five or six printed pages. Others are frightened by the large number of possible (but unlikely) disasters that might happen, and so refuse to participate. Of course, unlikely disasters sometimes happen. When they do, lawsuits follow and the consent forms become yet longer and more detailed.

CONFIDENTIALITY

Ethical problems do not disappear once a study has been cleared by the review board, has obtained consent from its subjects, and has actually collected data about the subjects. It is important to protect the subjects' privacy by keeping all data about individuals confidential. The report of an opinion poll may say what percent of the 1500 respondents felt that legal immigration should be reduced. It may not report what *you* said about this or any other issue.

Notice that confidentiality is not the same as **anonymity.** Anonymity means that subjects are anonymous—their names are not known even to the director of the study. Anonymity causes severe problems because

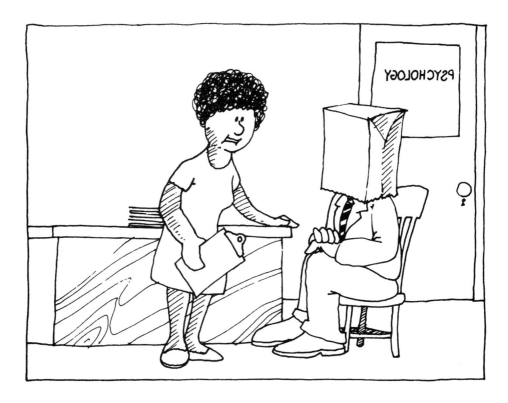

"I realize the participants in this study are to be anonymous, but you're going to have to expose your eyes."

it prevents knowing who responded to a survey and who did not. This means that no follow-up work can be done to increase the response rate. Anonymity is usually possible only in mail surveys, because responses can be mailed in without any identification. Anonymity is therefore quite rare in statistical studies.

Any breach of confidentiality is a serious violation of data ethics. The best practice is to separate the identity of the subjects from the rest of the data at once. Sample surveys, for example, use the identification only to check on who did or did not respond. In an era of advanced technology, however, it is no longer enough to be sure that each individual set of data protects people's privacy. The government, for example, maintains a vast amount of data on citizens in many separate data bases—census responses, tax returns, Social Security information, data from surveys such as the Current Population Survey, and so on.

Many of these data bases can be searched by computers for statistical studies. A clever computer search of several data bases might be able, by combining information, to identify you and learn a great deal about you even if your name and other identification have been removed from the data available for search. Privacy and confidentiality of data is a hot issue among statisticians in the computer age.[19]

> **EXAMPLE 22. Use of government data bases.** Citizens are required to give information to the government. Think of tax returns and Social Security contributions. The government needs these data for administrative purposes—to see if I paid the right amount of tax and how large a Social Security benefit I am owed when I retire. Some people feel that individuals should be able to forbid any other use of their data, even with all identification removed. This would prevent using government records to study, say, the ages, incomes, and household sizes of Social Security recipients. Such a study could well be vital to debates on reforming Social Security.

Example 22 illustrates the central dilemma of data ethics: *Where is the proper balance between protecting individuals and producing knowledge that will benefit other individuals in the future?* I would be happy if no one could access my tax and Social Security files, even with my name removed. Yet wise Social Security policy may depend on studies that use data about me. I might refuse to participate in a medical study when faced with a five-page list of possible complications. Yet many future patients may be helped by the results of the study, even if I am not helped. It is probably fair to say that in the past the protection of individuals did not get enough attention. Institutional review boards and informed consent have tilted the balance toward protecting subjects. Has the tilt impeded the growth of useful knowledge? Arguments over data ethics will surely continue.

SECTION 9 EXERCISES

Most of the exercises in this section pose issues for discussion. There are no right or wrong answers, but there are more or less thoughtful answers.

1.82 **The General Social Survey.** One of the most important non-government surveys in the United States is the General Social Survey conducted by the National Opinion Research Center at the University of Chicago. The General Social Survey regularly monitors public opinion on a wide variety of political and social issues. Interviews are conducted in person in the subject's home. Are a subject's responses to General Social Survey questions anonymous, confidential, or both? Explain your answer.

1.83 **Informed consent.** A researcher suspects that orthodox religious beliefs tend to be associated with an authoritarian personality. She prepares a questionnaire that measures authoritarian tendencies and also asks many religious questions. Write a description of the purpose of this research to be read by subjects in order to obtain their informed consent. You must balance the conflicting goals of not deceiving the subjects as to what the questionnaire will tell about them and of not biasing the sample by scaring off certain types of people.

1.84 **Institutional review boards.** Your college or university has an institutional review board that screens all studies that use human subjects. Get a copy of the document that describes this board.

(a) According to this document, what are the duties of the board?

(b) How are members of the board chosen? How many members are not scientists? How many members are not employees of the college? Do these members have some special expertise, or are they simply members of the "general public"?

1.85 **Institutional review boards.** Suppose that you attend a university whose institutional review board consists entirely of faculty. Now the board is about to add several members from outside the university. Do you think the new members should have some special expertise? In particular, do you favor including local clergy? Lawyers? Doctors? Can you offer the university any suggestions about choosing the outside members?

1.86 **Opinion polls.** The presidential election campaign is in full swing, and the candidates have hired polling organizations to take regular polls to find out what the voters think about the issues. What information should the pollsters be required to give out?

(a)   What does the standard of informed consent require the pollsters to tell potential respondents?

(b)   The standards accepted by polling organizations also require giving respondents the name and address of the organization that carries out the poll. Why do you think this is required?

(c)   The polling organization usually has a professional name such as "Samples Incorporated," so respondents don't know that the poll is being paid for by a political party or candidate. Would revealing the sponsor to respondents bias the poll? Should the sponsor always be announced whenever poll results are made public?

(d)   Some people think that the law should require that all poll results be made public. Otherwise, the possessors of poll results can use the information to their own advantage. They can act on the information, release only selected parts of it, or time the release for best effect. The candidate's organization replies that they are paying for the poll in order to gain information for their own use, not to amuse the public. Do you favor requiring complete disclosure of political poll results? What about other private surveys, such as market research surveys of consumer tastes?

1.87   **Census questions.** The federal government, unlike opinion polls or academic researchers, has the legal power to compel response to survey questions. The 1990 census long form, sent to 17% of all households, contained the following question:

H10.   Do you have COMPLETE plumbing facilities in this house or apartment; that is, (1) hot and cold piped water, (2) a flush toilet, and (3) a bathtub or shower?
O   *Yes, have all three facilities*
O   *No*

When a question about plumbing was first asked in the 1970 census, some members of Congress felt that this was an invasion of privacy and that the Census Bureau should be prohibited from asking such questions. The bureau replied that, as with all census data, no individual information would be released, only averages for various regions. Moreover, lack of plumbing is the best single measure of substandard housing, and the government needs this information to plan housing programs.

Do you feel that this question is proper? If not, when do you think that the government's need for information outweighs a citizen's wish to withhold personal facts? If you do think the plumbing question is proper, where does the citizen's right to withhold information begin?

1.88 **Charging for data?** Data produced by the government are often available free or at low cost to private users. For example, satellite weather data produced by the U.S. National Weather Service are available free to TV stations for their weather reports. *Opinion*: Government data should be available to all citizens at minimal cost.

European governments, on the other hand, charge TV stations for weather data. *Opinion*: The satellites are expensive, and the TV stations are making a profit from their weather services, so they should share the cost.

Which opinion do you support, and why?

1.89 Some common practices may appear to offer anonymity, while actually delivering only confidentiality. Market researchers often use mail surveys that do not ask the respondent's identity but contain hidden codes on the questionnaire that identify the respondent. Invisible ink coding and code numbers hidden under the flap of the return envelope are the usual techniques. A false claim of anonymity is clearly unethical. If only confidentiality is promised, is it also unethical to hide the identifying code, perhaps causing respondents to believe their replies are anonymous?

1.90 **Testing blood for the AIDS virus?** How can the government gauge the spread of infection with the AIDS virus? Plans for a national sample survey were dropped because the people most at risk of AIDS, especially intravenous drug users, often refused to participate. Instead, we rely on large-scale blood tests that do not require consent. For example, blood specimens are routinely drawn from newborn infants; in 30 states, these specimens are tested for the AIDS virus. Some hospitals, family doctors, and blood-testing laboratories also test random samples of blood specimens for AIDS. All of these tests are anonymous, and all use blood specimens drawn for other purposes. Put together, they monitor the extent of AIDS in the general population.

Here are two possible, and opposite, objections. On the one hand, some object to secret use of blood specimens drawn for other purposes, even though there is complete anonymity. On the other hand, some object to the anonymity because it prevents notifying the people whose blood tested positive for the AIDS virus. What do you think?

1.91 **Use of records without consent.** Does having an abortion affect the health of mother or child in any future live births? To study this question, the New York State Health Department sampled medical records in the state. It chose a sample of 21,000 women who had legal abortions and 27,000 women who gave birth to living children in the same year. The department used the medical records of these 48,000 women to examine their maternity patterns in the five years following the abortion or live birth. None of the women was contacted in person.

Unlike the AIDS studies in the previous example, the subjects were not anonymous. Locating the 48,000 women to obtain consent would make the study impossibly expensive. Was this study an invasion of privacy? Was the study justified by the value of the information gained? Should consent always be required to use existing data for statistical studies such as this? Explain your opinion.

1.92 **Consent, or not?** In what circumstances is collecting personal information without the subject's consent permitted? Consider the following cases in your discussion:

(a) A government agency takes a random sample of income tax returns to obtain information on the average income of people in different occupations. Only the incomes and occupations are recorded from the returns, not the names.

(b) A psychologist asks a random sample of students to fill out a questionnaire. She explains that their responses will be used to measure several personality traits so that she can study how these traits are related. The psychologist does not inform the students that one trait measured by the test is how prejudiced they are toward other races.

(c) A social psychologist attends public meetings of a religious group to study the behavior patterns of members.

(d) The social psychologist pretends to be converted to membership in a religious group and attends private meetings to study the behavior patterns of members.

1.93 **Surveys of youth.** Some people think that surveys of children and teen-agers on issues such as sexual behavior and religion invade the privacy of families. They point to questions such as:

How old were you when you had sexual intercourse for the first time?

This question appeared on a survey of at-risk teens carried out by the Centers for Disease Control and Prevention. Certainly, say advocates of family privacy, parents should give written permission before their children are asked such questions.

Surveys of young people are usually carried out in schools. They are first cleared by a review board. Then a notice to parents tells them about the survey's content and asks them to allow their children to participate. More than half the parents don't respond. Researchers say that telephoning parents and other follow-ups to get written permission would raise the cost from about $1 a student to $25 a student. Leaving out students whose parents don't return a permission slip would bias the surveys.

Is it OK to include all children whose parents do not actually refuse to give consent, as long as parents always have the opportunity to refuse? Does your opinion depend on the topic of the survey? If so, what topics would you not allow without consent? Does your opinion depend on the age of the children?

Notes

1. For basic information about the Current Population Survey, see the current edition of the Bureau of Labor Statistics' *Handbook of Methods.* The BLS web site, http://www.bls.gov, also has much information.
2. The General Social Survey is described in detail in J. A. Davis and T. W. Smith, *The NORC General Social Survey,* Sage Publications, 1992.
3. From an article by M. Kernan of the *Washington Post,* printed in the *Lafayette Journal and Courier,* April 19, 1976.
4. From the *New York Times,* August 21, 1989.
5. Based on M. R. Kagay, "90's, in poll: a good life amid old ills," *New York Times,* January 1, 1990.
6. You can find details on telephone surveys in R. M. Groves *et al.* (eds.) *Telephone Survey Methodology,* Wiley, 1988. The finding on women is reported in the *New York Times,* September 12, 1988. The 1984 campaign example is reported in P. E. Converse and M. W. Traugott, "Assessing the accuracy of polls and surveys," *Science,* Volume 234 (1986), pp. 1094–1098.
7. Information on the census undercount appears in a special section of the *Journal of the American Statistical Association,* Volume 88 (1993), pp. 1044–1166. The nonresponse rates cited in the example appear in E. P. Ericksen and T. K. DeFonso, "Beyond the net undercount: how to measure census error," *Chance,* Volume 6 (1993), pp. 38–43.
8. Results from a New York Times/CBS News Poll, reported in the *New York Times,* July 5, 1992.
9. From D. Goleman, "Pollsters enlist psychologists in quest for unbiased results," *New York Times,* September 7, 1993.
10. P. H. Lewis, "Technology" column, *New York Times,* May 29, 1995.
11. M. R. Kagay, "Poll on doubt of Holocaust is corrected," *New York Times,* July 8, 1994.
12. Recorded in *Public Opinion Quarterly,* Volume 38 (1974–1975), pp. 492–493.
13. From the article cited in Note 9.
14. These are the words of J. B. Lansing and J. N. Morgan, *Economic Survey Sampling,* Institute for Social Research, The University of Michigan, 1971.
15. Cited by D. Trewin, "Non-sampling errors in sample surveys," *CSIRO Division of Mathematics and Statistics Newsletter,* June 1977.
16. For details about the 1970 draft lottery, see S. E. Fienberg, "Randomization and social affairs: the 1970 draft lottery," *Science,* Volume 171 (1971), pp. 255–261.
17. It's even a little more complicated than I have described. For the details, see J. R. Rosenblatt and J. J. Filliben, "Randomization and the draft lottery," *Science,* Volume 171 (1971), pp. 306–308.

18. J. C. Bailar, III, "The real threats to the integrity of science," *The Chronicle of Higher Education,* April 21, 1995, pp. B1–B2.

19. You can find details about current policies and an exploration of the issues in G. T. Duncan, T. B. Jabine, and V. A. DeWolf (eds.), *Private Lives and Public Policies: Confidentiality and Accessibility of Government Services,* National Academy Press, 1993.

REVIEW EXERCISES

1.94 **Mail to a member of Congress.** You are on the staff of a member of Congress who is considering a controversial bill that provides for government-sponsored insurance for nursing home care. You report that 1128 letters have been received on the issue, of which 871 oppose the legislation. "I'm surprised that most of my constituents oppose the bill. I thought it would be quite popular," says the congresswoman. Are you convinced that a majority of the voters oppose the bill? State briefly how you would explain the statistical issue to the congresswoman.

1.95 A newspaper advertisement for *USA Today: The Television Show* once said:

> Should handgun control be tougher?
> You call the shots in a special call-in poll tonight.
> If yes, call 1-900-720-6181. If no, call 1-900-720-6182.
>
> Charge is 50 cents for the first minute.

Explain why this opinion poll is almost certainly biased.

1.96 A magazine for health foods and organic healing wants to establish that large doses of vitamins will improve health. The magazine asks readers who have regularly taken vitamins in large doses to write in, describing their experiences. Of the 2754 readers who reply, 93% report some benefit from taking vitamins.

Is the sample proportion 93% probably higher than, lower than, or about the same as the percent of all adults who would perceive some benefit from large vitamin intake? Why?

1.97 The noted scientist Dr. Iconu wanted to investigate attitudes toward television advertising among American college students. He decided to use a sample of 100 students. Students in freshman psychology (PSY 001) are required to serve as subjects for experimental work. Dr. Iconu obtained a class list for PSY 001 and chose a simple random sample of 100 of the 340 students on the list. He asked each of the 100 students in the sample the following question:

Do you agree or disagree that having commercials on TV is a fair price to pay for being able to watch it?

Of the 100 students in the sample, 82 marked "Agree." Dr. Iconu announced the result of his investigation by saying, "82% of American college students are in favor of TV commercials."

(a) What is the population in this example?

(b) What is the sampling frame in this example?

(c) Explain briefly why the sampling frame is or is not suitable.

(d) Discuss briefly the question Dr. Iconu asked. Is it a slanted question?

(e) Discuss briefly why Dr. Iconu's announced result is misleading.

(f) Dr. Iconu defended himself against criticism by pointing out that he had carefully selected a simple random sample from his sampling frame. Is this defense relevant? Why?

1.98 An advertising agency conducts a sample survey to see how adult women react to various adjectives that might be used to describe an automobile. The firm chooses 400 women from across the country. Each woman is read a list of adjectives, such as "elegant" and "prestigious." For each adjective, she is asked to indicate how desirable a car described this way seems to her. The possible responses are (1) highly desirable (2) somewhat desirable (3) neutral (4) not desirable. Of the women interviewed, 76% said that a car described as "elegant" was highly desirable.

(a)   What is the population in this sample survey?

(b)   Is the number 76% a parameter or a statistic?

(c)   The sample was a probability sample chosen using the Gallup poll's sampling procedure. Use Table 1-1 to make a 95% confidence statement about the reaction of women to the adjective "elegant."

1.99   A simple random sample of 1200 adult residents of California is asked whether they favor restricting imports of foreign cars. The margin of error for 95% confidence in this sample is ±1.4 percentage points. If the survey had selected an SRS of 1200 adults from the city of Chicago (population 3 million) rather than from the state of California (population 32 million), would the margin of error in the confidence statement be wider, about the same, or narrower?

1.100   **Do you jog?** A national opinion poll asked "Do you happen to jog?" Fifteen percent of the sample answered "Yes." A newspaper article reporting the poll said:

The latest findings are based on in-person interviews with 1540 adults, 18 and older, conducted in more than 300 scientifically selected locations across the country during the period June 7–10. The poll had a margin of error of plus or minus three percent.

(a)   Make a confidence statement based on the survey results.

(b)   Explain carefully what sources of error are included in the margin of error allowed by your confidence statement. Give two examples of possible sources of error that are not included in the margin of error.

1.101   **Prayer in the schools?** The New York Times/CBS News Poll conducts regular surveys of public opinion. One of the questions asked in a recent poll was "Do you favor an amendment to the Constitution that would permit organized prayer in public schools?" Sixty-six percent of the sample answered "Yes."

(a)   The article describing the poll says that it "is based on telephone interviews conducted from Sept. 13 to Sept. 18 with 1664

adults around the United States, excluding Alaska and Hawaii.
... the telephone numbers were formed by random digits, thus
permitting access to both listed and unlisted residential numbers." This sampling method excludes some groups of adults,
which may cause some bias in the poll results. What groups of
adults are excluded from the sample?

(b)   The article gives the margin of error as three percentage
points. Make a confidence statement about the percent of all
adults (excluding the groups you named in part (a)) who favor
a school prayer amendment.

(c)   The news article goes on to say that "The theoretical errors
do not take into account a margin of additional error resulting
from the various practical difficulties in taking any survey of
public opinion." List some of the "practical difficulties" that may
cause errors in addition to the $\pm 3\%$ margin of error.

1.102 A friend who knows no statistics has encountered some statistical
terms in reading for her psychology course. Explain each of these
terms in one or two simple sentences.

(a)   Simple random sample

(b)   95% confidence

(c)   Nonsampling error

(d)   Informed consent

WRITING PROJECTS

1.1   **The Current Population Survey.** The Current Population Survey (CPS) is the most important sample survey of the federal
government. The CPS provides monthly information on employment and unemployment. It also gathers information on many
other economic and social issues on a less frequent basis by
varying the questions asked each month.

Locate information about the design of the CPS. Here are
some suggested sources. The Bureau of Labor Statistics makes
information available via the Internet. The current World Wide

Web location is http://stats.bls.gov, and the CPS has its own home page at http://stats.bls.gov/cpshome.htm (these addresses may change in the future). The sampling design of the CPS is described in detail in the Bureau of Labor Statistics' *Handbook of Methods,* which is updated from time to time and can be found in most college libraries.

Using information from one of these sources, write a clear description of the sampling design used for the CPS. How does the CPS use ideas such as stratification and multistage design? Does the design employ ideas not discussed in this chapter?

1.2 **Design your own sample survey.** Choose an issue of current interest to students at your school. Prepare a short (no more than five questions) questionnaire to determine opinions on this issue. Choose a sample of about 25 students, administer your questionnaire, and write a brief description of your findings. Also write a short discussion of your experiences in designing and carrying out the survey.

(Although 25 students are too few for you to be statistically confident of your results, this project centers on the practical work of a survey. You must first identify a population; if it is not possible to reach a wider student population, use students enrolled in this course. Did the subjects find your questions clear? Did you write the questions so that it was easy to tabulate the responses? At the end, did you wish you had asked different questions?)

1.3 **Data ethics.** Several exercises in Section 9, Data Ethics, raise issues that deserve careful discussion. Choose one of the issues listed below and prepare an essay that makes clear both your opinion on the issue and the reasons that you hold this opinion.

▶ Testing blood for the AIDS virus (Exercise 1.90)

▶ Using medical records without consent (Exercise 1.91)

▶ Surveys of youth on sensitive issues (Exercise 1.93)

CHAPTER 2

EXPERIMENTS

Does taking aspirin regularly help protect people against heart attacks? The best way to find out is to actually give people aspirin regularly and note its effect. That is an *experiment*. We might also choose a sample of people and just ask them if they take aspirin regularly. That's not an experiment. The experiment, if we arrange it properly, will tell us much more about the effect of taking aspirin than will the sample survey. Why this is so, and how we should arrange the experiment, are the themes of this chapter.

▶ 1 WHY EXPERIMENT?

Observation and experiment

An **observational study** observes individuals and measures variables of interest but does not attempt to influence the responses. Sample surveys are an important kind of observational study. The purpose of an observational study is to describe some group or situation.

An **experiment** deliberately imposes some treatment on individuals in order to observe their responses. The purpose of an experiment is to study whether the treatment causes a change in the response.

EXAMPLE 1. Aspirin and heart attacks. Does taking aspirin regularly help protect people against heart attacks? The Physicians'

Health Study was a medical experiment that helped answer this question. Half of a group of 22,000 male doctors were chosen at random to take an aspirin every other day. The other half of the doctors took a *placebo,* a dummy pill that looked and tasted just like the aspirin but had no active ingredient. After several years, 239 of the placebo group but only 139 of the aspirin group had suffered heart attacks. This difference is large enough to give good evidence that taking aspirin does reduce heart attacks.[1]

EXAMPLE 2. Does studying a foreign language in high school increase students' verbal ability in English? Julie obtains lists of all seniors in her high school who did and did not study a foreign language. Then she compares their scores on a standard test of English reading and grammar given to all seniors. The average score of the students who studied a foreign language is much higher than the average score of those who did not.

Sorry: This observational study gives no evidence that studying another language builds skill in English. Students decide for themselves whether or not to elect a foreign language. Those who choose to study a language are mostly students who are already better at English than most students who avoid foreign languages. The difference in average test scores just shows that students who choose to study a language differ (on the average) from those who do not. We can't say whether studying languages *causes* this difference.

Examples 1 and 2 illustrate the big advantage of experiments over observational studies. *In principle, experiments can give good evidence for causation.* All the doctors in the Physicians' Health Study took a pill every other day, and all got the same schedule of checkups and information. The only difference was the content of the pill. When one group had many fewer heart attacks, we conclude that it was the content of the pill that made the difference. Julie's observational study—a census of all seniors in her high school—does a good job of describing differences between seniors who have studied foreign languages and those who have not. But she can't say anything about cause and effect.

If Julie could direct half of the students in her school to study a foreign language, and deny the other half any contact with languages other than English, she would have an experiment. That experiment isn't practical (or ethical), so we are left without a clear conclusion about whether foreign language study causes better English skills. The most intense controversies about "statistical evidence" occur when there is a strong relationship between two variables but experiments can't be done.

THE LANGUAGE OF EXPERIMENTS

Here is the vocabulary we need to discuss experiments in more detail.

The vocabulary of experiments

Units—the objects on which the experiment is done. When the units are human beings, they are called **subjects**.

Variable—a measured characteristic of a unit.

Response variable—a variable whose changes we wish to study; an outcome or result.

Explanatory variable—a variable that explains or causes changes in the response variables.

Treatment—any specific experimental condition applied to the units. If an experiment has several explanatory variables, a treatment is a combination of specific values of these variables.

In Example 1, the subjects are the 22,000 doctors, the explanatory variable is whether a subject took aspirin or a placebo, and the response variable is whether or not the subject had a heart attack. Example 2, although it is not an experiment, also has an explanatory variable (did the student study a foreign language?) and a response variable (score on the English test). *The active imposition of a treatment, not just the presence of explanatory and response variables, makes a study an experiment.* Here is an example of an experiment with two explanatory variables.

EXAMPLE 3. **The Physicians' Health Study.** In fact the Physicians' Health Study looked at the effects of two drugs: aspirin and beta carotene. The body converts beta carotene into vitamin A, which may help prevent some forms of cancer. Figure 2-1 shows how these two explanatory variables combine to form four treatments. On odd-numbered days, the subjects took a white tablet that contained either aspirin or a placebo. On even-numbered days, they took a red capsule containing either beta carotene or a placebo. There were also several response variables—the study looked for heart attacks, several kinds of cancer, and other medical outcomes.

It may be that taking both aspirin and beta carotene together has effects (either good or bad) that we could not predict from studying aspirin alone and beta carotene alone. These joint effects are called **interactions** between the two explanatory variables. Some subjects in the Physicians' Health Study took aspirin alone, some took beta carotene alone, some took both, and some took neither. Comparing

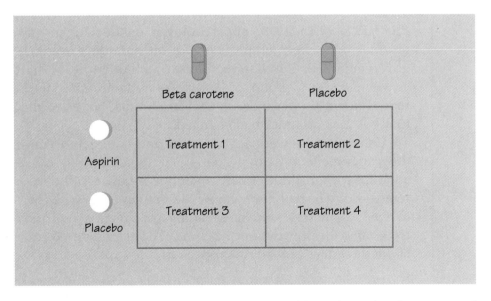

Figure 2-1 The treatments in the Physicians' Health Study. Each subject received one of these four drug combinations.

these four groups allowed the study to ask about interactions in addition to asking about the effects of taking aspirin and the effects of taking beta carotene.

Why experiment? We can study the effects of exactly the treatments we are interested in. Few people may choose on their own to take both aspirin and beta carotene, but an experiment can ask them to do so. We can study interactions among several explanatory variables. Most of all, we can hope to get good evidence for cause-and-effect relationships.

HOW TO EXPERIMENT BADLY

Julie's observational study of the effect of studying foreign languages (Example 2) failed because we cannot separate the effects of actually studying languages from the effects of being the kind of student who chooses to study languages. Badly designed experiments have the same weakness.

> EXAMPLE 4. In 1940, a psychologist conducted an experiment to study the effect of propaganda on attitude toward a foreign government. He administered a test of attitude toward the German government to a group of American students. After the students read German propaganda for several months, he tested them again to see if their attitudes had changed.
>
> Unfortunately, Germany attacked and conquered France while the experiment was in progress. The students did change their attitudes toward the German government between the test and the retest—but we shall never know how much of the change was due to the explanatory variable (reading propaganda) and how much to the historical events of that time. The data give no information about the effect of reading propaganda.

In both Example 2 and Example 4, the effect of the explanatory variable on the response is hopelessly mixed up with influences lurking in the background.

> **Confounding**
>
> A **lurking variable** is a variable that has an important effect on the relationship among the variables in a study but is not included among the variables studied.
>
> Two variables are **confounded** when their effects on a response variable cannot be distinguished from each other. The confounded variables may be either explanatory variables or lurking variables.

Experiments in the science laboratory often have a simple design: impose the treatment and see what happens. We can outline that design like this:

Treatment ⟶ Response

In the laboratory, we try to avoid confounding by rigorously controlling the environment of the experiment so that nothing except the experimental treatment influences the response. Once we get out of the laboratory, however, there are almost always lurking variables waiting to confound us. When our experimental units are people or animals rather than electrons or chemical compounds, confounding can happen even in the controlled environment of a laboratory or medical clinic. Here is an example:

> **EXAMPLE 5. Gastric freezing.** "Gastric freezing" is a clever treatment for stomach ulcers. The patient swallows a deflated balloon with tubes attached; then a refrigerated solution is pumped through the balloon for an hour. The idea is that cooling the stomach will reduce its production of acid and so relieve ulcers. An experiment reported in the *Journal of the American Medical Association* showed that gastric freezing did relieve ulcer pain. The design was:
>
> ### Gastric freezing ⟶ Reduced pain?
>
> The treatment was safe and easy and was widely used for several years.
>
>   Sorry once again: The patients' response to gastric freezing was confounded with the **placebo effect. A placebo** is a dummy treatment that can have no physical effect. Many patients respond

favorably to *any* treatment, even a placebo, presumably because of trust in the doctor and expectations of a cure. This response to a dummy treatment is the placebo effect.

A second experiment, done several years later, divided ulcer patients into two groups. One group was treated by gastric freezing as before. The other group received a placebo treatment in which the solution in the balloon was at body temperature rather than freezing. The results: 34% of the 82 patients in the treatment group improved, but so did 38% of the 78 patients in the placebo group. This and other properly designed experiments showed that gastric freezing was no better than a placebo, and its use was abandoned.[2]

Both observation and simple experiments often yield useless data because of confounding with lurking variables. It is hard to avoid

"I want to make one thing perfectly clear, Mr. Smith. The medication I prescribe will cure that run-down feeling."

confounding when only observation is possible. Experiments offer better possibilities—for example, both the Physicians' Health Study and the second gastric freezing experiment included a group of subjects who received only a placebo. This allows us to see whether the treatment being tested does better than a placebo and so has more than the placebo effect going for it. Aspirin passes this test, but gastric freezing does not.

SECTION 1 EXERCISES

2.1   **The gender gap.** There appears to be a "gender gap" in political party preference in the United States, with women more likely than men to prefer Democratic candidates. A political scientist selects a large sample of registered voters, both men and women. She asks each voter whether he or she voted for the Democratic or the Republican candidate in the last congressional election. Is this study an experiment? Why or why not? What are the explanatory and response variables?

2.2   **The effects of exercise.** Some people think that exercise raises the body's metabolic rate for as long as 12 to 24 hours, enabling us to continue to burn off fat after we end our workout. An exercise physiologist studying this effect asks subjects to walk briskly on a treadmill for several hours. He measures their metabolic rate before, immediately after, and 12 hours after the exercise. Is this study an experiment? Why or why not? What are the explanatory and response variables?

2.3   Before a new variety of frozen muffin is put on the market, it is subjected to extensive taste testing. People are asked to taste the new muffin and a competing brand, and to say which they prefer. (Both muffins are unidentified in the test.) Is this an observational study or an experiment? Explain your answer.

2.4   **Treating breast cancer.** What is the preferred treatment for breast cancer that is detected in its early stages? The most common treatment was once mastectomy (removal of the breast). It is now usual to remove the tumor and nearby lymph nodes, followed by radiation. To study whether these treatments differ in their effectiveness, a medical team examines the records of 25 large hospitals and compares the survival times after surgery of all women who have had either treatment.

(a)   What are the explanatory and response variables?

(b)   Explain carefully why this study is not an experiment.

(c)   Do you think this study will show whether a mastectomy causes longer average survival time? Explain your answer carefully.

2.5   An educator wants to compare the effectiveness of computer software that teaches reading with that of a standard reading curriculum. She tests the reading ability of a group of 60 fourth graders, then divides them into two classes of 30 students each. One class uses the computer regularly, while the other studies a standard curriculum. At the end of the year, she retests the students and compares the average increases in reading ability in the two classes.

(a)   Explain carefully why this study is an experiment.

(b)   What are the explanatory and response variables?

(c)   Do you think this study will show whether using the computer causes greater improvement in reading scores? Explain your answer carefully.

2.6   **Better corn?** New varieties of corn with altered amino acid content may have higher nutritional value than standard corn, which is low in the amino acid lysine. An experiment compares two new varieties, called opaque-2 and floury-2, with normal corn. The researchers mix corn-soybean meal diets using each type of corn at each of three protein levels: 12% protein, 16% protein, and 20% protein. They feed each diet to 10 one-day-old male chicks and record their weight gain after 21 days. The amount of weight the chicks gain is a measure of the nutritional value of their diet.

(a)   What are the experimental units and the response variable in this experiment?

(b) How many explanatory variables are there? How many treatments? Use a diagram like Figure 2-1 to describe the treatments.

(c) How many experimental units does the experiment require?

2.7 A chemical engineer is designing the production process for a new product. The chemical reaction that produces the product may have higher or lower yield, depending on the temperature and the stirring rate in the vessel in which the reaction takes place. The engineer decides to investigate the effects of combinations of two temperatures ($50°$ C and $60°$ C) and three stirring rates (60 rpm, 90 rpm, and 120 rpm) on the yield of the process. She will process two batches of the product at each combination of temperature and stirring rate.

(a) What are the experimental units and the response variable in this experiment?

(b) How many explanatory variables are there? How many treatments? Use a diagram like that in Figure 2-1 to describe the treatments.

(c) How many experimental units does the experiment require?

2.8 **The effect of public housing.** A study of the effect of living in public housing on family stability and other variables in poverty-level households was carried out as follows. The study team obtained a list of applicants accepted for public housing, together with a list of families who applied but were rejected by the housing authorities. A random sample was drawn from each list, and the two groups were observed for several years.

(a) Is this an experiment? Why or why not?

(b) What are the explanatory and response variables?

(c) Does this study contain confounding that may prevent valid conclusions on the effects of living in public housing? Explain.

2.9 **Safety of anesthetics.** The death rates of surgical patients are different for operations in which different anesthetics are used. An observational study found these death rates for four anesthetics:

Anesthetic	Halothane	Pentothal	Cyclopropane	Ether
Death rate	1.7%	1.7%	3.4%	1.9%

This is *not* good evidence that cyclopropane is more dangerous than the other anesthetics. Suggest some lurking variables that may be confounded with the choice of anesthetic in surgery, and that could explain the different death rates.[3]

2.10 **Nursing your baby.** An article in a women's magazine reported that women who nurse their babies feel warmer and more receptive toward the infants than mothers who bottle-feed. The author concluded that nursing has desirable effects on the mother's attitude toward the child. But women choose whether to nurse or bottle-feed. Explain why this fact makes any conclusion about cause and effect untrustworthy. Use the language of lurking variables and confounding in your explanation.

2.11 Last year only 10% of a group of adult men did not have a cold at some time during the winter. This year all the men in the group took 1 gram of vitamin C each day, and 20% had no colds. Explain why this result does not give good evidence that vitamin C prevents colds. In particular, describe some lurking variables that may have been confounded with taking vitamin C.

2.12 A state institutes a job training program for manufacturing workers who lose their jobs. After five years, the state reviews how well the program works. Critics claim that because the state's unemployment rate for manufacturing workers was 8% when the program began and 12% five years later, the program is ineffective.

(a)   Explain why higher unemployment does not necessarily mean that the training program failed. In particular, identify some lurking variables whose effect on unemployment may be confounded with the effect of the training program.

(b)   Briefly suggest how an experiment could have produced better information about the effects of the training program.

2.13 A college student thinks that drinking herbal tea will improve the health of nursing home patients. She and some friends visit a large nursing home regularly, serving herbal tea to the residents in one wing. Residents in the other wing are not visited. After six months, the first group had fewer days ill than the second. Are you convinced that herbal tea improves health? Explain your answer.

2.14 **The placebo effect.** A survey of physicians found that some doctors give a placebo to a patient who complains of pain for which the physician can find no cause. If the patient's pain improves, these doctors conclude that it had no physical basis. The medical school researchers who conducted the survey claimed that these doctors do not understand the placebo effect. Why?

▶ 2 RANDOMIZED COMPARATIVE EXPERIMENTS

The first goal in designing an experiment is to ensure that it will show us the effect of the explanatory variables on the response variables. The simple design

Experimental units $\longrightarrow$ **Treatment** $\longrightarrow$ **Response**

often fails to do this because of confounding. Here is an example of a properly designed experiment.

> EXAMPLE 6. **Sickle cell disease.** Sickle cell disease is an inherited disorder of the red blood cells that in the United States affects mostly blacks. It can cause severe pain and many complications. In 1992, the National Institutes of Health began a study of the drug hydroxyurea for treatment of sickle cell disease. The subjects were 300 adult patients who had had at least three episodes of pain from sickle cell disease in the previous year.
>
> Simply giving hydroxyurea to all 300 subjects would confound the effect of the medication with the placebo effect and other lurking variables such as the effect of knowing that you are a subject in an experiment. Instead, half of the subjects received hydroxyurea, and the other half received a placebo that looked and tasted the same. All subjects were treated exactly the same (same schedule of medical checkups, for example) except for the content of the medicine they took. Lurking variables therefore affected both groups equally and

should not cause any differences between their average responses. The placebo group is called a **control group** because it allows us to control the effects of lurking variables.

The two groups of subjects must be similar in all respects before they start taking the medication. Just as in sampling, the best way to avoid bias in choosing which subjects get hydroxyurea is to allow impersonal chance to make the choice. An SRS of 150 of the subjects formed the hydroxyurea group; the remaining 150 subjects made up the control group. Here is the experimental design in outline:

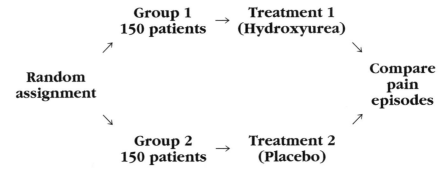

In January of 1995, the experiment was stopped ahead of schedule because the hydroxyurea group had only half as many pain episodes as the control group. This was compelling evidence that hydroxyurea is an effective treatment for sickle cell disease, good news indeed for those who suffer from this serious illness.[4]

Example 6 illustrates the simplest **randomized comparative experiment,** one that compares just two treatments. The diagram outlines the essential information about the design: random assignment; one group for each treatment; the number of subjects in each group (it is generally best to keep the groups similar in size); what treatment each group gets; and the response variable we compare. You know how to carry out the random assignment of subjects to groups. Label the 300 subjects 001 to 300, then read three-digit groups from the table of random digits (Table A) until you have chosen the 150 subjects for Group 1. The remaining 150 subjects form Group 2.

To compare more than two treatments, we can randomly assign the available experimental units or subjects to as many groups as there are treatments.

EXAMPLE 7. **Conserving energy.** Many utility companies have programs to encourage their customers to conserve energy. An electric company is considering placing electronic meters in households to show what the cost would be if the electricity use at that moment continued for a month. Will meters reduce electricity use? Would cheaper methods work almost as well? The company decides to design an experiment.

One cheaper approach is to give customers a chart and information about monitoring their electricity use. The experiment compares these two approaches (meter, chart) with each other and also with a control. The control group of customers receives no help in monitoring electricity use. The response variable is total electricity used in a year. The company finds 60 single-family residences in the same city willing to participate, so it assigns 20 residences at random to each of the three treatments. The outline of the design is:

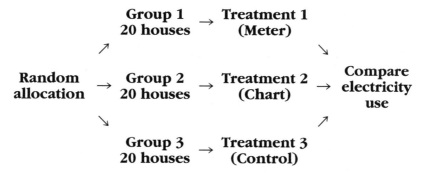

To carry out the random assignment, we label the 60 houses 01 to 60. We then enter Table A and select an SRS of 20 houses to receive the meters. Continuing in Table A, we select 20 more to receive charts. The remaining 20 form the control group.

THE LOGIC OF EXPERIMENTAL DESIGN

The randomized comparative experiment is one of the most important ideas in statistics. It is designed to allow us to draw cause-and-effect conclusions. Be sure you understand the logic:

▶ Randomization produces groups of experimental units that should be similar in all respects before we apply the treatments.

▶ Comparative design ensures that influences other than the experimental treatments operate equally on all groups.

▶ Therefore, differences in the response variable must be due to the effects of the treatments.

We use chance to choose the groups in order to eliminate any systematic bias in assigning the subjects to groups. In the sickle cell study, for example, a doctor might subconsciously assign the most seriously ill patients to the hydroxyurea group, hoping that the untested drug will help them. That would bias the experiment against hydroxyurea. Choosing an SRS of the subjects to be Group 1 gives everyone the same chance to be in either group. We expect the two groups to be similar in all respects—age, seriousness of illness, smoker or not, and so on. Chance tends to assign equal numbers of smokers to both groups, for example, even if we don't know which subjects are smokers.

Random assignment cannot *guarantee* that the groups are similar in all respects. There is no bias in choosing the groups, but we might by bad luck end up with all the smokers in one group. That could easily happen if we had only a few subjects. We would not trust the results of an experiment with only two subjects in each group. With 150 subjects in each group, it is likely that the smokers are nearly evenly divided between the groups. Just as we prefer large samples, we prefer experiments with many subjects. The outcomes of a randomized experiment, like the result of a random sample, depend on chance. In both settings, however, the laws of probability describe how much

---

Principles of experimental design

The basic principles of statistical design of experiments are:

1. Control the effects of lurking variables on the response, most simply by comparing several treatments.

2. Randomize—use impersonal chance to assign subjects to treatments.

3. Replicate—repeat the experiment on many subjects to reduce chance variation in the results.

chance variation is present. More subjects means that there is less chance variation among the treatment groups, and less chance variation in the outcomes of the experiment. "Use enough subjects" joins "compare several treatments" and "randomize" as a basic principle of statistical design of experiments.

STATISTICAL SIGNIFICANCE

The presence of chance variation requires us to look more closely at the logic of randomized comparative experiments. We cannot say that *any* difference in the average number of pain episodes between the hydroxyurea and control groups must be due to the effect of the drug. There will be some differences even if both treatments are the same, because there will always be some differences in the individuals who are our subjects. Even though randomization eliminates systematic differences between the groups, there will still be chance differences. We should insist on a difference in the responses so large that it is unlikely to happen just because of chance variation.

Statistical significance

An observed effect so large that it would rarely occur by chance is called **statistically significant**.

The difference between the average number of pain episodes for subjects in the hydroxyurea group and the average for the control group was "highly statistically significant." That means that a difference this large would almost never happen just by chance. We do indeed have strong evidence that hydroxyurea beats a placebo in helping sickle cell disease sufferers. You will often see the phrase "statistically significant" in reports of investigations in many fields of study. It tells you that the investigators found good evidence for the effect they were seeking.

SECTION 2 EXERCISES

2.15 You are testing a vaccine for use against a dangerous virus. You have available 10 rats (named below), which you will expose to the virus. Unprotected rats often die when infected.

Alfie	Chuck	Frank	Lyman	Polyphemus
Bernie	David	Harry	Mercedes	Zaffo

(a)   Outline the design of an experiment to test the effectiveness of the vaccine. Use the diagram in Example 6 as a model.

(b)   Use line 140 of Table A to carry out the random assignment.

(c)   Why would you prefer to use more rats in the experiment?

2.16   **Calcium and blood pressure.** Some medical researchers suspect that added calcium in the diet reduces blood pressure. You have available 40 men with high blood pressure who are willing to serve as subjects.

(a)   Outline an appropriate design for the experiment, taking the placebo effect into account. Use the diagram in Example 6 as a model.

(b)   The names of the subjects appear below. Use Table A beginning at line 111 to do the randomization required by your design. List the subjects to whom you will give the drug.

Abrams	Danielson	Gutierrez	Lippman	Rosen
Adamson	Durr	Howard	Martinez	Sugiwara
Afifi	Edwards	Hwang	McNeill	Thompson
Brown	Fluharty	Iselin	Morse	Travers
Cansico	Garcia	Janle	Ng	Turing
Chen	Gerson	Kaplan	Quinones	Ullmann
Cortez	Green	Kim	Rivera	Williams
Curzakis	Gupta	Lattimore	Roberts	Wong

2.17   **Do charts help investors?** Some investment advisers believe that charts of past trends in the prices of securities can help predict future prices. Most economists disagree. In an experiment to examine the effects of using charts, business students trade (hypothetically) a foreign currency at computer screens. There are 20 student subjects, named for convenience A, B, C,..., T. Their goal is to make as much money as possible, and the best performances are rewarded with small prizes. The student traders have the price history of the foreign currency in dollars in their computers. They may or may not also be given software that charts past trends. Describe a design to test whether the chart software helps traders make more money. Use Table A to carry out the randomization required by your design.

2.18 A college allows students to choose either classroom or self-paced instruction in a basic economics course. The college wants to compare the effectiveness of self-paced and regular instruction. A professor proposes giving the same final exam to all students in both versions of the course and comparing the average score of those who took the self-paced option with the average score of students in regular sections.

(a) Explain why confounding makes the results of that study worthless.

(b) Given 30 students who are willing to use either regular or self-paced instruction, outline an experimental design to compare the two methods of instruction. Then use Table A starting at line 108 to carry out the randomization.

2.19 **Medical records as data.** A large study used records from Canada's national health care system to compare the effectiveness of two ways to treat prostate disease. The two treatments are traditional surgery and a new method that does not require surgery. The records described many patients whose doctors had chosen each method. The study found that patients treated by the new method were significantly more likely to die within 8 years.[5]

(a) Further study of the data showed that this conclusion was wrong. The extra deaths among patients who got the new method could be explained by lurking variables. What lurking variables might be confounded with a doctor's choice of surgical or nonsurgical treatment?

(b) You have 300 prostate patients who are willing to serve as subjects in an experiment to compare the two methods. Use a diagram to outline the design of a randomized comparative experiment.

(c) Use Table A to choose *only the first 5* subjects for the surgery group. Be sure to explain how you used the table.

2.20   **Does child care attract employees?** Will providing child care for employees make a company more attractive to women, even those who are unmarried? You are designing an experiment to answer this question. You prepare recruiting material for two fictitious companies, both in similar businesses in the same location. Company A's brochure does not mention child care. There are two versions of Company B's material, identical except that one describes the company's on-site child-care facility. Your subjects are 150 unmarried women who are college seniors seeking employment. Each subject will read recruiting material for both companies and choose the one she would prefer to work for. You will give each version of Company B's brochure to half the women. You suspect that a higher percentage of those who read the description that includes child care will choose Company B.

(a)   Outline the design of the experiment. Be sure to identify the response variable.

(b)   Use Table A to choose *only the first 5* subjects to receive the brochure that mentions child care. Be sure to explain how you used the table.

2.21   You wish to compare three treatments for effectiveness in preventing flu: (1) a flu vaccine, (2) 1 gram of vitamin C per day, and (3) a placebo taken daily. Describe how you would use 600 volunteer subjects in a designed experiment to compare these treatments. (Do not actually do any randomization, but do include a diagram showing your design. Be sure to identify your response variable.)

2.22   **Raising turkeys.** Turkeys raised commercially for food are often fed the antibiotic salinomycin to prevent infections from spreading among the birds. Salinomycin can damage the birds' internal organs, especially the pancreas. A researcher believes that adding vitamin E to the diet may prevent injury. He wants to explore the effects of three levels of vitamin E added to the diet of turkeys along with the usual dose of salinomycin. There are 30 turkeys available for the study. At the end of the study, the birds will be killed and each pancreas will be examined under a microscope.

Outline the design of a randomized comparative experiment in this setting. Then label the 30 turkeys 00 to 29 and use Table A beginning at line 125 to carry out the random assignment required by your design.

2.23 To demonstrate how randomization reduces confounding, consider the following situation: A nutrition experimenter intends to compare the weight gain of newly weaned male rats fed diet A with that of rats fed diet B. To do this, she will feed each diet to 10 rats. She has available 10 rats of genetic strain 1 and 10 of strain 2. Strain 1 is more vigorous, so if the 10 rats of strain 1 were fed diet A, the effects of strain and diet would be confounded, and the experiment would be biased in favor of diet A.

(a) Label the rats 00, 01, ..., 19. Use Table A to assign 10 rats to diet A. Do this four times, using different parts of the table, and write down the four groups assigned to diet A.

(b) Unknown to you, the rats labeled 00, 02, 04, 06, 08, 10, 12, 14, 16, and 18 are the 10 strain 1 rats. How many of these rats were in each of the four diet A groups that you generated? What was the average number of strain 1 rats assigned to diet A?

2.24 Comparison alone does not make a study an experiment. Example 2 in Section 1 is an observational study that compares two groups, but it is not an experiment. Describe clearly the difference between a comparative observational study and a comparative experiment. What advantages do comparative experiments have over comparative observational studies?

2.25 You read in a newspaper that "nonphysical treatments such as meditation and prayer have been shown to be effective in controlled scientific studies for such ailments as high blood pressure, insomnia, ulcers, and asthma." Explain in simple language what the article means by "controlled scientific studies" and why such studies can show that meditation and prayer are effective treatments for some medical problems.

2.26 **Calcium and blood pressure.** A randomized comparative experiment examines whether a calcium supplement in the diet reduces the blood pressure of healthy men. The subjects receive

Heart Treatment's Value Doubted

By Michael Specter
Washington Post Staff Writer

In a study with far-reaching implications for the routine treatment of heart attack patients, researchers have found that the immediate use of special balloons to force open clogged arteries is unnecessary in the vast majority of cases if the patient is treated with a clot-dissolving drug.

Most heart specialists have assumed that the balloon treatment, an expensive and increasingly popular procedure called balloon angioplasty, should routinely follow the use of drugs, such as TPA, that dissolve the blood clots that cause most heart attacks.

But in a study of 3,262 heart attack patients that is expected to transform the standards for treatment of the nation's leading killer, researchers at medical centers across the country found the extra measures were rarely needed. Half of the randomly selected patients were treated solely with the clot-dissolving drug TPA, while the other half were given TPA followed by angioplasty.

The results, reported in today's issue of The New England Journal of Medicine, showed that for most people adding angioplasty was no better than relying on the less complicated and less costly drug treatment.

Only 10 percent of the group assigned solely to drug treatment died or had another heart attack in the six weeks following treatment. By comparison, 11 percent of the group assigned to receive the combination drug and angioplasty treatment suffered the same fate. The difference could have occurred by chance. But the fact that angioplasty did not prove to be the better option was a shock even to some of the nation's most renowned heart experts.

"There is no question that it bucks the trend," said Eugene Braunwald, professor of medicine at Harvard Medical School and the study chairman for the research project. "It will spare many patients unnecessary surgery and reduce the cost of medical care greatly."

either a calcium supplement or a placebo for 12 weeks. The researchers conclude that "the blood pressure of the calcium group was significantly lower than that of the placebo group."

"Significant" in this conclusion means statistically significant. Explain what statistically significant means in the context of this experiment, as if you were speaking to a doctor who knows no statistics.

2.27   The financial aid office of a university asks a sample of students about their employment and earnings. The report says that "for academic year earnings, a statistically significant difference was found between the sexes, with men earning more on the average. No significant difference was found between the earnings of black and white students." Explain both of these conclusions, for the effects of sex and of race on average earnings, in language understandable to someone who knows no statistics.

2.28   A psychologist reports that "in our sample, ethnocentrism was significantly higher among church attenders than among nonattenders." Explain what this means in language understandable to someone who knows no statistics. Do not use the word "significance" in your answer.

2.29   **Understanding the news.** The article from which the excerpts on the facing page were taken appeared in the *Washington Post* of March 9, 1989. Describe the design of the experiment in as much detail as the article allows. Include the treatments, the response variable, and the design outline. Does the design of the experiment give you confidence in its result?

2.30   **Understanding the news.** The excerpts on the next page are from an Associated Press dispatch that appeared in newspapers on December 27, 1995.[6]

(a)   What treatments did this experiment compare? What response variables are mentioned?

(b)   What important information about the design of the experiment appears in the article?

(c)   Is any important information omitted? If so, what?

(d)   Based on a combination of the information in the article and your knowledge of sound experimental design, make an outline that shows how you think the experiment was designed.

For fat men, losing weight even smarter than exercise

The best exercise for your heart is pushing yourself away from the table, a new study suggests.

Losing weight by itself works better than aerobic exercise by itself in reducing the risk of heart disease, a study of fat men found.

Dr. Leslie I. Katzel and his colleagues at the University of Maryland School of Medicine studied 111 men, ages 46 to 60, who were sedentary and obese—that is,

20 percent to 60 percent overweight—but were otherwise healthy.

The men were divided into groups; one pursued weight loss without exercise; the second exercised without trying to lose weight; and the third neither exercised nor dieted.

Researchers were surprised to find weight loss clearly produced more benefits than exercise alone:

- Levels of "good" cholesterol improved 13 percent in the weight-loss group, vs. virtually no change in the exercise group.

- Blood pressure dropped 8 percent in the weight-loss group, compared with a 2 to 3 percent drop among exercisers.

"Based on these results, we feel that if you're overweight, you really need to lose weight to decrease your chances of developing heart disease," Katzel said.

The results were expected to be the same among women, who are being studied, Katzel said.

▶ 3 EXPERIMENTS IN PRACTICE

Probability samples are a big idea, but they don't solve all the difficulties of sampling in practice. Randomized comparative experiments are also a big idea, but they don't solve all the difficulties of experimenting. A sampler must know exactly what information she wants and must compose questions that extract that information from her sample. An experimenter must know exactly what treatments and responses he wants information about, and he must construct the apparatus needed to apply the treatments and measure the responses. This is what psychologists or medical researchers or engineers mean when they talk about "designing an experiment." We are concerned with the *statistical* side of designing experiments, ideas that apply to experiments in psychology, medicine, engineering, and other areas as well. Even at this general level, you should understand that practical problems can prevent an experiment from giving useful data.

HIDDEN BIAS

The logic of a randomized comparative experiment assumes that all of the subjects are treated alike except for the treatments that the experiment is designed to compare. Any other unequal treatment can cause bias.

> EXAMPLE 8. **Double-blind experiments.** Subjects in a medical experiment are not told whether they are receiving a standard drug, a new drug, or a placebo. Knowing that they are getting "just a placebo" would no doubt reduce their expectation of success and bias the experiment in favor of the other treatments. If doctors and other medical personnel know that a subject is getting "just a placebo," they may also expect less than if they know the subject is receiving a promising experimental drug. The doctors' expectations can subconsciously change the way they interact with patients and even the way they diagnose a patient's condition. In a **double-blind experiment,** both the subjects and everyone who works with them are kept in the dark about which treatment each subject is receiving. Until the study ends and the results are in, only the study's statistician knows for sure.

> EXAMPLE 9. **Lab rats and rabbits.** Rats and rabbits, specially bred to be uniform in their inherited characteristics, are the subjects in many experiments. It turns out that animals, like people, are quite sensitive to how they are treated. This creates some amusing opportunities for hidden bias.
>
> Studies of the nutritional value of a new breakfast cereal compare the weight gains of young rats fed the new product and rats fed a standard diet. The rats are randomly assigned to diets and are

"Dr. Burns, are you sure this is what the statisticians call a double-blind experiment?"

housed in large racks of cages. It turns out that rats in upper cages grow a bit faster than rats in bottom cages. If the experimenters put rats fed the new product at the top and those fed the standard diet below, the experiment is biased in favor of the new product. Solution: Assign the rats to cages at random.[7]

Another study looked at the effects of human affection on the cholesterol level of rabbits. All of the rabbit subjects ate the same diet. Some (chosen at random) were regularly removed from their cages to have their furry heads scratched by friendly people. The rabbits who received affection had lower cholesterol. So affection for some but not other rabbits could bias an experiment in which the rabbits' cholesterol level is a response variable.

REFUSALS, NONADHERERS, AND DROPOUTS

Sample surveys suffer from nonresponse due to failure to contact some people selected for the sample and the refusal of others to participate. Experiments with human subjects suffer from similar problems.

> **EXAMPLE 10. Minorities in medical experiments.** Refusal to participate is a serious problem for medical experiments on treatments for serious diseases such as cancer. As in the case of samples, bias can result if those who refuse are systematically different from those who cooperate. An article in the *New York Times* said:[8]
>
> *Patients are often reluctant to serve as "human guinea pigs," even though they may stand to benefit directly from the study. Many patients fear being randomized into the placebo group. A study of African-Americans, Hispanic people, and American Indians, who have very low rates of participation in clinical trials, revealed a lack of information and a basic "mistrust of white people" who generally run the trials.*

Subjects who participate but don't follow the experimental treatment, called **nonadherers,** can also cause bias. AIDS patients who participate in trials of a new drug sometimes take other treatments on their own, for example. What is more, some AIDS subjects have their medication tested and drop out or add other medications if they were not assigned to the new drug. This may bias the trial against the new drug.

Experiments that continue over an extended period of time also suffer **dropouts,** subjects who begin the experiment but do not complete it. If the reasons for dropping out are unrelated to the experimental treatments, no harm is done other than reducing the number of subjects. If subjects drop out because of their reaction to one of the treatments, some bias can result.

EXAMPLE 11. Dropouts in a medical study. Exercise 2.30 on page 115 describes an experiment that showed that weight loss beats exercise for improving factors such as high cholesterol and high blood pressure that put men at risk of a heart attack. The study began with 170 subjects, who were randomly assigned to a weight-loss program, an exercise program, or a control group. By the end of the study a year later, 29 of the 73 men in the weight-loss group had dropped out, along with 22 of the 71 in the exercise group and 8 of the 26 in the control group. The study's conclusions are based on the 111 of the original 170 subjects who completed their assigned treatment.

The dropouts in the exercise group were mostly due to the time commitment required to come to the exercise program three times a week. Many of the dropouts in the other two groups, however, were unhappy at being assigned to a program that included no exercise. Because the percents who dropped out were similar in all three groups, the study authors believe that the dropouts did not cause a strong bias.[9]

WERE THE TREATMENTS REALISTIC?

Medical experiments test real therapies on real patients. In many other settings, experimental treatments can only imitate the situations we want to study. This is true in particular of many experiments on human behavior.

EXAMPLE 12. Studying frustration. A psychologist wants to study the effects of failure and frustration on the relationships between members of a work team. She forms a team of students, brings them to the psychology laboratory, and has them play a game that requires teamwork. The game is rigged so that they lose regularly. The psychologist observes the students through a one-way window and notes the changes in their behavior during an evening of game-playing.

Playing a game in a laboratory for small stakes, knowing that the session will soon be over, is a long way from working for months developing a new product that never works right and is finally abandoned by your company. Does the behavior of the students in the lab tell us much about the behavior of the team whose product failed?

Psychologists do their best to devise clever experiments that avoid artificial settings like that of Example 12. When they study behavior, lack of realism remains a barrier to firm conclusions. What is worse, subjects in any experiment may change their normal behavior because they know they are being studied.

> **EXAMPLE 13. The Hawthorne effect.** In the 1920s, the Hawthorne Works of the Western Electric Company tried to discover by experiment what changes in working conditions would improve the productivity of their workers. They found that *any* change made while the workers knew that a study was going on increased productivity. More lighting helped, but so did less lighting. The fact that people change their behavior when they know they are being studied is now called the *Hawthorne effect*.

The Hawthorne effect makes it hard to find treatments that will reliably change human behavior when people are not being studied. It is another strong argument for comparative experiments, in which the Hawthorne effect at least acts equally on all the treatment groups.

HOW FAR CAN WE GENERALIZE OUR CONCLUSIONS?

A well-designed experiment tells us whether the treatments caused a change in the response for this particular group of subjects. Experimenters usually want to draw conclusions about a much larger group of people or things. We just noted that they may also want to draw conclusions about the effects of treatments that are not exactly like those applied in the actual experiment. How far the conclusions of an experiment generalize is often unclear. Example 12 shows the problems: the psychologist wants conclusions about adult workers (not just students) faced with genuine failure (not just losing a game in a laboratory). Here are some examples with more impact on public policy.

> **EXAMPLE 14. Center brake lights.** Cars sold in the United States since 1986 have been required to have high center brake lights in addition to the usual two brake lights at the rear of the vehicle. This safety requirement was justified by randomized comparative experiments with fleets of rental and business cars. The experiments showed that the third brake light reduced rear-end collisions by as much as 50%.
>
> After almost a decade in actual use, the Insurance Institute found only a 5% reduction in rear-end collisions, helpful but much less than the experiments predicted. What happened? Most cars did not have the extra brake light when the experiments were carried out, so it caught the eye of following drivers. Now that almost all cars have the third light, they no longer capture attention. The experimental conclusions did not generalize as well as safety experts hoped.

> **EXAMPLE 15. Diet and cancer.** Substances that cause cancer should not appear in our food. We don't want to experiment on people to learn what substances cause cancer, so we experiment on rats instead. The rats are specially bred to have more tumors than humans do. They are fed large doses of the test chemical for most of their natural life, about two years. These randomized comparative experiments tell us whether a substance fed in *high doses* causes cancer in *rats*. What we want to know is whether *low doses* cause cancer in *people*. Many scientists now think that rat results overstate the risk to people.

THE DELANEY AMENDMENT

Sec. 409 (c) (3) (A). No additive shall be deemed to be safe if it is found to induce cancer when ingested by man or animal, or if it is found, after tests which are appropriate for the evaluation of the safety of food additives, to induce cancer in man or animal.

–Federal Food, Drug and Cosmetic Act, 1958

Federal law long made the problem worse. The Delaney amendment prohibited the presence in food of any amount of an additive that causes cancer in animals. Current technology can detect in food minute traces of farm pesticides that (in high doses) cause cancer in rats. The traces in our food are so small that most scientists think they pose very little risk. In 1996, Congress replaced the Delaney amendment with a "negligible risk" standard. The rules change, but the issue remains. Stay tuned.[10]

When experiments are not fully realistic, statistical analysis of the experimental data cannot tell us how far the results will generalize. Experimenters generalizing from rats to people must argue based on their understanding of how rats and people function. Other experts may disagree. This is one reason why a single experiment is rarely completely convincing, despite the compelling logic of experimental design. The true scope of a new finding must usually be explored by a number of experiments in various settings.

A convincing case that an experiment is sufficiently realistic to produce useful information is based not on statistics but on the experimenter's knowledge of the subject matter of the experiment. The attention to detail required to avoid hidden bias also rests on subject matter knowledge. Good experiments combine statistical principles with understanding of a specific field of study.

SECTION 3 EXERCISES

2.31 **The Physicians' Health Study.** The article in the *New England Journal of Medicine* that presents the final results of the Physicians' Health Study begins with these words: "The Physicians' Health Study is a randomized, double-blind, placebo-controlled trial designed to determine whether low-dose aspirin (325 mg every other day) decreases cardiovascular mortality and whether beta carotene reduces the incidence of cancer." Doctors are expected to understand this. Explain to a doctor who knows no statistics what "randomized," "double-blind," and "placebo-controlled" mean.

2.32   Fizz Laboratories, a pharmaceutical company, has developed a new pain-relief medication. Sixty patients suffering from arthritis and needing pain relief are available. Each patient will be treated and asked an hour later, "About what percentage of pain relief did you experience?"

(a)   Why should Fizz not simply administer the new drug and record the patients' responses?

(b)   Design an experiment to compare the drug's effectiveness with that of aspirin and that of a placebo.

(c)   Should patients be told which drug they are receiving? How would this knowledge probably affect their reactions?

(d)   If patients are not told which treatment they are receiving, the experiment is single-blind. Should this experiment be double-blind? Explain.

2.33   **The effects of meditation?** An experiment that was publicized as showing that a meditation technique lowered the anxiety level of subjects was conducted as follows: the experimenter interviewed the subjects and assessed their levels of anxiety. The subjects then learned how to meditate and did so regularly for a month. The experimenter re-interviewed them at the end of the month and assessed whether their anxiety levels had decreased or not.

(a)   There was no control group in this experiment. Why is this a blunder? What lurking variables may be confounded with the effect of meditation?

(b)   The experimenter who diagnosed the effect of the treatment knew that the subjects had been meditating. Explain how this knowledge could bias the experimental conclusions.

(c)   Briefly discuss a proper experimental design, with controls and blind diagnosis, to assess the effect of meditation on anxiety level.

2.34   **Taste tests.** Taste tests ask subjects to compare the taste of two food products, such as Pepsi and Coke, and say which they prefer. There is no separate control group—each subject serves as his or her own control by tasting both products.

(a)   Randomization remains important in a taste test. How should randomization be used?

(b)   How does the idea of blindness apply in a taste test?

2.35   **Dealing with dropouts**. Experiments with human subjects that continue over a long period of time face the problem of dropouts and nonadherers. In an account of a large medical experiment, we read the following:[11]

For this complex problem, we followed the generally accepted practice of comparing the mortality experience of the originally randomized groups, and of not eliminating dropouts or non-adherers from the analysis. This practice is conservative in that it dilutes whatever treatment effects, beneficial or adverse, are present.

Subjects who drop out are in effect counted as "living" when the group death rates are compared. Explain why this practice reduces the observed effect of a treatment relative to a placebo.

2.36   A number of clinical trials have studied whether reducing blood cholesterol using either drugs or diet reduces heart attacks. The first researchers followed their subjects for five to seven years. In order to see results in this relatively short time, the subjects were chosen from the group at greatest risk, middle-aged men with high cholesterol levels or existing heart disease. The experiments generally showed that reducing blood cholesterol does decrease the risk of a heart attack. Some doctors questioned whether these experimental results applied to many of their patients. Why?

2.37   For what population do you think the experimental conclusions are valid in each of the following cases? (You cannot be definite about this question without expert knowledge, so your answers will be partly guesses.)

(a)   The Physicians' Health Study (Example 1 on page 94).

(b)   The energy conservation experiment of Example 7 (page 107).

(c)   The behavioral experiment of Example 12 (page 120).

▶ 4 OTHER EXPERIMENTAL DESIGNS

The experimental designs we have met all have the same pattern: divide the subjects at random into as many groups as there are treatments, then apply each treatment to one of the groups. These are *completely randomized* designs.

Completely randomized design

In a **completely randomized** experimental design, all the experimental units are allocated at random among all the treatments.

A completely randomized design can have any number of explanatory variables. The Physicians' Health Study, for example, had two variables: aspirin or placebo and beta carotene or placebo. Figure 2-1 on page 97 shows how these variables form four treatments. The completely randomized design assigns one-fourth of the 22,000 subjects to each of these treatments. Once the layout of treatments is set, the randomization needed for a completely randomized design is tedious but straightforward.

BLOCK DESIGNS

Completely randomized designs are the simplest statistical designs for experiments. They are the analog of simple random samples. In fact, each treatment group is an SRS drawn from the available subjects. Completely randomized designs illustrate clearly the principles of control, randomization, and replication. However, just as in sampling, more elaborate statistical designs are often superior. In particular, matching the subjects in various ways can produce more precise results than simple randomization.

> **EXAMPLE 16. Men, women, and advertising.** Women and men respond differently to advertising. An experiment to compare the effectiveness of three television commercials for the same product will want to look separately at the reactions of men and women, as well as assess the overall response to the ads.

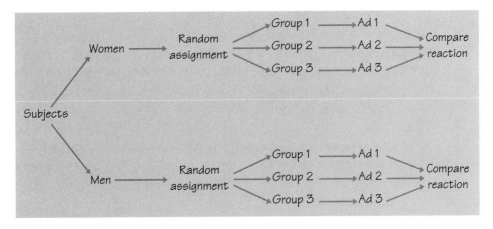

Figure 2-2 A block design to compare the effectiveness of three TV advertisements. Male and female subjects form two blocks.

A completely randomized design considers all subjects, both men and women, as a single pool. The randomization assigns subjects to three treatment groups without regard to their sex. This ignores the differences between men and women. A better design considers women and men separately. Randomly assign the women to three groups, one to view each commercial. Then separately assign the men at random to three groups. Figure 2-2 outlines this improved design.

The design of Figure 2-2 uses the principles of comparison, randomization, and replication. However, the randomization is not complete (all subjects randomly assigned to treatment groups). It is restricted to operate only within groups of similar subjects. The groups are called *blocks,* and the design is called a *block design.*

Block design

A **block** is a group of experimental units or subjects that are similar in ways that are expected to affect the response to the treatments. In a **block design**, the random assignment of units to treatments is carried out separately within each block.

Blocks are another form of *control*. They control the effects of some lurking variables (such as the sex of the subjects in Example 16) by bringing those variables into the experiment to form the blocks. Block designs are similar to stratified designs for sampling. Blocks and strata both group similar units. We use two different names only because the idea developed separately for sampling and experiments. Blocks allow us to draw separate conclusions about each block; for example, about men and women in Example 16. Blocking also allows more precise overall conclusions, because the systematic differences between men and women can be removed when we study the overall effects of the three commercials.

MATCHED PAIRS DESIGNS

A special type of block design is the **matched pairs design.** Matched pairs designs compare two treatments. Each block consists of just two units, as closely matched as possible. These units are assigned at random to the treatments by tossing a coin or reading odd and even digits from Table A. Alternatively, each block in a matched pairs design may consist of just one subject who gets both treatments one after the other. Each subject then serves as his or her own control.

> **EXAMPLE 17. Right hand versus left hand.** Is the right hand of right-handed people generally stronger than the left? A student designs an experiment to study this question. She fastens an ordinary bathroom scale to a shelf five feet from the floor, with the end of the scale projecting out from the shelf. Subjects squeeze the scale between their thumb beneath it and their fingers on top. The scale reading in pounds measures hand strength.
>
> A completely randomized design would require two groups of subjects, one for each hand. Surely it is more natural to have each subject use both hands, so that we have a direct right-left comparison. This is a matched pairs design. The name reminds us that instead of comparing two groups we make comparisons within matched pairs of observations, in this case the two hands of the same person.

What about randomization in this experiment? We can't assign subjects at random to groups. Instead, we choose at random which hand each subject tries first. A subject may gain confidence or learn better how to grasp the scale after the first try. Randomizing the order prevents this learning from being confounded with right versus left hand.

SECTION 4 EXERCISES

2.38 Suppose that you have 20 subjects to assign to the treatments for the Physicians' Health Study (Figure 2-1 on page 97). Call the subjects A, B, C, ..., T for convenience. Use Table A starting at line 128 to do the randomization.

2.39 A food products company is preparing a new cake mix for marketing. It is important that the taste of the cake not be changed by small variations in baking time or temperature. In an experiment, cakes made from the mix are baked at 300, 320, and 340°F, and for 1 hour and 1 hour and 15 minutes. Ten cakes are baked at each combination of temperature and time. A panel of tasters scores each cake for texture and taste.

(a) What are the explanatory variables and the response variables for this experiment?

(b) Make a diagram like Figure 2-1 on page 97 to describe the treatments. How many treatments are there? How many units (cakes) are needed?

(c) Explain why it is a bad idea to bake all 10 cakes for one treatment at once, then bake the 10 cakes for the second treatment, and so on. Instead, the experimenters will bake the cakes in a random order determined by the randomization in your design.

2.40 **Calcium and blood pressure.** You are participating in the design of a medical experiment to investigate whether a calcium supplement in the diet will reduce the blood pressure of middle-aged men. Preliminary work suggests that calcium may be effective and that the effect may be greater for black men than for white men.

(a)   Outline the design of an appropriate experiment.

(b)   Explain in plain language the advantage of using larger groups of subjects.

2.41   You suspect that a drug affects the coordination of subjects. The drug can be administered in three ways: orally, by injection under the skin, or by injection into a vein. The effect of the drug may depend on the method of administration as well as on the dose administered. You therefore wish to study the effects of two explanatory variables: dose (at two levels) and type of administration (by the three methods mentioned). The response variable is the score of the subjects on a standard test of coordination. Ninety subjects are available.

(a)   List the treatments that can be formed from the two explanatory variables.

(b)   Describe an appropriate completely randomized design. (Just outline the design; don't do any randomization.)

(c)   You could study the effect of dose by performing an experiment comparing two doses for one method of administration. You could separately study the effect of administration by comparing the three methods for one dose. What advantages does the experiment you designed in (a) have over these two experiments together?

2.42   Twenty overweight females have agreed to participate in a study of the effectiveness of four weight-loss treatments. Call the treatments A, B, C, and D. The researcher first calculates how overweight each subject is by comparing her actual weight with her "ideal" weight. The subjects and their excess weights in pounds are:

Birnbaum	35	Hernandez	25	Moses	25	Smith	29
Brown	34	Jackson	33	Nevesky	39	Stall	33
Brunk	30	Kendall	28	Obrach	30	Tran	35
Cruz	34	Loren	32	Rodriguez	30	Wilansky	42
Deng	24	Mann	28	Santiago	27	Williams	22

The response variable is the weight lost after eight weeks of treatment. Because a subject's excess weight will influence the response, we will use a block design.

(a) Arrange the subjects in order of increasing excess weight. Form 5 blocks of 4 subjects each by grouping the 4 least overweight, then the next 4, and so on.

(b) Use Table A to randomly assign the 4 subjects in each block to the 4 weight-loss treatments. Be sure to explain exactly how you used the table. This block design matches the 4 groups as closely as possible in excess weight at the start of the study.

2.43 In Exercise 23 on page 113, a nutritionist had 10 rats of each of two genetic strains. We can control the effect of genetic strain by treating the strains as blocks and randomly assigning five rats of each strain to diet A. The remaining five rats of each strain receive diet B. Outline the design. Then use Table A beginning at line 111 to do the randomization.

2.44 An agronomist wishes to compare the yield of five corn varieties. The field in which the experiment will be carried out increases in fertility from north to south. The agronomist therefore divides the field into 30 plots of equal size, arranged in six east-west rows of five plots each, and employs a block design with the rows of plots as the blocks. Identify the experimental units, the treatments, and the blocks. Outline the arrangement of the block design, but do not actually do the randomization.

2.45 Do consumers prefer the taste of Pepsi or Coke in a blind test in which neither cola is identified? Describe briefly the design of a matched pairs experiment to investigate this question.

2.46 **The culture of Mexican Americans.** There are several psychological tests that measure the extent to which Mexican Americans are oriented toward Mexican/Spanish or Anglo/English culture. Two such tests are the Bicultural Inventory (BI) and the Acculturation Rating Scale for Mexican Americans (ARSMA). To study the correlation between the scores on these two tests, researchers will give both tests to a group of 22 Mexican Americans.

(a) Briefly describe a matched pairs design for this study. In particular, how will you use randomization in your design?

(b) You have an alphabetized list of the subjects (call them A to V). Carry out the randomization required by your design and report the result.

2.47 **Do charts help investors?** Some investment advisors believe
that charts of past trends in the prices of securities can help predict future prices. Most economists disagree. In an experiment
to examine the effects of using charts, business students trade
(hypothetically) a foreign currency at computer screens. There
are 20 student subjects available, named for convenience A, B,
C, ..., T. Their goal is to make as much money as possible,
and the best performances are rewarded with small prizes. The
student traders have the price history of the foreign currency in
dollars in their computers. They may or may not also be given
software that charts past trends. Describe *two* designs for this experiment, a completely randomized design and a matched pairs
design in which each student serves as his or her own control.
In both cases, carry out the randomization required by your
design.

▶ 5 ISSUES: PUBLIC POLICY EXPERIMENTS

The effectiveness of a new fertilizer or a new drug is always tested
by a randomized comparative experiment—and for good reason. A
well-designed experiment can provide clearer answers than any other
method of study. What about testing the effectiveness of a new welfare
program or health insurance system or preschool education program?
Public policy decisions in these areas have usually been based on much
supposition and little knowledge. It is tempting to try an experiment in
the hope of finding clear answers.

Recent years have seen many statistically designed experiments to
test changes in public policy. Several of these experiments provided
support for changes in the welfare system that supports poor families.
Here is an example that, on the whole, illustrates the difficulties of
public policy experiments better than it displays their strengths.

> EXAMPLE 18. **The New Jersey Income-Maintenance Experi-
> ment.** This early experiment on welfare policy began in 1968. It
> examined the effects of replacing fixed welfare payments that end
> as soon as the family earns more than a small amount of other

income with a sliding scale of payments that decrease gradually as the family earns more. Supporters of the new policy argued that it gives more incentive to work because welfare payments don't disappear as soon as the recipient finds a minimum-wage job. Opponents said the new program would be too expensive.

The experimental units were households chosen at random from those that met the experiment's criteria. This initial random sampling helped ensure that the households in the experiment were typical of the larger population of welfare households. In the end, 1357 households participated.

There were two explanatory variables, the guaranteed minimum income level and the rate at which welfare payments decreased as earnings went up. Eight combinations of these variables were tried, with the current welfare system as a control, giving nine treatments in all. The many response variables included measures of family income, employment, and family stability. The 1357 families that participated were divided into three blocks of 400 to 500 families each, depending on their recent income level. The block design was used because recent income helps predict future income.[12]

The New Jersey researchers faced abundant practical difficulties. They had to contend with dark stairwells, vicious dogs, hostile local politicians, and militant community groups. Moreover, 10% of the dwelling units in the initial sample were empty, 19% of the remaining households were never at home (in five tries), and 18% refused to speak to the interviewer in four visits. An initial sample of 48,000 households yielded 27,000 that were interviewed. Of these, only 3124 met the detailed criteria for participation in the experiment. Another 425 families had vanished by the time interviewers returned to invite them to join the experiment, and others refused, leaving the final 1357.

The New Jersey welfare experiment spent considerable time and money finding the 1357 subject families even before the experiment proper could begin. The final results were not available until nine years after the start of the study. By this time the political climate had changed. The study findings had no effect on welfare policy.

Disadvantages of policy experiments

Public policy experiments often face severe **practical difficulties** that hinder good experimental practice.

Such experiments are often quite **expensive** and **take a long time to complete.**

Experimental results have had **little influence** on major questions of public policy, where politics dominates.

Not all policy experiments have so little to show for their work. Here is an example in which experiments *did* help change welfare policy.

> EXAMPLE 19. **The Baltimore Options Program.** Another round of welfare experiments began in 1982. The goal this time was to compare the existing welfare system with a new program that did not change welfare payments but offered basic education, job training, unpaid work experience, and job search help. The population of interest was welfare recipients who were able-bodied single parents (90% of them women) with no children less than six years old.
>
> The experimental design was simpler than that of the New Jersey study. Subjects were simply assigned at random to the existing system or the new program. The practical difficulties were also reduced because participation in the program was required as a condition of receiving welfare payments. This was possible because all subjects received their regular welfare payments, with or without the extra preparation for employment. There were 1362 subjects in the experimental group and 1395 subjects in the control group.[13]

Even the simpler Baltimore Options Program suffered its share of practical problems. Many subjects had legitimate reasons not to participate in the training and work programs, and many left the program through marriage or the birth of another child. As a result, only 45% of the subjects in the experimental group actually participated in one of the new activities within a year of the start of the study.

Nonetheless, the Baltimore experiment provided good evidence that the new program did increase the employment and earnings of welfare recipients. By the third year, the earnings of the experimental group were almost 25% higher than in the control group. The costs of the new training programs were balanced by reduced welfare payments to those who found work and by the taxes they paid. Partly as a result of evidence produced by the Baltimore Options Program and other welfare experiments, requirements for participation in job training and placement programs became a common part of welfare programs.

> **Advantage of policy experiments**
>
> A randomized comparative experiment can provide clear evidence that changes in policy actually **cause** changes in outcomes.

Welfare is a large, expensive, and complex issue. Many contradictory changes in welfare policy have been proposed, often in such broad terms that it is hard to decide exactly what proposals an experiment should compare. Experiments on public policy succeed more easily when the questions are smaller and the treatments to be compared are clearer. For example, experiments have helped change police response to domestic violence calls and have shown the ineffectiveness of most programs intended to change the behavior of people convicted of drunk driving.

Public policy decisions affect us all, either directly or indirectly by changing our society and spending our taxes. Most policy debates involve lots of words and some data that may or may not be relevant. Experiments promise—and sometimes deliver—specific answers to specific questions about the effects of proposed new policies.

SECTION 5 EXERCISES

2.48 **Daytime running lights.** Canada requires that cars be equipped with "daytime running lights," headlights that automatically come on at a low level when the car is started. Some manufacturers are now equipping cars sold in the U.S. with running lights. Will running lights reduce accidents by making cars more visible?

(a) Briefly discuss the design of an experiment to help answer this question. In particular, what response variables will you examine?

(b) Example 14 on page 122 discusses center brake lights. What cautions do you draw from that example that apply to an experiment on the effects of running lights?

2.49   **Treating drunk drivers.** Once a person has been convicted of drunk driving, one purpose of court-mandated treatment or punishment is to prevent future offenses of the same kind. Suggest three different treatments that a court might require. Then outline the design of an experiment to compare their effectiveness. Be sure to specify the response variables you will measure.

2.50   **Reducing domestic violence.** The usual police response to domestic violence cases was to calm the situation and to warn the offender, but not to arrest him unless use of a weapon or other circumstances required an arrest. The first experiments on police response to domestic violence suggested that arrest reduces future incidents. This evidence, along with pressure from women's groups, has changed police policy in many cities, where people accused of domestic violence are now arrested.

Outline the design of an experiment to compare "warn and release" with "arrest and hold." What are your response variables? How will you help the police on the scene do the randomization required?

2.51   **Speed kills?** The elimination of the national 55 miles per hour speed limit in 1995 provoked much debate about whether lowering highway speed limits reduces traffic fatalities.

(a) States now set their own speed limits. Can we answer the question by comparing traffic deaths in states with lower limits (like Massachusetts) and states with higher limits (like Montana)? Explain your answer.

(b) Proponents of lower speed limits point out that there were fewer deaths after the national speed limit was lowered to 55 miles per hour in the 1970s. Is this convincing? Explain your answer.

(c) How would you design an experiment to compare speed limits of 55, 65, and 75 miles per hour? In addition to an outline of the statistical design, describe the physical working of the experiment. What kind of roads would you use? Would you change speed limits on the same roads over time, or assign different speed limits on similar roads at the same time?

2.52 Another of the early experimental trials of new welfare systems was conducted in Gary, Indiana. The statistical design of the Gary experiment divided families into two blocks, those with and those without an employable adult male. Families were randomly assigned to treatments separately within each block. Why do you think these blocks were chosen?

2.53 **Controlling health care spending**. It is often suggested that people will spend less on health care if their health insurance requires them to pay some part of the cost of medical treatment themselves. One experiment on this issue asked if the percent of medical costs that are paid by health insurance has an effect either on the amount of medical care that people use or on their health. The treatments were four insurance plans. Each plan paid all medical costs above a ceiling. Below the ceiling, the plans paid 100%, 75%, 50%, or 5% of costs incurred.[14]

(a) Outline the design of a randomized comparative experiment suitable for this study. Will you use blocking?

(b) Describe briefly the practical and ethical difficulties that might arise in such an experiment.

2.54 Choose an issue of public policy that you feel might be clarified by an experiment. Briefly discuss the statistical design of the experiment you are suggesting. What are the treatments? What are the response variables? Should blocking be used?

▶ 6 ISSUES: ETHICS AND EXPERIMENTS

All studies that collect data about people should be guided by the basic principles of data ethics described in Section 1.9: screening in advance by an institutional review board, informed consent of the subjects, and confidentiality of all data on individuals. Experiments with human subjects, which actually do something to people, pose more complex ethical problems than do observational studies. Institutional review boards and the writers of informed consent forms devote particular care to experiments. In this section, we will look at some more specific issues about experiments with human subjects, first in medical studies and then in the behavioral and social sciences.

CLINICAL TRIALS

Experiments that study the effectiveness of medical treatments on actual patients are called *clinical trials*. Clinical trials are the gold standard for evidence that new treatments really do work.

> **EXAMPLE 20. Mammary artery ligation.** Medical treatments that are introduced without the firm evidence of effectiveness that only a randomized comparative experiment can give may be ineffective and even dangerous. Consider the case of mammary artery ligation.[15] This surgical treatment for angina, the severe pain caused by inadequate blood supply to the heart, was so popular in the 1950s that it was the subject of an article in *Readers' Digest*. The surgeon opened the patient's chest and tied off the internal mammary arteries in the hope of forcing more blood to flow to the heart by other routes, thus relieving the patient's angina.
>
> In 1958 and 1959, some skeptical researchers finally conducted a randomized comparative double-blind experiment. They compared mammary artery ligation with a "placebo operation" in which the surgeon did not tie off the arteries. Result: some improvement in angina for both groups, but no difference between them. The claims for mammary artery ligation had been based entirely on the placebo effect. Surgeons abandoned mammary artery ligation at once. Thanks to a properly designed experiment, angina sufferers no longer undergo chest surgery for no good reason.

It is tempting to try to bypass randomized comparative experiments, which are expensive, time-consuming, and can present difficult ethical problems. Why not use available medical records to compare various treatments? Several Canadian provinces have provincial health plans that keep records on all patients in the province. In the United States, large Health Maintenance Organizations have similarly comprehensive records on many thousands of patients. We could use these records, for example, to compare the progress of patients whose cancer was treated by chemotherapy with those who received surgery. You can see the flaw in this suggestion: when doctors decide who gets surgery and who gets chemotherapy, they take into account many facts about the patient that don't appear in the records. The records very likely compare the results of giving the two treatments to two quite different groups of patients, resulting in a bias whose direction and strength we don't know. As a leading medical statistician, Richard Peto of Oxford University, says, investing in large studies of medical records "is worse than just destroying the money, because it gives the illusion of information."[16]

Your understanding of the power of randomized comparative experiments is an important preliminary to considering ethical problems. It sets one side of the balance between protecting subjects and gaining new knowledge by reminding us that there are no substitutes for well-designed experiments if we are to find more effective treatments for our diseases and infirmities. Very few medical treatments have such dramatic benefits that experiments are not needed. Medicine advances in small steps: the death rate for the new operation is 10% lower than for the old one, or patients given the new drug live 5% longer than those given the old. Only a randomized comparative experiment can give clear evidence for or against such moderate gains. The dilemma is that the benefit of medical advances goes mainly to future patients, while the risk falls on the current patients who serve as subjects. Here are some current ethical issues concerning clinical trials, with the arguments on both sides.

> **EXAMPLE 21. Placebo controls?** You are testing a new drug. Is it ethical to give a placebo to a control group if an effective drug already exists?
>
> *Yes:* The placebo gives a true baseline for the effectiveness of the new drug. There are three groups: new drug, best existing drug, and placebo. Every clinical trial is a bit different, and not even genuinely effective treatments work in every setting. The placebo control helps us see if the study is flawed so that even the best existing drug does not beat the placebo. So placebo controls are ethical except for life-threatening conditions.

No: It isn't ethical to deliberately give patients an inferior treatment. We don't know whether the new drug is better than the existing drug, so it is ethical to give both in order to find out. If past trials showed that the existing drug is better than a placebo, it is no longer right to give patients a placebo. (If the existing drug is an older one that did not undergo proper clinical trials, a placebo is ethical.)[17]

EXAMPLE 22. Fair representation? Many large clinical trials concerned with preventing or treating heart attacks have used only middle-aged men as subjects. This is true, for example, of the Physicians' Health Study. Should trials on issues such as heart attacks that affect both sexes use subjects of just one sex?

No: This is a matter of simple fairness. The Physicians' Health Study showed that taking aspirin helps prevent heart attacks in men. How do we know whether aspirin also helps women if no women were among the subjects? Congress has in fact now passed a law requiring that clinical trials be designed to allow tests of whether the response to the treatments differs by sex or minority status.

Yes: The new law is a triumph of politics over medical and statistical sense. Although the strength of the response may vary somewhat between men and women, a treatment that helps men will almost surely also help women. As one expert says, "The biology that unites us all is usually stronger than the gender, ethnic, and political differences that divide us."[18] To study prevention of heart attacks without spending vast amounts of money, we need subjects who are likely to have heart attacks and are not likely to die of other causes during the study. That, not a wish to discriminate against women, is why middle-age men are the subjects of choice. About one in 5 men has a heart attack before age 65, but only about one in 17 women. Adding enough women to measure the difference between aspirin and a placebo would greatly increase the expense, thus reducing the number of studies we can afford, and would almost surely give little additional information.

EXAMPLE 23. AIDS research: speed versus thoroughness. Patients with fatal diseases such as AIDS and cancer want treatment that will help them now. Researchers want to find the best treatments for all future patients. Both groups want better treatments, but the difference in their time horizons causes extreme tension, particularly in the case of AIDS. Only 10% of AIDS patients survive more than five years. Should we largely bypass the traditional slow-but-sure process of requiring randomized clinical trials before new AIDS drugs are approved for general use?

Yes: Patients with a fatal disease should have immediate access to any drug that *might* help them, even before its effectiveness has been clearly shown.

No: Experience shows that most new drugs will be ineffective or have dangerous side effects. Without careful experiments, we just don't know. Why release drugs that will mostly do more harm than good?

Example 23 raises a particularly difficult issue. Both sides have good arguments. Much has been done to bridge the gap. Since 1989, experimental AIDS drugs have been made available to patients not enrolled in the clinical trials that are testing the effectiveness of these drugs. The process of testing drugs and approving them for use has been accelerated in several ways. For example, rather than waiting to see whether an experimental drug actually prolongs the life of AIDS patients, a clinical trial may examine only whether the drug increases the count of the blood cells that the AIDS virus attacks.

Neither side is satisfied. AIDS activists are disappointed that no really effective treatment or vaccine for AIDS has emerged. Researchers worry that the search for a treatment has been slowed by the changes in the traditional system for testing new drugs. When experimental drugs are freely available outside of clinical trials, the trials may give inaccurate results. Subjects often take any available drug along with their randomly assigned treatment. They may have their treatment drug tested (destroying the blindness of the experiment) and drop out if they are not getting the new drug. The subjects' behavior is understandable; after all, they are dying. Yet, as Anthony Fauci, the head of AIDS research at the National Institutes of Health, says, the purpose of clinical trials "is not to deliver therapy. It's to answer a scientific question so that the drug can be available for everybody once you've established safety and efficacy."[19] Here we meet in acute form the conflict between helping the subjects now and getting firm knowledge to help others in the future.

BEHAVIORAL AND SOCIAL SCIENCE EXPERIMENTS

When we move from medicine to the social and behavioral sciences, the direct risks to experimental subjects are less acute, but so is the value to the subjects of the knowledge gained. Consider, for example, the experiments conducted by psychologists in their study of human behavior.

"I'm doing a little study on the effects of emotional stress. Now, just take the axe from my assistant."

EXAMPLE 24. **Group pressure.** Stanley Milgram of Yale conducted a famous experiment "to see if a person will perform acts under group pressure that he would not have performed in the absence of social inducement."[20]

The subject arrives with three others who (unknown to the subject) are stooges of the experimenter. The experimenter explains that he is studying the effects of punishment on learning. The Learner (one of the stooges) is strapped into an electric chair, mentioning in passing that he has a mild heart condition. The subject and the other two stooges are Teachers. They sit in front of a panel with switches with labels ranging from "Slight Shock" to "Danger: Severe Shock." The experimenter instructs them to shock the Learner whenever he fails in a memory learning task. How badly the Learner is shocked is up to the Teachers and will be the *lowest* level suggested by any Teacher on that trial.

All is rigged. The Learner answers incorrectly on 30 of the 40 trials. The two Teacher-stooges call for a higher shock level at each failure. As the shocks increase, the Learner protests, asks to be let out, shouts, complains of heart trouble, and finally screams in pain. (The "shocks" are phony and the Learner is an actor, but the subject does not know this.) What will the subject do? He can keep the shock at the lowest level, but the two Teacher-stooges are pressuring him to torture the Learner more for each failure.

What the subject most often does is give in to the pressure. "While the experiment yields wide variation in performance, a substantial number of subjects submitted readily to pressure applied to them by the confederates." Milgram noted that in questioning after the experiment, the subjects often admit that they acted against their own principles and are upset by what they did. "The subject was then dehoaxed carefully and had a friendly reconciliation with the victim."[21]

This experiment was acceptable in the early 1960s, when Milgram did his work. Psychologists felt that it gave insight into the power of group pressure to coerce individuals into acts that they would not otherwise commit. Now, however, such a study would be considered clearly unethical, despite the lack of any possible physical harm to the subjects. The likely embarrassment and possible emotional effects on the subjects would lead any institutional review board to reject such a study.

Milgram's experiment illustrates the difficulties facing those who plan and review behavioral studies.

▶ What should we protect subjects from when physical harm is unlikely? Possible emotional harm? Emotional harm if a subject is already disturbed? Undignified situations? Bad taste?

▶ How informed must informed consent be? Behavioral experiments often rely on keeping subjects unaware of the true purpose of the study. Some, like Milgram's, actively deceive subjects. Subjects are asked to consent on the basis of vague information. They receive full information only after the experiment.

Here is an example in which the subjects get no information and give no consent. They don't even know that an experiment may be sending them to jail for the night.

> **EXAMPLE 25. Domestic violence.** How should police respond to domestic violence calls? In the past, the usual practice was to remove the offender and order him to stay out of the household overnight. Police were reluctant to make arrests because the victims rarely pressed charges. Women's groups argued that arresting offenders would help prevent future violence even if no charges were filed. Is there evidence that arrest will reduce future offenses? That's a question that experiments have tried to answer.
>
> A typical domestic violence experiment compares two treatments: arrest the suspect and hold him overnight, or warn the suspect and release him. When police officers reach the scene of a domestic violence call, they calm the participants and investigate. Weapons or death threats require an arrest. If the facts permit an arrest but do not require it, an officer radios headquarters for instructions. The person on duty opens the next envelope in a file prepared in advance by a statistician. The envelopes contain the treatments in random order. The police either arrest the suspect or warn and release him, depending on the contents of the envelope.

The researchers then watch police records and visit the victim to see if the domestic violence reoccurs.

Such experiments appear to show that arresting domestic violence suspects does reduce their future violent behavior. As a result of this evidence, arrest has become the common police response to domestic violence in many cities.

Because there is no informed consent, the ethical rules that govern clinical trials and most social science studies would forbid the experiments described in Example 25. The studies have been allowed because, in the words of one domestic violence researcher, "These people became subjects by committing acts that allow the police to arrest them. You don't need consent to arrest someone."

SECTION 6 EXERCISES

There are no "right" answers to these exercises, but there are more and less thoughtful answers. I wish mainly to invite you to think about the ethics of experimentation.

2.55 **Mammary artery ligation.** Example 20 mentions a "placebo operation" that was compared with mammary artery ligation to show that this popular surgery for angina was in fact no more effective than a placebo. The placebo operation required that the surgeon open the patient's chest and then *not* tie off the mammary arteries. Current ethical standards would not permit this placebo operation, because it exposes patients to some risk but offers no benefits. Yet without the experiment, many patients might still be undergoing mammary artery ligation, a worthless procedure. Do you agree that the placebo operation should not be allowed?

2.56 Medical records do not give trustworthy data on the effectiveness of treatments. Nevertheless, the records of thousands of patients who have received a treatment can give valuable information about side effects that were not picked up by the clinical trials carried out before the treatment was approved for general use. Explain why this is so.

2.57 **How common is AIDS?** Researchers from Yale, working with medical teams in Tanzania, wanted to know how common infection with the AIDS virus is among pregnant women in that African country. To do this, they planned to test blood samples drawn from pregnant women.

Yale's institutional review board insisted that the researchers get the informed consent of each woman and tell her the results of the test. This is the usual procedure in developed nations. The Tanzanian government did not want to tell the women why blood was drawn or tell them the test results. The government wanted to avoid panic if many people turned out to have an incurable disease for which the country's underdeveloped medical system could not provide care. The study was canceled. Do you think that Yale was right to apply its usual standards for protecting subjects?[22]

2.58 **Testing vaccines for the AIDS virus.** AIDS researchers are now deciding whether to begin large-scale trials of possible vaccines against the AIDS virus. The subjects will no doubt be chosen from groups that are at high risk of AIDS. Should strong efforts be made to persuade subjects to reduce high-risk behaviors such as unprotected sex and sharing needles to inject drugs? That is the ethical thing to do, but changing the subjects' behavior could greatly reduce their chance of exposure to the AIDS virus, and that would make it hard to see if a vaccine works. Once again the welfare of the subjects stands in the way of obtaining information that might save many other lives. What would you do?

2.59 **Testing vaccines for the AIDS virus.** Infection with the AIDS virus is more common in some African cities than anywhere else in the world. This makes these cities natural places to test vaccines that may protect against AIDS. However, a successful vaccine may be so expensive to make and administer that few Africans would be able to afford it. Is it ethical to test a vaccine in Africa if its benefits would go mainly to richer parts of the world?

No: If Africans are used as subjects, the countries and companies doing the testing should agree to supply any successful vaccine at an affordable cost in the places where the tests take place.

Yes: It is enough to give the vaccine to all the actual subjects, including the placebo group, if the test shows that the vaccine works. All of the subjects will benefit from better medical care than they would otherwise receive, whether or not the vaccine works. Because we can't predict what a vaccine will cost, insisting on large-scale, low-cost distribution may prevent testing in Africa.

What do you think?

2.60 **Informed consent.** The subjects in the New Jersey Income-Maintenance Experiment (Example 18 on page 132) were given a complete explanation of the purpose of the study and of the workings of the treatment to which they were assigned. They were not told that there were other treatments that would have paid them more (or less), or that the luck of randomization had determined the income they would receive. Do you agree or disagree that the information given is adequate for informed consent? Explain your opinion.

2.61 **Deception of subjects.** A psychologist conducts the following experiment: a team of subjects plays a game of skill against a computer for money rewards. Unknown to the subjects, one team member is a stooge whose stupidity causes the team to lose regularly. The experimenter observes the subjects through one-way glass. Her intent is to study the behavior of the subjects toward the stupid team member.

This experiment involves no risk to the subjects and is intended simply to create the kind of situation that might occur in any pickup basketball game. To create the situation, the subjects are deceived. Is this deception ethically objectionable? Explain your position.

2.62 **Enticement of subjects.** A psychologist conducts the following experiment: she measures the attitude of subjects toward cheating, then has them play a game rigged so that winning without cheating is impossible. The computer that organizes the game also records—unknown to the subjects—whether or not they cheat. Then attitude toward cheating is retested.

Subjects who cheat tend to change their attitudes to find cheating more acceptable. Those who resist the temptation to cheat tend to condemn cheating more strongly on the second test of attitude. These results confirm the psychologist's theory.

Unlike the experiment of the previous exercise, this experiment entices subjects to engage in behavior (cheating) that probably contradicts their own standards of behavior. The subjects are led to believe that they can cheat secretly when in fact they are observed. Is this experiment ethically objectionable? Explain your position.

2.63   **Morality, good taste, and public money.** Congress has often objected to spending public money on projects that seem "indecent" or "in bad taste." The arts are most often affected, but science can also be a target. Congress once refused to fund an experiment to study the effect of marijuana on sexual response. Like all government-supported research, the proposed study had been reviewed by a panel of scientists, both for scientific value and for risk to the subjects. The journal *Science* reported:[23]

Dr. Harris B. Rubin and his colleagues at the Southern Illinois Medical School proposed to exhibit pornographic films to people who had smoked marihuana and to measure the response with sensors attached to the penis.

Marihuana, sex, pornographic films—all in one package, priced at $120,000. The senators smothered the hot potato with a ketchup of colorful oratory and mixed metaphors.

"I am firmly convinced we can do without this combination of red ink, 'blue' movies, and Acapulco 'gold,'" Senator John McClellan of Arkansas opined in a persiflage of purple prose . . . .

The research community is up in arms because of political interference with the integrity of the peer review process.

(a)   Assume that no physical or psychological harm can come to the volunteer subjects. You might still object to the experiment on grounds of "decency" or "good taste." If you were a member of a review panel, would you veto the experiment on such grounds? Explain.

(b)   Suppose we concede that a free society should permit any legal experiment with volunteer subjects. It is a further step to say that any such experiment is entitled to government funding if the usual review procedure finds it scientifically worthwhile. If you were a member of Congress, would you ever refuse to pay for an experiment on grounds of "decency" or "good taste"?

2.64   **Prison experiments.** It was once common to carry out medical experiments on prison inmates who gave their consent in return for good behavior credit. Now we worry that prisoners are not really free to refuse, and the law forbids medical tests in prisons. Do you agree that all such tests, even an experiment to see if vitamin C helps prevent colds and flu, should be banned from prisons because of the difficulty of obtaining truly voluntary consent?

2.65   **Students as subjects.** Students taking Psychology 001 are required to serve as experimental subjects. Students in Psychology 002 are not required to serve, but they are given extra credit if they do so. Students in Psychology 003 are required either to sign up as subjects or to write a term paper. Serving as an experimental subject may be educational, but current ethical standards frown on using "dependent subjects" such as prisoners or charity medical patients. Students are certainly somewhat dependent on their teachers. Do you object to any of these course policies? If so, which ones, and why?

2.66   **Popular pressure.** Patients with serious diseases such as AIDS and cancer sometimes become convinced that an untested treatment will help their condition. For example, the substance Laetrile received abundant publicity in the 1970s as a treatment for cancer. Like hundreds of thousands of other chemical compounds, Laetrile had earlier been tested to see if it showed antitumor activity in animals. It didn't. Because no known drug fights cancer in people but not in animals, the medical community branded Laetrile as worthless. Advocates of Laetrile wanted the government to sponsor a clinical trial on human cancer patients.

It is usually considered unethical to test a drug on people without some promise, based on animal trials, that it is safe and effective. What is more, Laetrile may have toxic side effects. Would you approve a clinical trial of Laetrile if many people were using it anyway and wanted it tested?

2.67   **Equal treatment.** Researchers on aging proposed to investigate the effect of supplemental health services on the quality of life of older persons. Eligible patients on the rolls of a large medical clinic were to be randomly assigned to treatment and control groups. The treatment group would be offered hearing aids, dentures, transportation, and other services not available without charge to the control group. The review board felt that providing these services to some but not other persons in the same institution raised ethical questions. Do you agree?

2.68   **Animal welfare.** Many people are concerned about the ethics of experimentation with living animals. Some go so far as to regard any animal experiments as unethical, regardless of the benefits to human beings. Briefly explain your position on each of these uses of animal subjects:

(a)   Military doctors use goats that have been deliberately shot (while completely anesthetized) to study and teach the treatment of combat wounds. There is no equally effective way to prepare doctors to treat human wounds.

(b)   Several states have passed legislation that would end the practice of using cats and dogs from pounds in medical research. Instead, the animals will be killed at the pounds.

(c)   The cancer-causing potential of chemicals is assessed by exposing lab rats to high concentrations. The rats are bred for this specific purpose. (Would your opinion differ if dogs or monkeys were used?)

2.69   Examples 21, 22, and 23 describe three current ethical issues in clinical trials. Choose one of these issues and write a brief statement of your own opinion. Be sure to explain the reasons for your opinion. (Each example gives a reference where you can find more information if you wish.)

NOTES

1. Steering Committee of the Physicians' Health Study Research group, "Final report on the aspirin component of the ongoing Physicians' Health Study," *New England Journal of Medicine,* Volume 321 (1989), pp. 129–135.
2. L. L. Miao, "Gastric freezing: an example of the evaluation of medical therapy by randomized clinical trials," in J. P. Bunker, B. A. Barnes, and F. Mosteller (eds.), *Costs, Risks, and Benefits of Surgery,* Oxford University Press, 1977, pp. 198–211.
3. From L. E. Moses and F. Mosteller, "Safety of anesthetics," in J. M. Tanur et al. (eds.), *Statistics: A Guide to the Unknown,* 3rd edition, Wadsworth, 1989, pp. 15–24.
4. W. E. Leary, "Sickle cell trial called success, halted early," *New York Times,* January 31, 1995. More detail appears in a "clinical alert" posted by the National Institutes of Health at many Internet medical information sites on January 30, 1995.
5. Based on C. Anderson, "Measuring what works in health care," *Science,* Volume 263 (1994), pp. 1080–1082.
6. More detail about this study appears in L. I. Katzel et al., "Effects of weight loss vs. aerobic training on risk factors for coronary disease in healthy, obese, middle-aged and older men: a randomized controlled trial," *Journal of the American Medical Association,* Volume 274 (1995), pp. 1915–1921.
7. E. Street and M. B. Carroll, "Preliminary evaluation of a new food product," in J. M. Tanur et al. (eds.), *Statistics: A Guide to the Unknown,* 3rd edition, Wadsworth, 1989, pp. 161–169.
8. J. E. Brody, "Personal health" column, *New York Times,* November 16, 1994.
9. See the article cited in Note 6.
10. The EPA's dilemma is described in K. Schneider, "A trace of pesticide, an accepted risk," *New York Times,* February 7, 1993. The lack of realism of high-dose rodent trials and of the Delaney amendment are discussed in numerous editorials by P. H. Abelson in *Science,* including "Diet and cancer in humans and rodents," Volume 255 (1992), p. 141; "Risk assessments of low-level exposures," Volume 265 (1994),p. 1507; and "Flaws in risk assessment," *Science,* Volume 270 (1995), p. 215. *Science,* Volume 266 (1994), pp. 1141–1145 carries several letters commenting on the second editorial cited.
11. J. Cornfield, "The university group diabetes program," *Journal of the American Medical Association,* Volume 217 (1971), pp. 1676–1687.
12. For a full account of this policy experiment, see D. Kershaw and J. Fair, *The New Jersey Income-Maintenance Experiment, Volume I,* Academic Press, 1976.

13. See D. Friedlander et al., *Maryland: Final Report on the Employment Initiatives Evaluation,* Manpower Demonstration Research Corporation, 1985, and D. Friedlander, *Supplemental Report on the Baltimore Options Program,* Manpower Demonstration Research Corporation, 1987.

14. Such an experiment is described in J. P. Newhouse, "A health insurance experiment," in J. M. Tanur et al. (eds.), *Statistics: A Guide to the Unknown,* 3rd edition, Wadsworth, 1989, pp. 31–40.

15. E. M. Barsamian, "The rise and fall of internal mammary artery ligation," in J. P. Bunker, B. A. Barnes, and F. Mosteller (eds.), *Costs, Risks, and Benefits of Surgery,* Oxford University Press, 1977, pp. 212–220.

16. Quoted in the *Science* article cited in Note 5. A specific example and references to others appear in S. W. Wen, R. Hernandez, and C. D. Naylor, "Pitfalls in nonrandomized outcomes studies: the case of incidental appendectomy with open cholecystectomy," *Journal of the American Medical Association,* Volume 274 (1995), pp. 1687–1691.

17. See D. Clery, "Use of placebo controls in clinical trials disputed," *Science,* Volume 267 (1995), pp. 25–26.

18. J. Wittes, "Subgroup representation in randomized clinical trials," in *Clinical Trials and Statistics: Proceedings of a Symposium,* National Academy Press, 1993, pp. 15–21.

19. Quoted in J. Palca, "AIDS drug trials enter new age," *Science,* Volume 246 (1989), pp. 19–21.

20. S. Milgram, "Group pressure and action against a person," *Journal of Abnormal and Social Psychology,* Volume 69 (1964), pp. 137–143.

21. Postexperimental attitudes and the final quotation are in S. Milgram, "Liberating effects of group pressure," *Journal of Personality and Social Psychology,* Volume 1 (1965), pp. 127–134.

22. From a news item in *Science,* Volume 250 (1990), p. 199.

23. From a news item in *Science,* Volume 192 (1976), p. 1086.

24. C. S. Fuchs et al., "Alcohol consumption and mortality among women," *New England Journal of Medicine,* Volume 332 (1995), pp. 1245–1250.

25. The study is described in G. Kolata, "New study finds vitamins are not cancer preventers," *New York Times,* July 21, 1994. Look in the *Journal of the American Medical Association* of the same date for the details.

REVIEW EXERCISES

2.70 TV harms children? Observational studies suggest that children who watch many hours of television get lower grades in school and are more likely to commit crimes than those who watch less TV. Explain clearly why these studies do *not* show that watching TV *causes* these harmful effects. In particular, suggest some lurking variables that may be confounded with heavy TV viewing.

2.71 **Left-handers die younger?** A well-publicized study claimed to show that right-handed people lived almost a decade longer on the average than left-handers. The study surveyed a large number of deaths, recorded the ages of the deceased people, and asked their survivors if the deceased had been right- or left-handed. The conclusion is wrong. Early in the century, it was common to force natural left-handers to use their right hand. More recently, this practice ended and left-handers were allowed to be left-handed. Explain how this change in social practice created a bias that explains the study's wrong conclusion.

2.72 In a study of the relationship between physical fitness and personality, middle-aged college faculty who have volunteered for an exercise program are divided into low-fitness and high-fitness groups on the basis of a physical examination. All subjects then take the Cattell Sixteen Personality Factors Questionnaire, a 187-item multiple-choice test often used by psychologists, and the results for the two groups are compared. Is this study an experiment? Explain your answer.

2.73 **Running changes your mood?** A study of the effect of running on personality involved 231 men who ran about 20 miles each week. The runners were given the Cattell Sixteen Personality Factors Questionnaire. A news report (*New York Times,* February 15, 1988) said, "The researchers found statistically significant personality differences between the runners and the 30-year-old male population as a whole." A headline on the article said, "Research has shown that running can alter one's moods." Explain carefully, to someone who knows no statistics, why the headline is misleading.

2.74 **A little alcohol is good for you.** The Nurses Health Study has queried a sample of over 100,000 female registered nurses every two years since 1976. Beginning in 1980, the study asked questions about diet, including alcohol consumption. After looking at all deaths among these nurses through May of 1992, the researchers concluded that "As compared with nondrinkers and heavy drinkers, light-to-moderate drinkers had a significantly lower risk of death."[24]

The word "significantly" in this conclusion has a technical meaning. Explain to someone who knows no statistics what "significant" means in a statistical study.

2.75 **Do antioxidants prevent cancer?** People who eat lots of fruits and vegetables have lower rates of colon cancer than those who eat little of these foods. Fruits and vegetables are rich in "antioxidants" such as vitamins A, C, and E. Will taking antioxidants help prevent colon cancer? A clinical trial studied this question with 864 people who were at risk of colon cancer. The subjects were divided into four groups: daily beta carotene, daily vitamin C and E, all three vitamins every day, and daily placebo. After four years, the researchers were surprised to find no significant difference in colon cancer among the groups.[25]

(a)   What are the explanatory and response variables in this experiment?

(b)   Outline the design of the experiment. Use your judgment in choosing the group sizes.

(c)   Assign labels to the 864 subjects and use Table A starting at line 118 to choose the *first 5* subjects for the beta carotene group.

(d)   The study was double-blind. What does this mean?

(e)   What does "no significant difference" mean in describing the outcome of the study?

(f)   Suggest some lurking variables that could explain why people who eat lots of fruit and vegetables have lower rates of colon cancer. The experiment suggests that these variables, rather than the antioxidants, may be responsible for the observed benefits of fruits and vegetables.

2.76 You are in charge of a writing course for college freshmen in which 1200 students are enrolled. You wish to learn if the students write better essays when they are required to use computer word processing than when they write and revise essays by hand. Describe the design of an experiment to determine whether word processing results in better essays. Be sure to include a description of the treatments and the response variable or variables. Also comment on issues related to "blinding" the scoring of the responses.

2.77 The design of controls and instruments determines how easily people can use them. A student investigates this effect by asking right-handed students to turn a knob (with their right hands) that moves a pointer by screw action. There are two identical instruments. The knob turns clockwise on one, and counterclockwise on the other. The response variable is the number of seconds required to move the pointer a fixed distance. Thirty right-handed students are available to serve as subjects. You have been asked to lay out the statistical design of this experiment. Describe your design, and carry out any randomization that is required.

2.78 **Speeding the mail?** Is the number of days a letter takes to reach another city affected by the time of day it is mailed and whether or not the Zip code is used? Describe briefly the design of an experiment to investigate this question. Be sure to specify the treatments exactly and to tell how you will handle lurking variables such as the day of the week on which the letter is mailed.

2.79 **Taste testing.** Do consumers prefer the taste of a cheeseburger from McDonald's or from Burger King in a blind test in which neither burger is identified? Describe briefly the design of a matched pairs experiment to investigate this question.

WRITING PROJECTS

2.1 **Medical experiments in the news.** Articles in the press often describe medical findings based on an experiment. The conclusion of the Physicians' Health Study that taking aspirin regularly helps prevent heart attacks is an example. Find an article in a newspaper or magazine that deals with a recent medical experiment. Describe the purpose and design of the experiment. What were the conclusions of the study, and how well grounded do you think they are?

Add a brief critique of the news article's presentation. Does the article mention a control group? Does it mention random assignment of the subjects? Often the article will cite the journal in which the full account of the study appears, such as the *Journal of the American Medical Association* or the *New England Journal of Medicine*. You may be able to find the journal in your library (look for the same date as the news report) and use it to help critique the news presentation.

2.2   **Design your own experiment.** Exercises 2.78 and 2.79 illustrate the use of statistically designed experiments to answer questions that arise in everyday life. Select a question of interest to you that an experiment might answer and briefly discuss the design of an appropriate experiment. Comment on practical difficulties that your experiment may face, and explain how you will deal with them.

2.3   **Experiments in ecology.** Ecologists have traditionally used careful observation in the field to study such topics as why animal populations decline and how different species interact. The ability of controlled experiments to yield clear conclusions about cause and effect based on rigorous statistics has made experiments increasingly popular in ecology. For example, an experimenter may divide a pond by a barrier and remove a species from one side but not from the other in order to see the effect of this species on other species in the pond. Some ecologists say that the trend has gone too far—that experimental results often don't generalize to large-scale natural systems.

You can find a full discussion in a series of articles in *Science*, Volume 269 (1995), pp. 313–315 and 324–331. Read these articles. Then write a short essay on the use of experiments in ecology. Include some examples of ecological experiments, an account of the advantages of experiments, and an account of the criticisms aimed at too much reliance on experiments.

CHAPTER 3

MEASUREMENT

Statistics deals with numbers. Planning the production of data through a sample or an experiment does not by itself produce numbers. Once we have our sample respondents or our experimental subjects, we must still *measure* whatever characteristics interest us. Here is the definition:

Measurement

We **measure** a property of a person or thing when we assign a number to represent the property. We usually use an **instrument** to make a measurement.

If I want to measure the length of my bed, I use a tape measure (that's the instrument). The measurement is a length in inches or in centimeters. If I want to measure a student's readiness for college, I might have the student take the Scholastic Assessment Test (SAT). The SAT exam form is the instrument. The measurement is a score in points, somewhere between 400 and 1600 if I combine the verbal and mathematics sections of the SAT.

Measuring length is more straightforward than measuring college readiness. That is because we understand length better than we understand what makes a student ready for college work. As we discuss measurement, we will introduce ideas in the simple setting of measuring physical properties such as length and weight. Then we will move to the less simple task of measuring properties like employment and readiness for college.

▶ 1 MEASUREMENTS VALID AND INVALID

In Chapters 1 and 2, we called any measured characteristic a *variable*. Measurement is the process of turning vague concepts like length or employment status into precisely defined variables. Using a tape measure to turn the idea of "length" into a number is straightforward. Here are some examples in which measuring is both complex and controversial.

> **EXAMPLE 1. Measuring unemployment.** The Bureau of Labor Statistics (BLS) produces data on employment and unemployment. Figure 3-1 is a graph of the nation's unemployment rate that appeared on the front page of the BLS monthly news release on the employment situation in 1994.[1] Much to the relief of politicians, the percent of workers without jobs had been dropping since mid-1992.
>
> There is a gap in the graph in January 1994. Unemployment seems to jump up that month. It's a measurement problem—the BLS

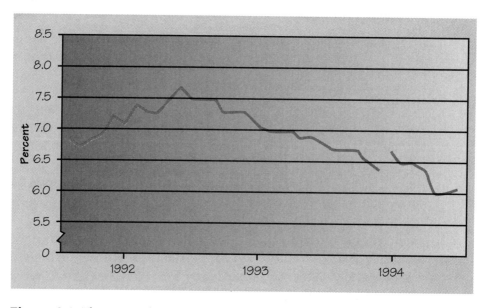

Figure 3-1 The unemployment rate from August 1991 to July 1994. The gap shows the effect of a change in how the government measures unemployment.

changed the way it measures unemployment. The unemployment rate would have been 6.3% under the old system. It was 6.7% under the new system. The politicians were not happy.

EXAMPLE 2. Measuring television viewing. Nielsen Media Research places "people meters" similar to TV remote controls in a sample of homes to measure how many people and what kind of people watch a television show. The ratings determine how much advertisers will pay, and eventually whether the show stays on the air. The TV networks think that Nielsen's electronic gadgets favor younger viewers who often watch cable channels. Perhaps another method would give the networks higher ratings. And so we read in the news:

> *After years of complaining about inaccuracies in the Nielsen ratings, the three biggest television networks announced yesterday that they were underwriting a multimillion-dollar research effort to find an alternative system.*[2]

EXAMPLE 3. Measuring readiness for college. Most colleges base admission in part on scores on standardized tests, especially the SAT. Many other factors play a role in getting into a selective college—it helps to be an athlete, a musician, the child of alumni, or to be from a background underrepresented in the student body. That last factor means that in most places it helps to be black or Hispanic. When the admissions process is done, selective colleges end up with diverse student bodies in which blacks and Hispanics have on the average lower SAT scores than whites and Asians. At UCLA, for example, average SAT scores are 951 for blacks, 1186 for whites, and 1182 for Asians.[3] Opponents of affirmative action say this is unfair to whites and Asians. Supporters often say that the SAT is biased against minorities. Does the SAT really measure readiness for college work? Are SAT scores "fair" measures for students from different backgrounds? To give informed answers, we must know more about measurement in general.

VALIDITY

Let's cut through the argument over using the SAT in college admissions. We will just measure the height in inches of all applicants and accept the tallest. Bad idea, you say. Why? Because height has nothing to do with being prepared for college. In more formal language, height is not a *valid* measure of a student's academic background.

Valid measurement

A variable is a **valid** measure of a property if it is relevant or appropriate as a representation of that property.

EXAMPLE 4. Measuring highway safety. The 1970s saw the completion of most of the Interstate highway system, an emphasis on better safety features in new cars, and a reduction in the national speed limit to 55 miles per hour. Did driving become safer? Here's one measure: the number of motor vehicle deaths was 52,600 in 1970 and 51,091 in 1980. That's a small change. We might think that driving was no safer in 1980 than in 1970.

However, the number of motor vehicles registered in the United States grew from 108 million in 1970 to 156 million in 1980. The count of deaths does not take account of the fact that more people did more driving in 1980 than in 1970. It is an *invalid* measure of highway safety.

A valid measure of the dangers of driving is a *rate*, the number of deaths divided by the number of miles driven. This rate takes into account the total miles driven by all vehicles. In 1980, vehicles drove 1,527,000,000,000 miles in the U.S. Because this number is so large, it is usual to give the death rate per 100 million miles driven. For 1980, this rate is

$$\frac{\text{Motor vehicle deaths}}{\text{100s of millions of miles driven}} = \frac{51,091}{15,270}$$

$$= 3.3$$

The death rate fell from 4.7 deaths per 100 million miles in 1970 to 3.3 in 1980. That 30% drop tells us that driving did get safer.

> ### Rates and counts
>
> Often a **rate** (a fraction, proportion, or percent) at which something occurs is a more valid measure than a simple **count** of occurrences.

The Bureau of Labor Statistics reports the count of unemployed people, but the most important measure is the national unemployment rate. People who are not available for work (retired people, for example, or students who do not want to work while in school) should not be counted as unemployed just because they don't have a job. To be unemployed, a person must first be in the labor force. That is, they must be available for work and looking for work. The unemployment rate is

$$\text{Unemployment rate} \ = \ \frac{\text{Number of people unemployed}}{\text{Number of people in the labor force}}$$

"Unemployed? Not me, I'm out of the labor force."

The BLS estimates the unemployment rate based on interviews with the sample in the monthly Current Population Survey. The interviewer can't simply ask "Are you in the labor force?" and "Are you employed?" Both need precise definition. What must a person do to show that they are looking for work? How many hours per week must someone work to be employed full time? What about workers who are on strike? The BLS has detailed definitions that deal with these and many other issues. The interviewer must ask many detailed questions that allow the BLS to classify a person as employed, unemployed, or not in the labor force. Changing these details can change the unemployment rate. That's what happened in January 1994. After several years of planning, the BLS introduced computer-assisted interviewing (Example 15 of Chapter 1, page 45) and thoroughly revised the questions asked. The new measuring process gives a slightly higher unemployment rate. The BLS says that the meaning of "unemployed" did not change, and that the new interview measures unemployment more accurately. The change does, however, make it hard to directly compare unemployment rates before and after January 1994.

Details matter

In measuring complex properties, different methods can give different results even when all the methods are valid.

In some cases, such as economic and social data, having a measure that remains the same over time is important for ease of interpretation.

EXAMPLE 5. **How many people have AIDS?** The number of new AIDS cases reported by the Centers for Disease Control and Prevention jumped from 43,672 in 1991 and 45,472 in 1992 to 103,691 in 1993. Was the AIDS epidemic surging? No. In 1993, the Centers expanded the definition of what it means to have AIDS. The number of cases would have remained stable under the old definition. The new definition may be medically more accurate, but the data could easily mislead an unwary reader. When you see a big jump up or down in a series of data, you should suspect that there was a change in measurement.

VALIDITY FOR WHAT PURPOSE?

Using height to measure readiness for college, and even using counts when rates are needed, are examples of clearly invalid measures. The tougher questions concern measures that are neither clearly invalid nor obviously valid. We can clarify the issues by asking "valid for what purpose?"

> **EXAMPLE 6. Achievement tests.** When you take a subject-matter examination, you hope that the exam will cover the announced material. A statistics exam that covers the main points of your course syllabus is a valid measure of how much you know about the course material. The College Board, which administers the SAT, also offers achievement tests in a variety of disciplines (including an advanced placement exam in statistics). These achievement tests are not very controversial. Experts can judge whether they examine their announced content. That is a question of **content validity**.

> **EXAMPLE 7. IQ tests.** Psychologists would like to measure aspects of the human personality that can't be observed directly, such as "intelligence" or "degree of authoritarian personality." Does an IQ test measure intelligence? Some psychologists say "Yes" rather loudly. There is such a thing as general intelligence, they argue, and the various standard IQ tests do measure it, though not perfectly. Other experts say "No" equally loudly. There is no single intelligence, just a variety of mental abilities that no one instrument can measure.
>
> The disagreement over the validity of IQ tests is rooted in disagreement about the nature of intelligence. If we can't agree on exactly what intelligence is, we can't agree on how to measure it. Whether an instrument like an IQ test validly measures a complex property like intelligence is a question of **construct validity** in the language of psychologists. Construct validity is difficult to establish and often controversial.

> **EXAMPLE 8. The SAT again.** "Readiness for college academic work," like intelligence, is a vague concept that probably combines inborn intelligence (whatever we decide that is), learned knowledge, study and test-taking skills, and motivation to work at academic

subjects. We cannot claim that the SAT has construct validity as a measure of readiness for college.

Instead, we ask a simpler and more easily answered question: Do SAT scores help predict students' success in college? Success in college is a clear concept, measured by whether students graduate and by their college grades. Students with high SAT scores are more likely to graduate and earn (on the average) higher grades than students with low SAT scores. We say that SAT scores have *predictive validity* as measures of readiness for college. This is the only kind of validity that data can assess directly.

Predictive validity

A measurement of a property has **predictive validity** if it can be used to predict success on tasks that are related to the property to be measured.

Predictive validity is the clearest and most useful form of validity from the statistical viewpoint. "Do SAT scores help predict college grades?" is a much clearer question than "Do IQ test scores measure intelligence?" However, predictive validity is not a yes-or-no idea. We must ask *how accurately* SAT scores predict college grades. There are statistical ways to describe "how accurately," and we will meet them later. Here's the big picture.

SAT scores do predict college grades. They usually predict college grades somewhat better than high school grades do. Moreover, SAT scores predict grades for blacks just as accurately as they do for whites. That blacks have (on the average) lower SAT scores reflects the fact that they have (again on the average) less money and fewer educational opportunities before college. That is, the SAT exams are not biased against blacks—they just report the consequences of the disadvantaged position of many blacks in American society, a position that hurts readiness for college work. Nonetheless, *SAT scores do not predict success in college very accurately.* Students with the same SAT scores often have very different college grades. Motivation and study habits matter a lot. Selective colleges like UCLA (Example 3) are justified in paying some attention to SAT scores, but they are also justified in looking beyond SAT scores for the motivation that can bring success to students with weaker academic preparation. Although both sides in the debate over affirmative action in college admissions point to the SAT, the debate is not really about the numbers. It is about how colleges should use all the information they have in deciding whom to admit, and also about the goals colleges should have in forming their entering classes.

SECTION 1 EXERCISES

3.1 **Measuring length of hair.** You are studying the relationship between political attitudes and length of hair among male college students. You will measure political attitudes with a standard questionnaire. How will you measure length of hair? Give precise instructions that an assistant could follow. Include a description of the measuring instrument that your assistant is to use.

3.2 **Measuring physical fitness.** You want to measure the "physical fitness" of college students. Give an example of a clearly invalid way to measure fitness. Then briefly describe a measurement process that you think is valid.

3.3 **Measuring intelligence.** "Intelligence" means something like "general problem-solving ability." Explain why is it *not* valid to measure intelligence by a test that asks questions such as

Who wrote Romeo and Juliet?

or

Who won the last soccer World Cup?

3.4 Customers returned 36 coats to Sears this holiday season, and only 12 to La Boutique Classique next door. Sears sold 1100 coats this season, while La Boutique sold 200.

(a) Sears had a greater number of coats returned. Why does this not show that Sears' coat customers were less satisfied than those of La Boutique?

(b) What is the rate of returns (percent of coats returned) at the two stores?

3.5 **Bicycle safety.** Bicycle riding has become more popular. Is it also getting safer? In 1988, 24,800,000 people rode a bicycle at least six times and 949 people were killed in bicycle accidents. In 1992, there were 54,632,000 riders and 903 bicycle fatalities.

(a)   Compare the death rates for the two years. What do you conclude?

(b)   Although the data come from the same government publication, we suspect that some change in measurement took place between 1988 and 1992 that makes the data for the two years not directly comparable. Why should we be suspicious?

3.6 **Highway safety.** Example 4 discusses measuring highway safety. We might divide each year's motor vehicle deaths by the number of vehicles registered that year to get the rate of deaths per vehicle registered. This may be a better measure than the simple count of deaths, but it is inferior to deaths per miles driven as a measure of highway safety. Explain why.

3.7 **Fighting cancer.** Congress wants the medical establishment to show that progress is being made in fighting cancer. Some variables that might be used are:

(a)   Total deaths from cancer. These have risen sharply over time, from 331,000 in 1970 to 505,000 in 1980 to 531,000 in 1993.

(b)   The percent of all Americans who die from cancer. The percent of deaths due to cancer has also risen steadily, from 17.2% in 1970 to 20.9% in 1980 to 23.4% in 1993.

(c)   The percent of cancer patients who survive for five years from the time the disease was discovered. These rates are rising slowly. For whites, the 5-year survival rate was 50.3% in the 1974 to 1976 period and 55.5% in 1983 to 1989.

Discuss the validity of each of these variables as a measure of the effectiveness of cancer treatment. In particular, explain why both (a) and (b) could increase even if treatment is getting more effective, and why (c) could increase even if treatment is getting less effective.

3.8 Politicians often choose the details of the statistics they cite in a way that favors their cause. In 1994, George Pataki unseated longtime governor Mario Cuomo to become governor of New

York State. Pataki attacked Cuomo for New York's high taxes. In fact, New York's state taxes, measured as a percentage of personal income, were 22nd from the top among the 50 states. When local taxes are included, however, New York ranked second to Alaska. Which measure of tax burden did Pataki use in his campaign? Why?

3.9 **Testing job applicants.** The law requires that tests given to job applicants must be shown to be directly job-related. The Labor Department believes that an employment test called the General Aptitude Test Battery (GATB) is valid for a broad range of jobs. As in the case of the SATs, blacks and Hispanics get lower average scores on the GATB than do whites. Describe briefly what must be done to establish that the GATB has predictive validity as a measure of future performance on the job.

3.10 **Testing job applicants.** A news article reported a study of how well a job performance test given before hiring predicted the subsequent job performance of 1400 government technicians. Such tests are often accused of being biased against minority groups. A psychologist commenting on the results of the study said, "Six years later, we found that belief wrong, if you define bias as meaning the scores are unrealistically low in relation to performance on the job." (*New York Times*, July 27, 1973.)

Explain in what way a test that is "biased against minority groups" would lack predictive validity.

3.11 **TV violence.** You want to study the effect of violent television programs on antisocial behavior in young children.

(a) Define (that is, tell how to measure) the explanatory and response variables. You have many possible choices of variables— just be sure that the ones you choose are valid and clearly defined.

(b) Do you think that an experiment is practically and morally possible? If so, briefly describe the design of an experiment for this study.

(c) Briefly discuss the design of an observational study that avoids the ethical difficulties of an experiment. Will confounding with other variables threaten the validity of your conclusions about the effect of TV violence on child behavior?

▶ 2 MEASUREMENTS ACCURATE AND INACCURATE

Using a bathroom scale to measure your weight is valid. But if the scale has marks only at ten-pound intervals, forcing you to guess single pounds by eye, your measurements will not be very accurate. On the other hand, some measurements are very accurate but are not valid. Here is an example.

> **EXAMPLE 9. Do big skulls house smart brains?** In the mid-nineteenth century, it was thought that measuring the volume of a human skull would measure the intelligence of the skull's owner. It was difficult to measure a skull's volume accurately, even after it was no longer attached to its owner. Paul Broca, a professor of surgery, showed that filling a skull with small lead shot, then pouring out the shot and weighing it, gave quite accurate measurements of the skull's volume. These accurate measurements do not, however, give a valid measure of intelligence. Skull volume turned out to have no relation to intelligence or achievement.

We have met some basic concepts for dealing with the validity of measurements. Now we must deal with accuracy.

UNBIASEDNESS AND RELIABILITY

Like validity, accuracy of measurement is clearest for physical properties such as length and weight. That's because to talk about the accuracy of a measurement we compare the measurement with the *true value* of the property. We think we understand what the true weight of a person is, so it is easy to talk about more or less accurate measurements of that weight. We will therefore use physical measurements to introduce the basic ideas. Accuracy has two aspects, *lack of bias* and *reliability*. Both describe what would happen if we repeat the measurement many times.

> ### Unbiased and reliable measurements
>
> A measurement process is **unbiased** if it does not systematically overstate or understate the true value of the variable.
>
> A measurement process is **reliable** if repeated measurements on the same individual give the same (or approximately the same) results.

EXAMPLE 10. Bias and reliability. A bathroom scale that always gives a weight five pounds too low is biased but reliable. This scale measures systematically too low, not just on one try but on repeated tries. That's bias. Nevertheless, it gives close to the same weight every time. A person who weighs 160 pounds always gets a scale reading close to 155 pounds. That's reliability.

A scale that is erratic, so that you get different weights by stepping off the scale and back on again, is unreliable. It may or may not also be biased. If the erratic results of repeat weighings of the same person center about the true weight, some low and some high but not systematically low or systematically high, the scale has small bias.

Unbiasedness and reliability are separate aspects of accurate measurement. An accurate measurement process has both small bias and high reliability.

IMPROVING RELIABILITY

Research laboratories use scales that are more accurate than your bathroom scale, but even the best scales have some bias and are not perfectly reliable. A laboratory calibrates its scales by weighing "mass standards" whose true weight is known very accurately. The National Institute of Standards and Technology (NIST) will sell you any of the mass standards shown in the photo. The set in the foreground, a set of weights covering 1 milligram to 1 kilogram, costs $4657. The cost reflects the care NIST takes to produce accurate weights. The mass standards you can buy from NIST are accurate because they have been

compared with super-standard weights kept by NIST. These weights in turn have been compared with the International Prototype Kilogram, which lives in a guarded vault in Paris and indirectly determines all the world's measures of weight.

> **EXAMPLE 11. Measuring weight accurately.** NIST, as the nation's guardian of physical measurements, measures weight accurately—but not perfectly accurately. Here are the results of 11 NIST measurements of the weight (in grams) of the mass standard named NB 10:[4]
>
> | | | |
> |---|---|---|
> | 9.9995992 | 9.9995985 | 9.9995947 |
> | 9.9996008 | 9.9995978 | 9.9996027 |
> | 9.9995925 | 9.9995929 | 9.9996006 |
> | 9.9995988 | 9.9996014 | |

You see that the 11 measurements do vary. There is no such thing as an absolutely accurate measurement. The average (mean) of these measurements is 9.9995982 grams. This average is a more reliable estimate of the true mass than a single measurement. In fact, NIST says it is 95% confident that the true weight of NB 10 is within ± 0.0000023 gram of this value. NB 10 is supposed to weigh 10 grams, but it is light by about the weight of a grain of salt.

When NIST wants to get a more reliable result, it repeats its measurements and uses the average of several weighings. Students doing a lab assignment are often advised to do the same thing, though 11 repeats would be a bit much for a chemistry lab.

Use averages to improve reliability

No measuring process is perfectly reliable. The average of several repeated measurements of the same individual is more reliable (less variable) than a single measurement.

REDUCING BIAS

Unfortunately, there is no similarly straightforward way to reduce the bias of measurements. Bias depends on how good the measuring instrument is. To reduce the bias, you need a better instrument.

> **EXAMPLE 12. Measuring unemployment again.** Measuring unemployment is also "measurement." The concepts of bias and reliability apply here just as they do to measuring the mass of NB 10.
>
> The Bureau of Labor Statistics checks the *reliability* of its measurements of unemployment by having supervisors reinterview about 5% of the sample. This is repeated measurement on the same individual, just as when NB 10 is weighed several times.
>
> The BLS attacks *bias* by improving its instrument. That's what happened in January 1994, when the Current Population Survey was given its biggest overhaul in more than 50 years. The old system for measuring unemployment, for example, underestimated unemployment among women because the detailed procedures had not kept up with changing patterns of women's work. The new measurement system corrected that bias—and raised the reported rate of unemployment.[5]

Bias and reliability are tough issues for behavioral and social scientists. Bias is hard to assess because we can't say what the true value of "college readiness" or "authoritarian personality" or "ego strength" is. Reliability should be easier, because it requires only that repeated measurements give about the same result. Unfortunately, we can't assess the reliability of the SAT by measuring the same students several times—taking the test once teaches a student how to do better the next time. The fact that people, unlike NB 10, learn while being measured makes measurement a more complex topic for psychologists than for engineers.

STATISTICS IN MEASUREMENT

Bias and reliability in measurement are similar to bias and precision in sampling. Bias is systematic error that pulls results away from the truth in the same direction on every try. Reliability and precision both refer to the repeatability of results—do we get about the same answer on every try? If we have no bias, we can make a confidence statement that says how accurate our results will be if we make many tries. You can read a 95% confidence statement from NIST about the weight of NB 10 in Example 11. So similar statistical ideas apply to studying sampling and measurement. You might even think of NIST's 11 measurements of the weight of NB 10 as a sample from the population of all the measurements they would get if they kept trying forever.

Sampling and measurement differ in what is being repeated. Bias and precision in sampling refer to what would happen if we drew many samples from the same population. Bias and reliability in measurement refer to what would happen if we measured the same individual many times.

SECTION 2 EXERCISES

3.12   Give your own example of a measurement process that is valid but has large bias. Then give your own example of a measurement process that is invalid but highly reliable.

3.13   It is hard to measure "intelligence." Let's do it the easy way: measure height in inches, and call the result "intelligence." Not only is this method easy, it gives the same number every time we repeat the measurement on the same person. Is this measurement process reliable? Is it valid?

3.14 **Measuring authoritarian personality?** A psychologist claims that a standard psychological test for "authoritarian personality" does not really measure an aspect of the subject's personality, but instead measures other factors such as religious beliefs. Is she attacking the *accuracy* or the *validity* of the test? Explain your answer.

3.15 Use a ruler to mark off a piece of stiff paper in inches (mark only full inches, no fractions) as shown here:

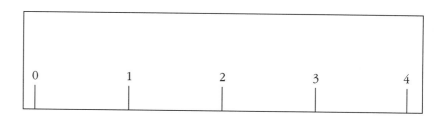

Measure the line below with your instrument, recording your answer to a hundredth of an inch (such as 2.23 inches or 2.39 inches).

_____

To make this measurement, you must estimate what portion of the distance between the 2- and 3-inch marks the line extends. Careful measurements usually involve an uncertainty such as this. We have merely magnified it by using an instrument divided into inches only.

(a) What is the result of your measurement?

(b) Ask four other people to measure the line with your instrument and record their results. Average the five measurements. What margin of error do you think a measurement with your instrument has? (That is, how reliable is it?)

(c) Suppose that someone measures the line by placing its left end at the end of the instrument instead of at the 0 mark. This causes bias. Explain why. Is the reliability of the measurement also affected, or not?

Comment: If you collect the results of the entire class for part (a) and display them in a graph like Figure 1-2 in Chapter 1, you will have a picture of the reliability of this measurement similar to the graph used to present the precision of a sample statistic.

3.16   Take a 1-foot ruler and measure the length of one wall of a room to the nearest inch. Do this five times, recording your answers in feet and inches. What is your average result? Now use a tape measure at least as long as the wall to get a more accurate measurement. Did your ruler measurements show bias? (For example, were they almost all too long?)

3.17   **An experiment on bias.** Let's study bias in an intuitive measurement. Here is a drawing of a tilted container:

Reproduce this drawing on 10 sheets of paper. Choose 10 people, 5 men and 5 women. Explain that the drawing represents a tilted container of water. Ask each subject to draw the water level when the container is full of water.

The correct level is horizontal (straight back from the lower lip of the container). Many people make large errors in estimating the level. Use a protractor to measure the angle of each subject's error. Were your subjects systematically wrong in the same direction? How large was the average error? Was there a clear difference between the average errors made by men and women?

3.18 **Measuring pulse rate.** All the members of a physical education class are asked to measure their pulse rate as they sit in the classroom. The students use a variety of methods. Method 1: count heart beats for 6 seconds and multiply by 10 to get beats per minute. Method 2: count heart beats for 30 seconds and multiply by 2 to get beats per minute. Which method is more reliable? Why? Is either method clearly more biased than the other? Why?

3.19 **Measuring pulse rate.** One student in the class of the previous exercise proposes a third method: starting exactly on a heart beat, measure the time needed for 50 beats and convert this time into beats per minute. This method is more accurate than either of the two methods mentioned in the previous exercise. Why?

3.20 **Measuring crime.** Crime data make headlines. We measure the amount of crime by the number of crimes committed or (better) by crime rates (crimes per 100,000 population). The FBI publishes data on crime in the United States based on reports to local police departments. The National Crime Victimization Survey publishes data based on a national probability sample of 50,000 households. The victim survey shows about two and a half times as many crimes as the FBI report. Explain why the FBI reports have a large downward bias for many types of crime. (Here is a case in which bias in producing data leads to bias in measurement.)

3.21 Example 11 says that NIST is 95% confident that the true weight of the mass standard NB 10 is within ± 0.0000023 gram of 9.9995982 grams. Explain in simple language what "95% confident" means in this setting.

▶ 3 SCALES OF MEASUREMENT

We are accustomed to treating numbers with some respect. They seem so solid and precise. Now that we know about bias in data production and the problems of validity and accuracy in measurement, we will surely ask that data earn our respect by demonstrating that they are carefully produced and soundly measured. There is more to think about: *not all numbers carry the same amount of information.* That is the message of this section.

Consider, for example, my employment status. I am either employed, unemployed, or not in the labor force. We might represent this by a variable having value 0 if I am not in the labor force, value 1 if I am unemployed, and value 2 if I am employed. Now, 2 is twice as much as 1. And 2 inches is twice as much as 1 inch—but an employment status of 2 is *not* twice an employment status of 1. The numbers used to code employment status are just category labels disguised as numbers. We could use the labels A, B, and C in place of numbers. (Labeling is still measurement. It is only convenience that leads us to require that measurements be numbers.)

Not all numbers resulting from measurement carry information, such as "twice as much," that we naturally associate with numbers. What we can do with data depends on how much information the numbers carry.

Types of measurement scales

A measurement of a property has a **nominal scale** if the measurement tells only into *what class* a unit falls with respect to the property.

The measurement has an **ordinal scale** if it also tells when one individual has *more of* the property than does another individual.

The measurement has an **interval scale** if the numbers also tell us that one individual *differs by a certain amount* of the property from another individual.

The measurement has a **ratio scale** if in addition the numbers tell us that one individual has *so many times as much* of the property as does another individual.

NOMINAL SCALE

Measurements in a *nominal scale* place units in categories, nothing more. Properties such as race, sex, and employment status are measured in a nominal scale. We can code the sex of a person by

0 – female
1 – male

or by

0 – male
1 – female.

Which numbers we assign makes no difference. The value of this variable just labels the sex of the person. It makes no sense to do arithmetic with these labels—we can't calculate the "average sex" of the 1500 people who respond to an opinion poll.

ORDINAL SCALE

In an *ordinal scale*, the order of numbers is meaningful. If a committee ranks 10 fellowship candidates from 10 (strongest) to 1 (weakest), the candidate ranked 8 is better than the candidate ranked 6—not just different (as a nominal scale would tell us), but better. However, the usual arithmetic is not meaningful; 8 is not twice as good as 4, and the difference in quality between the candidates ranked 8 and 6 need not be the same as between the candidates ranked 6 and 4. Only the order of the values is meaningful.

Ordinal scales appear when social scientists measure properties such as "authoritarian personality" by giving a test on which a subject can score, say, between 0 and 100 points. If the test is valid as a measure of this property, then Esther who scores 80 is more authoritarian than Lydia who scores 60. If Jane scores 40, however, we can probably not conclude that Esther is "twice as authoritarian" as Jane. Nor can we say that "the difference in authoritarianism between Esther and Lydia is the same as between Lydia and Jane" just because their scores differ by 20 in each case. Whether a particular test has an ordinal scale or actually does carry information about differences and ratios we leave for psychologists to discuss. Many tests have ordinal scales.

INTERVAL AND RATIO SCALES

With *interval and ratio scales* we reach the kind of measurement most familiar to us. These are *measurements made on a scale of equal units*, such as height in centimeters, reaction time in seconds, or temperature in degrees Celsius. Arithmetic such as finding differences is meaningful when these scales are used. A cockroach 4 centimeters long is 2 centimeters longer than one 2 centimeters long.

There is a rather fine distinction between interval and ratio scales. A cockroach 4 centimeters long is twice as long as one 2 centimeters long; length in centimeters has a ratio scale. When the temperature is 40°C, however, it is *not* twice as hot as when it is 20°C. Temperature in degrees Celsius has an interval scale, not a ratio scale. Another way of expressing the difference is that *ratio scales have a meaningful zero*. A length of 0 centimeters is "no length," and a time of 0 seconds is "no time." But a temperature of 0°C is just the freezing point of water, not "no heat."

We will not distinguish between interval and ratio scales of measurement. We will use "interval/ratio scale" for any scale of equal units such as centimeters or degrees Celsius. However, it is important (and usually easy) to distinguish an interval/ratio scale from an ordinal scale (objects are ordered in some way) or a nominal scale (objects are put into categories).

The discussion of measurement scales reminds us that not all numbers carry the same kind of information. "Paula finished first in the 1500 meter race" is ordinal. "Paula ran 1500 meters in 3 minutes 58 seconds" is interval/ratio. They give different information about Paula's performance.

One concluding comment: *The scale of a measurement depends mainly on the measuring process, not on the property measured.* The weight of a carton of eggs measured in grams has an interval/ratio scale. But if I label the carton as one of small, medium, large, or extra large, I have measured the weight in an ordinal scale. If a standard test of authoritarian personality has an ordinal scale, this does not mean that it is impossible to measure authoritarian personality on an interval/ratio scale, only that this test does not do so.

SECTION 3 EXERCISES

3.22 Identify the scale of each of the following variables as nominal, ordinal, or interval/ratio:

(a) The concentration of DDT in a sample of milk, in milligrams per liter.

(b) The species of each insect found in a sample plot of ground.

(c) A subject's response to the following personality test question: It is natural for people of one race to want to live away from people of other races.

Strongly agree
Agree
Undecided
Disagree
Strongly disagree

(d)   The pressure in pounds per square inch required to crack a specimen of copper tubing.

3.23   Identify the scale of each of the following variables as nominal, ordinal, or interval/ratio:

(a)   The position of the Chicago Bulls in the standings of the Central Division of the National Basketball Association (1st, 2nd, 3rd, 4th, 5th, 6th, 7th, or 8th).

(b)   The reaction time of a subject, in milliseconds, after exposure to a stimulus.

(c)   The score of a student on an examination in this statistics course.

(d)   A person's occupation as classified by the Bureau of Labor Statistics (managerial and professional, technical, sales, and so on).

3.24   A company database contains the following information about each employee: age, date hired, sex (male or female), ethnic group (Asian, black, Hispanic, etc.), job category (clerical, management, technical, etc.), yearly salary. In what type of scale is each of these six variables measured?

3.25   The Gallup poll from time to time asks its sample how highly they regard political candidates. The respondents are asked to rate each candidate on a 10-point scale that runs from "highly favorable" down to "highly unfavorable." Dwight Eisenhower received one of the highest ratings, with 65% of the voters giving him a "highly favorable" rating during the 1956 campaign. The lowest rating belongs to Barry Goldwater, who was rated "highly favorable" by only 15% of the voters in 1964. What type of scale does Gallup use to measure the personal enthusiasm of a voter toward a candidate? Explain your answer.

3.26   What type of scale is illustrated by the numbers on the shirts of a basketball team?

3.27   What type of scale is illustrated by house address numbers along a typical city street?

3.28   The 1990 census long form, given to a sample of 17% of all households, asked, "In what state or foreign country was this person born?" Another question asked "If this person is a female – How many babies has she ever had, not counting stillbirths?" What type of measurement scale is used in each of these two questions?

▶ 4 ISSUES: HOW NUMBERS CAN TRICK US

Political rhetoric, advertising claims, debate on public issues—we are assailed daily by numbers employed to prove a point or to buttress an argument. Those who use data to argue a cause want to support the cause, not necessarily to employ numbers carefully or even honestly. You now know quite a bit about how to produce reliable data, and about the pitfalls that await even careful statisticians. Let's look at how bad data, or good data wrongly used, can trick the unwary.

Senator Bilbo mails registered voters in his state a questionnaire and announces that the results show strong support for his positions. Voluntary response makes the data production suspect. Moreover, the senator may have asked slanted questions, so that the measurement is also suspect. We know by now that we should always ask:

▶ How were the data produced?

▶ Exactly what was measured?

Most of the examples in this section fall under the heading of misleading uses of properly produced data, so that asking exactly what was measured is the data detective's key question.

THE TRUTH, BUT NOT THE WHOLE TRUTH

The most common way to mislead with data is to cite correct numbers that don't quite mean what they appear to say. The numbers were not made up, so the fact that the information is a bit incomplete may be an innocent oversight. Here are some examples. You decide how innocent they are.

"Sure your patients have 50% fewer cavities. That's because they have 50% fewer teeth!"

EXAMPLE 13. Snow! Snow! Snow! Crested Butte attracts skiers by advertising that it has the highest average snowfall of any ski town in Colorado. That's true. But skiers want snow on the ski slopes, not in the town—and many other Colorado resorts get more snow on the slopes.[6]

EXAMPLE 14. We attract really good students. Colleges know that many prospective students look at popular guidebooks to decide where to apply for admission. The guidebooks print information supplied by the colleges themselves. Surely no college would simply lie about, say, the average SAT score of its entering students. But we do want our scores to look good. How about leaving out the scores of our international and remedial students? Northeastern University did this, making the average SAT score of its freshman class 50 points higher than if all students were included. If we admit economically disadvantaged students under a special program sponsored by the state, surely no one will complain if we leave their SAT scores out of our average? New York University did this.[7]

EXAMPLE 15. **We don't really make that much.** Are doctors overpaid? The American Medical Association (AMA) has long issued annual reports giving the median income of doctors in private practice. (The median is the typical income—half of doctors earn more than the median and half earn less.) After the median income reached $177,400 in 1992, the AMA stopped releasing the data. In 1994, the association announced that it would again release income data, but would lump doctors in private practice with doctors still in training and those who work for the government in order to bring the median down. "Now the physician looks less like he's gouging America," said an AMA spokesperson.[8]

BEWARE OF GOOD CAUSES

Good causes seem to attract bad statistics. Activists are so convinced of the rightness of their cause that a little fudging of the numbers seems harmless. Anyone who questions the numbers seems to be attacking the good cause, so bad numbers are endlessly repeated.

EXAMPLE 16. **How many homeless?** For several years after the plight of the homeless became a recognized good cause, the press regularly said that there were two or three million homeless people in the United States. These numbers were simply made up by the activist Mitch Snyder, who thought—correctly—that big numbers would help attract attention to his good cause. It is hard to count homeless people, but careful block-by-block surveys in several cities suggest that Snyder's numbers were between five and ten times larger than the truth.

A similar dispute surrounds the question of what kind of people are homeless. Activists suggest that they are mostly ordinary people who have fallen on hard times. Many careful studies make it clear that around three-fourths of homeless people are seriously disturbed by mental illness, substance abuse, or other conditions. This knowledge helps design suitable programs to help the homeless. Effective compassion requires accurate information.[9]

EXAMPLE 17. The multicultural labor force. The makeup of the American work force is changing. Women now have careers as a matter of course, and immigration is increasing the presence of minority groups in the workplace. Recognizing and adapting to this change is surely a good cause as well as an economic necessity. Countless articles and statements in support of the good cause have repeated the claim that in the final years of this century, "only 15 percent of the new entrants to the labor force will be native white males."

This seems implausible. A look at the *Statistical Abstract of the United States* shows that non-Hispanic white males make up more than 35% of the age group reaching working age. The 15% number is not made up, however. It first appeared on the first page of a report done for the Department of Labor.[10] A bit deeper in the report we discover that the 15% figure applies to "net new entrants" to the work force. That means new workers *after subtracting those who die or retire.* Because most workers who die or retire are older white men, subtracting them reduces the percent of white males among new workers. A company that hires 3 new white male workers and sees 3 other white male workers retire that same year has hired no white males in the eyes of the "net new entrants" count. White males are reduced to about 15% net, even though the actual percent of white men among new workers is about 35%. It isn't clear why anyone would be interested in "net new entrants." Perhaps the report's authors thought that "only 15% white males" would make a good headline.

IMPLAUSIBLE NUMBERS

As Example 17 illustrates, you can often detect dubious numbers simply because they don't seem plausible. Sometimes you can check an implausible number against data in reliable sources such as the *Statistical*

Abstract. Sometimes, as the next example illustrates, you can do a calculation to show that a number can't be right.

> **EXAMPLE 18. The abundant melon field.** The very respectable journal *Science*, in an article on insects that attack plants, mentioned a California field that produces 750,000 melons per acre. A reader responded, "I learned as a farm boy that an acre covers 43,560 square feet, so this remarkable field produces about 17 melons per square foot. If these are cantaloupes, with each fruit covering about 1 square foot, I guess they must grow in a stack 17 deep."[11] Here is the calculation the reader did:
>
> $$\text{melons per square foot} = \frac{\text{melons per acre}}{\text{square feet per acre}}$$
> $$= \frac{750,000}{43,560} = 17.2$$
>
> The editor, a bit embarrassed, replied that the correct figure was about 11,000 melons per acre.

ARE THE NUMBERS CONSISTENT WITH EACH OTHER?

If the numbers in a presentation don't agree with each other, something is wrong. This is the question of *internal consistency.* A little attention to consistency can do wonders. Here is part of an article dealing with a cancer researcher at the Sloan-Kettering Institute who was accused of committing the ultimate scientific sin, falsifying data.

> **EXAMPLE 19. Fake data.** "One thing he did manage to finish was a summary paper dealing with the Minnesota mouse experiments. ...That paper, cleared at SKI and accepted by the *Journal of Experimental Medicine*, contains a statistical table that is erroneous in such an elementary way that a bright grammar school pupil could catch the flaw. It lists 6 sets of 20 animals each, with the percentages of successful takes. Although any percentage of 20 has to be a multiple of 5, the percentages that Summerlin recorded were 53, 58, 63, 46, 48, and 67."[12]

ARE THE NUMBERS TOO GOOD TO BE TRUE?

In Example 19, lack of internal consistency led to the suspicion that the data were phony. *Too much precision or regularity* can lead to the same suspicion, as when a student's lab report contains data that are exactly as the theory predicts. The laboratory instructor knows that the accuracy of the equipment and the student's laboratory technique are not good enough to give such perfect results. He suspects that the student made them up. Here is an example drawn from another account of fraud in medical research.

> **EXAMPLE 20. More fake data.** "...Lasker had been asked to write a letter of support. But in reading two of Slutsky's papers side by side, he suspected that the same 'control' animals had been used in both without mention of the fact in either. Identical data points appeared in both articles, but...the actual number of animals cited in each case was different. This suggested at best a sloppy approach to the facts. Almost immediately after being asked about the statistical discrepancies, Slutsky resigned and left San Diego."[13]

In this case, suspicious regularity (identical data points) combined with inconsistency (different numbers of animals) led a careful reader to suspect fraud.

IS THE ARITHMETIC RIGHT?

Conclusions that are wrong or just incomprehensible are often the result of plain old-fashioned blunders. Rates and percentages cause particular trouble. Sometimes you can straighten out the numbers by repairing the arithmetic. This is again a matter of internal consistency.

> **EXAMPLE 21.** A writer in *Science* (Volume 192 (1976), p. 1081) stated in 1976 that "people over 65, now numbering 10 million, will number 30 million by the year 2000, and will constitute an unprecedented 25 percent of the population." Such explosive growth

of the elderly—tripling in a quarter century to become a fourth of the population—would profoundly change any society.

Let's check the arithmetic. Thirty million is 25% of 120 million, because

$$\frac{30}{120} = 0.25 = 25\%$$

The U.S. population is already more than twice 120 million, so 30 million elderly is nowhere near 25% of the population. Something is wrong with the writer's figures.

Thus alerted, we can check the *Statistical Abstract* to learn the truth. In 1975, there were 22.4 million persons over 65, not 10 million. The projection of 30 million by the year 2000 is correct, but that is only 11% or 12% of the projected population for that year. The explosive growth of the elderly vanishes in the light cast by accurate statistics.

EXAMPLE 22. Counts and rates again. The Bureau of Labor Statistics' monthly report on unemployment once noted that the unemployment rate was 6.1% for whites and 14.5% for blacks. The *New York Times* explained the situation as follows:

> *The bureau also reported that the ratio of black to white jobless rates "continued its recent updrift to the unusually high level of 2.4 to 1 in August," meaning that 2.4 black workers were without jobs for every unemployed white worker.*[14]

Check that 14.5% is 2.4 times as great as 6.1%. The BLS was correct in stating that the ratio of black to white jobless rates was 2.4 to 1. But the *Times'* interpretation was completely wrong. Because blacks make up only about 11% of the labor force, there are many fewer jobless blacks than jobless whites even though the percent of blacks who are unemployed is higher than the percent of whites who are without jobs. The *Times* confused percent unemployed with actual counts of the number of unemployed workers.

Calculating the percent increase or decrease in some variable is a common source of arithmetic errors. The percent change in a quantity is found by

$$\text{percent change } = \frac{\text{amount of change}}{\text{starting value}} \times 100$$

> **EXAMPLE 23.** Last year the price of gold rose from $300 an ounce to $450 an ounce. This was an increase of 50%, because
>
> $$\frac{\text{increase}}{\text{starting value}} = \frac{\$150}{\$300} = 0.5 = 50\%$$
>
> This year, the gold price drops by 50%. What is an ounce now worth? Well, the amount of the decrease is 50% of $450, or $225. So the price of an ounce of gold is now $450 less $225, or $225.

An increase of 50%, followed by a decrease of 50%, does *not* bring us back to the original value. To give another example, an increase of 100% means that the quantity in question has doubled, since the amount of the increase is 100% of the original value. But a decrease of 100% means that the quantity is now zero—it has lost 100% of its value, and 100% is all there is.

WHEN YOU SEE A NUMBER, STOP AND THINK

The aim of statistics is to provide insight by means of numbers. Numbers are most likely to yield their insights to those who examine them closely. Pay attention to voluntary response samples and to confounding. Ask exactly what a number measures. Look for internal inconsistencies and check that the arithmetic is correct. Compare implausible numbers with numbers you know are right. If you form the habit of looking at numbers closely, your friends will soon think that you are brilliant. They might even be right.

SECTION 4 EXERCISES

3.29   **Advertising pain killers.** An advertisement for the pain reliever Tylenol was headlined: "Why Doctors Recommend Tylenol

More Than All Leading Aspirin Brands Combined." The makers of Bayer Aspirin, in a reply headlined "Makers of Tylenol, Shame On You!" accused Tylenol of misleading by giving the truth but not the whole truth. You be the detective: How is Tylenol's claim misleading even if true?

3.30 **Advertising pain killers.** Anacin was long advertised as containing "more of the ingredient doctors recommend most." Another over-the-counter pain reliever claimed that "doctors specify Bufferin most" over other "leading brands." Both advertising claims were literally true; the Federal Trade Commission found them both misleading. Explain why. (Hint: What is the active pain reliever in both Anacin and Bufferin?)

3.31 **Are incomes up or down?** Has the income of Americans gone down in recent decades? Here are some data that feature in the debate over this question. After adjusting for the effects of inflation, the median money income of U.S. households fell from $34,200 in 1973 to $30,500 in 1983 and $31,200 in 1993. That looks bad. Per capita money income (total income divided by total number of people), on the other hand, increased from $13,500 in 1973 to $13,850 in 1983 and $15,777 in 1993. That looks better. All of these numbers come from the Bureau of Labor Statistics, so they are relatively trustworthy.

A household consists of all people living together at the same address. Explain how changes in American households over the past decades can cause median household income to drop even when income per person is going up.

3.32 **Trash at sea?** In a report on the problem of vacation cruise ships polluting the sea by dumping garbage overboard, *Condé Nast Traveler* magazine (June 1992) said:

On a seven-day cruise, a medium-size ship (about 1,000 passengers) might accumulate 222,000 coffee cups, 72,000 soda cans, 40,000 beer cans and bottles, and 11,000 wine bottles.

Are these numbers plausible? Do some arithmetic to back up your conclusion. Suppose, for example, that the crew is as large as the passenger list. How many cups of coffee must each person drink every day?

3.33   **Battered women?** A letter to the editor of the *New York Times* complained about a *Times* editorial that said "an American woman is beaten by her husband or boyfriend every 15 seconds." The writer of the letter claimed that "at that rate, 21 million women would be beaten by their husbands or boyfriends every year. That is simply not the case." He cited the National Crime Victimization Survey, which estimated 56,000 cases of violence against women by their husbands and 198,000 by boyfriends or former boyfriends. The survey showed 2.2 million assaults against women in all, most by strangers or someone the woman knew who was not her past or present husband or boyfriend.[15]

(a)   First do the arithmetic. Every 15 seconds is 4 per minute. At that rate, how many beatings would take place in an hour? In a day? In a year? Is the letter-writer's arithmetic correct?

(b)   Is the letter-writer correct to claim that the *Times* overstated the number of cases of domestic violence against women?

3.34   **Don't dare to drive?** A university sends a monthly newsletter on health to its employees. A recent issue included a column called "What is the Chance" that said:

Chance that you'll die in a car accident this year: 1 in 75.

There are about 265 million people in the United States. About 40,000 people die each year from motor vehicle accidents. What is the chance a typical person will die in a motor vehicle accident this year?

3.35   **How many miles of highways?** *Organic Gardening* magazine (July 1983) once said that "the U.S. Interstate Highway System spans 3.9 million miles and is wearing out 50% faster than it can be fixed. Continuous road deterioration adds $7 billion yearly in fuel costs to motorists." The distance from the east coast to the west coast of the United States is about 3000 miles. How many separate highways across the continent would be needed to account for 3.9 million miles of roads? What do you conclude about the number of miles in the Interstate system?

3.36   *Organic Gardening* (March 1983), describing how to improve your garden's soil, said, "Since a 6-inch layer of mineral soil in a 100-square-foot plot weighs about 45,000 pounds, adding 230 pounds of compost will give you an instant 5% organic matter." What percent of 45,000 is 230? Does the error lie in the arithmetic, or is that 45,000 pounds too heavy?

3.37   **No eligible men?** A news report quotes a sociologist as saying that for every 233 unmarried women in their 40s in the United States, there are only 100 unmarried men in their 40s. These numbers point to an unpleasant social situation for women of that age. Are the numbers plausible?

3.38   **Boating safety.** Data on accidents in recreational boating in the *Statistical Abstract* show that the number of deaths has dropped from 1418 in 1970 and 1360 in 1980 to 865 in 1990. However, the number of injuries reported grew from 780 in 1970 to 2650 in 1980 and 3822 in 1990. Why are there so few injuries in these government data relative to the number of deaths? Which count (deaths or injuries) is probably more accurate? Why might the injury count rise when deaths did not?

3.39   **Fake data?** The late English psychologist Cyril Burt was known for his studies of the IQ scores of identical twins who were raised apart. The high correlation between the IQs of separated twins in Burt's studies pointed to heredity as a major factor in IQ. ("Correlation" is a measure of the connection or association between two variables. We will become better acquainted with it in Chapter 5.) Burt wrote several accounts of his work, each reporting on more pairs of twins. Here are his reported correlations as he published them:

Date of publication	Twins reared apart	Twins reared together
1955	0.771 (21 pairs)	0.944 (83 pairs)
1958	0.771 ("over 30" pairs)	0.944 (no count)
1966	0.771 (53 pairs)	0.944 (95 pairs)

What is suspicious here? (Further investigation made it almost certain that Burt fabricated at least some of his data.)

3.40   Here is a quotation from a book review in a scientific journal:

> ... *a set of 20 studies with 57 percent reporting significant results,
> of which 42 percent agree on one conclusion while the remaining
> 15 percent favor another conclusion, often the opposite one.*[16]

Do the numbers given in this quotation make sense? Can you
decide how many of the 20 studies agreed on "one conclusion,"
how many favored another conclusion, and how many did not
report significant results?

3.41   **The stock market crash of 1987.** On "Black Monday," October
19, 1987, the Dow Jones Industrial Average of common stock
prices fell from 2244 to 1736. What percent decrease was this?

3.42   A news report noted that Ford Motor Company had decided
to offer its new models with fancier equipment at considerably
higher prices. As an example, the base price of Ford's F-150 XLT
pickup truck rose from $20,315 for the 1996 model to $24,405
for the restyled 1997 model. What percent increase is this?

3.43   In 1982, the Census Bureau gave a simple test of literacy in
English to a random sample of 3400 people. The *New York
Times* (April 21, 1986) printed some of the questions under the
headline "113% of Adults in U.S. Failed This Test." Why is the
percent in the headline clearly wrong?

3.44   **We don't lose your baggage.** Continental Airlines once advertised that it had "decreased lost baggage by 100% in the past six
months." Do you believe this claim?

3.45   A new mouth rinse claimed to "reduce plaque on teeth by
300%." Explain carefully why it is impossible to reduce anything
by 300%.

3.46   The question-and-answer column of a campus newspaper was
asked what percent of the campus was "Greek." The answer
given was that "the figures for the fall semester are approximately
13 percent for the girls and 15–18 percent for the guys, which
produces a 'Greek' figure of approximately 28–31 percent of the
undergraduates at Purdue" (*Purdue Exponent*, September 21,
1977). Discuss the campus newspaper's arithmetic.

3.47 **Smoking and health.** Below is a table from *Smoking and Health Now*, a report of the British Royal College of Physicians. It shows the number and percent of deaths among men age 35 and over that are due to the chief diseases related to smoking. One of the entries in the table is incorrect, and an erratum slip was inserted to correct it. Which entry is wrong, and what is the correct value?

	Lung cancer	Chronic bronchitis	Coronary heart disease	All causes
Number	26,973	24,976	85,892	312,537
Percent	8.6%	8.0%	2.75%	100%

3.48 Find in a newspaper or magazine an example of one of the following. Explain in detail the statistical shortcomings of your example.

Omission of essential information

Lack of internal consistency

Implausible numbers

Faulty arithmetic

Notes

1. Bureau of Labor Statistics, *The Unemployment Situation, July 1994*, document USDL 94-382, August 5, 1994.
2. E. Kolbert, "Nielsen TV ratings face a challenge by 3 networks," *New York Times*, February 4, 1994.
3. C. Shea, "Under UCLA's elaborate system, race makes a difference," *Chronicle of Higher Education*, April 28, 1995, pp. A12–A14. The article is part of a special section titled "Affirmative Action on the Line." The SAT scores given predate the recent "recentering" that raises the scores.
4. Data from H. H. Ku, "Statistical concepts in metrology," in H. H. Ku (ed.), *Precision Measurement and Calibration*, National Bureau of Standards Special Publication 300, 1969.
5. Some details appear in R. D. Hershey, Jr., "Jobless rate underestimated, U.S. says, citing survey bias," *New York Times*, November 17, 1993.
6. R. J. Newan, "Snow job on the slopes," *US News & World Report*, December 17, 1994, pp. 62–65.

7. "Colleges inflate SATs and graduation rates in popular guidebooks," *Wall Street Journal*, April 5, 1995.

8. "New math used to clean up doctors' image," Associated Press dispatch appearing in the *New York Times*, June 17, 1994.

9. The single best source of information is C. Jencks, *The Homeless*, Harvard University Press, 1994.

10. W. B. Johnston, et al., *Workforce 2000: Work and Workers for the 21st Century*, The Hudson Institute, 1987.

11. Letter by J. L. Hoffman, *Science*, February 7, 1992, p. 665. The original article appeared on December 6, 1991.

12. Quoted from B. Yuncker, "The strange case of the painted mice," *Saturday Review/World*, November 30, 1974, p. 53.

13. Quoted from E. Marshall, "San Diego's tough stand on research fraud," *Science*, Volume 234 (1986), pp. 534–535.

14. From the *New York Times*, September 3, 1977.

15. The letter, by L. Jarvik, appeared in the *New York Times*, May 4, 1993. The editorial, "Muggings in the Kitchen," appeared on April 23, 1993.

16. From *Science*, Volume 189 (1975), p. 373.

17. From the *Lafayette Journal and Courier*, October 23, 1988.

REVIEW EXERCISES

3.49 You are writing an article for a consumer magazine based on a survey of the magazine's readers that asked about the reliability of their household appliances. Of 13,376 readers who reported owning Brand A dishwashers, 2942 required a service call during the past year. Only 192 service calls were reported by the 480 readers who owned Brand B dishwashers. Describe an appropriate variable to measure the reliability of a make of dishwasher, and compute the values of this variable for Brand A and for Brand B. Which brand is more reliable?

3.50 A friend tells you, "In American History, 20 students failed. Only 11 students failed Russian History. That American History prof is a tougher grader than the Russian History teacher." Explain why the conclusion may not be true. What additional information would you need to compare the courses?

3.51 **Testing job applicants.** A company used to give IQ tests to all job applicants. This is now illegal, because IQ is not related to the performance of workers in all the company's jobs. Does the reason for the policy change involve the *reliability*, the *bias*, or the *validity* of IQ tests as a measure of future job performance? Explain your answer.

3.52 A sociologist measures many variables on each of a sample of urban households. In what type of scale is each of the following variables measured?

(a) Annual household income in dollars.

(b) The race of the householder.

(c) The educational level of the householder (elementary school, 1–3 years of high school, completed high school, and so on).

3.53 **Capital punishment.** Here are data on the number of convicted criminals who were executed between 1977 and 1993 in several states, as well as the 1990 population of those states.

State	Population (thousands)	Executions
Alabama	4,040	10
Florida	12,938	32
Georgia	6,478	17
Nevada	1,202	5
Texas	16,986	71
Virginia	6,189	14

Texas and Florida lead in number of executions. Because these are large states, we might expect them to have many executions. Find the *rate* of executions for each of these states, in executions per million population. Because population is given in thousands, you can find the rate per million as

$$\text{rate per million} = \frac{\text{executions}}{\text{population in thousands}} \times 1000$$

Which state has the highest number of executions relative to its population? Are Florida and Texas still high by this measure?

3.54 **Airport delays.** An article in a midwestern newspaper about flight delays at major airports said:

*According to a Gannett News Service study of U.S. airlines' perfor-
mance during the past five months, Chicago's O'Hare Field scheduled 114,370 flights. Nearly 10 percent, 1,136, were canceled.*[17]

Check the newspaper's arithmetic. What percent of scheduled
flights from O'Hare were actually canceled?

3.55 **Housing costs.** A newspaper story on housing costs (*Lafayette
Journal and Courier*, September 21, 1977) noted that in 1975
the median price of a new house was $39,300 and the median
family income was $13,991. (Half of all families earn less than
the median income and half earn more.) The writer then claimed
that "the ratio of housing prices to income is much lower today
than it was in 1900 (2.8 percent in 1975 vs. 9.8 percent in 1900)."

I wish that I could buy a new house for 2.8% of my income.
Where did that 2.8 come from? What is the correct expression
for $39,300 as a percent of $13,991?

3.56 **Deer in the suburbs.** Westchester County is a suburban area
covering 438 square miles immediately north of New York City.
Fine Gardening magazine (September/October 1989, p. 76)
claimed that the county is home to 800,000 deer. Do a calculation that shows this claim to be implausible.

WRITING PROJECTS

3.1 **Measuring race.** "Race" is not a clearly defined biological idea,
but it is very important socially and politically. The 1990 Census
asked people to classify themselves as one of "White," "Black or
Negro," "Indian (Amer.)," "Eskimo," "Aleut," or "Asian or Pacific
Islander." The last category had 10 subcategories. If none of
these labels fit, you are "Other race." "Spanish/Hispanic origin"
was a separate question, because Hispanics can be of any race.

Now the Census Bureau is considering what to do for the
2000 Census. Arab-Americans want a "Middle Eastern" category.
Some Hawaiians want to be moved from "Asian or Pacific
Islander" to "Native American" (with American Indians) or to
their own category. Many of the growing number of Americans

whose parents are of different races would like a "Multiracial" category. Some black groups oppose that, however, fearing that fewer people would call themselves "Black."

Write a brief discussion about measuring race for government purposes. Why do the census figures on race matter? Do you agree that people should be allowed to say what race they are? What categories should the Census Bureau offer? If you prefer a "Multiracial" category, exactly how will you define it? (For example, is it enough that not all grandparents have the same race, or must the parents be of different races? Both "Korean" and "Vietnamese" are subcategories of Asian on the 1990 census form—so are children of a union between a Korean and a Vietnamese parent multiracial?)

3.2 **Measuring income.** What is the "income" of a household? Household income may determine eligibility for government programs that assist "low-income" people. Income statistics also have political effects. Political conservatives often argue that government data overstate how many people are poor because the data include only money income, leaving out the value of food stamps and subsidized housing. Political liberals reply that the government should measure money income so it can see how many people need help.

You are on the staff of a member of Congress who is considering new welfare legislation. Write an exact definition of "income" for the purpose of determining which households are eligible for welfare. A short essay will be needed. Will you include nonmoney income such as the value of food stamps or subsidized housing? Will you allow deductions for the cost of child care needed to permit the parent to work? What about assets that are worth a lot but do not produce income, such as a house?

"Tonight, we're going to let the statistics speak for themselves."

PART II

ORGANIZING DATA

Data, like words, speak clearly only when they are organized. Also like words, data speak more effectively when well organized than when poorly organized. Again like words, data can obscure a subject by their quantity, requiring a brief summary to highlight essential facts. The second of statistics' three domains is the organizing, summarizing, and presenting of data.

Data are recorded in many forms: completed questionnaires, laboratory notebooks, bits stored in a computer. Our task is to digest such sets of raw data, to organize and summarize them for human use. A few general principles can help us understand data. *The first step in data analysis is to display the data in a graph.* The human eye and mind can see more in a graph than any computer, so we begin by taking advantage of our ability to see and understand. When looking at a graph, avoid getting lost in the details. *Look first for an overall pattern in the data, and then for any striking exceptions to that pattern.* Numerical calculations can help us describe specific aspects of data, such as the average value of a variable. Another principle of data analysis is therefore to *move from graphs to well-chosen numerical descriptions.* What we see in the graph will help us choose numerical summaries.

Summarizing and presenting a large body of facts offers to ignorance or malice ample opportunity for distortion. This is no less (but also no more) the case when the facts to be summarized are numbers rather than words. We shall therefore take due note of the traps into which the ignorant fall and the tools of duplicity that the malicious use. Those who picture statistics as primarily a piece of the liar's art concentrate on the part of statistics that deals with summarizing and presenting data. I claim that misleading summaries and selective presentations go back to that after-the-apple conversation between Adam, Eve, and God. Don't blame statistics. Remember the saying "Figures don't lie, but liars figure," and beware.

CHAPTER 4

DESCRIBING DISTRIBUTIONS

What mental picture does the word *statistics* call to mind? Very likely, images of tables crowded with numbers and graphs zigging up or zagging down. That's not a bad picture: statistics deals with data, and we use tables and graphs to present data. Random sampling, randomized comparative experiments, and valid measurement produce good data. Now we must see what the data say.

There is an art to presenting complex data clearly—in tables, in graphs, and by summary numerical descriptions. This chapter introduces the art of describing data, concentrating on data about just one variable.

Distributions

The **distribution** of a variable tells us what values the variable takes and how often each value occurs.

▶ 1 DISPLAYING DATA

We use data, like words, to communicate facts and to support conclusions. Like words, data must be well organized if they are to communicate clearly. All but the smallest sets of data must be summarized, boiled down into simpler form, if they are to be at all clear. We summarize data in tables or, for greater impact, in graphs. The tables we refer to when seeking information are almost always summary tables that organize and greatly condense the original data.

DATA TABLES

I invite you to look at the *Statistical Abstract of the United States,* an annual volume packed with every variety of numerical information. Has the number of private elementary and secondary schools grown over time? What about minority enrollments in these schools? How many college degrees were given in each of the past several years, and how were these degrees divided among fields of study and by the age, race, and sex of the students? All this and more can be found in the education section of the *Statistical Abstract.* The tables that present this information are summary tables. We don't want to see information on every college degree individually, only the counts in categories of interest to us.

> **EXAMPLE 1. What makes a clear table?** Table 4-1 presents data on the marital status of women age 18 and over. This table illustrates some good practices for data tables. It is clearly *labeled* so that we can see the subject of the data at once. The main heading describes the general subject of the data and gives the date because these data will change over time. Labels within the table identify the variables and state the *units* in which they are measured. Notice, for example, that the counts are in thousands of women. The *source* of the data appears at the foot of the table.

TABLE 4-1	Marital status of adult women, 1994	
Marital status	Count (thousands)	Percent
Single	19,458	19.7
Married	58,113	58.8
Widowed	11,073	11.2
Divorced	10,120	10.2
TOTAL	98,765	100.0

SOURCE: *Statistical Abstract of the United States,* 1995.

One of our first acts in organizing a set of data is often to count how often each value occurs. Table 4-1 presents the counts of women in each marital status. **Rates** (percents or proportions) are often clearer than counts—it is more helpful to hear that 19.7% of adult women have never been married than to hear that there are 19,458,000 such women. The percents also appear in Table 4-1. The two columns of Table 4-1 present the distribution of the variable "marital status" in two alternate forms. Each column gives information about what values the variable takes and how often it takes each value.

> **EXAMPLE 2. Roundoff errors.** Did you check Table 4-1 for internal consistency? The total number of women should be
>
> $$19,458 + 58,113 + 11,073 + 10,120 = 98,764 \text{ (thousands)}$$
>
> The table gives the total as 98,765. What happened? Each entry is rounded to the nearest thousand. The rounded entries don't quite add to the total, which is rounded separately. Such *roundoff errors* will be with us from now on as we do more arithmetic.

PIE CHARTS AND BAR GRAPHS

Marital status has a nominal scale—this variable just places people into categories. To picture the distribution in a graph, we might use a **pie chart.** Figure 4-1 is a pie chart of the marital status of adult women. Pie charts show how a whole is divided into parts. To make a pie chart, first draw a circle. The circle represents the whole, in this case all adult women. Wedges within the circle represent the parts, with the angle spanned by each wedge in proportion to the size of that part. For example, 19.7% of adult women are single. Because there are 360 degrees in a circle, the "single" wedge spans an angle of

$$0.197 \times 360 = 71 \text{ degrees}$$

Pie charts force us to see that the parts do make a whole. But because angles are harder to compare than lengths, a pie chart is not a good way to compare the sizes of the various parts of the whole.

Figure 4-2 is a **bar graph** of the same data. The height of each bar shows the percent of women with the marital status marked at the bar's base. The bar graph makes it clear that more women are widowed than divorced—the "widowed" bar is taller. Because it is hard to see this

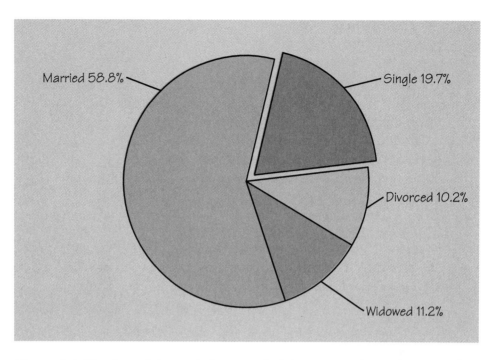

Figure 4-1 Pie chart of the marital status of adult women. The angle spanned by each wedge shows the percent of women having that marital status.

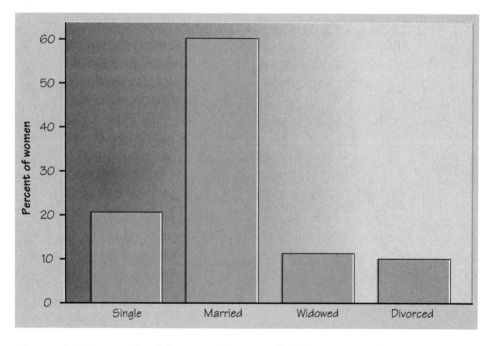

Figure 4-2 Bar graph of the marital status of adult women. The height of each bar shows the percent of women having that marital status.

from the wedges in the pie chart, we added labels to the wedges that include the actual percents. The bar graph is also easier to draw than the pie chart unless a computer is doing the drawing for you.

> **Displaying distributions**
>
> To display the distribution of a variable measured in a nominal scale, use a pie chart or a bar graph.

Pie charts emphasize how individual counts or percents are related to the whole. Bar graphs emphasize how quantities compare with one another. Although both graphs can display the distribution of a nominal variable, bar graphs are more generally useful. A bar graph, unlike a pie chart, can compare quantities that are not part of a whole.

EXAMPLE 3. **High taxes?** Figure 4-3 compares the level of taxation in seven large democratic nations. The height of the bars shows the

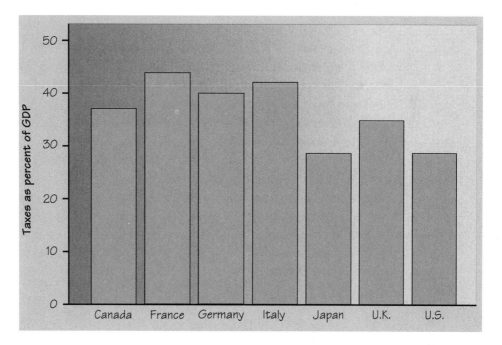

Figure 4-3 Bar graph comparing the level of taxes in seven large democracies. The height of each bar shows the percent of that nation's gross domestic product that is taken in taxes.

percent of each nation's gross domestic product (GDP, the total value of all goods and services produced) that is taken in taxes. Americans accustomed to complaining about high taxes may be surprised to see that the United States ties Japan for the lowest taxes, at 29.4% of GDP.

BEWARE THE PICTOGRAM

Bar graphs compare several quantities by comparing the heights of bars that represent the quantities. Our eyes, however, react to the *area* of the bars rather than to their height. When all bars have the same width, the area (width × height) varies in proportion to the height and our eyes receive the right impression. When you draw a bar graph, make the bars equally wide. Artistically speaking, bar graphs are a bit dull. It is tempting to replace the bars with pictures for greater eye appeal.

EXAMPLE 4. **A misleading graph.** Figure 4-4 is a *pictogram*. It is a bar graph in which pictures replace the bars. The graph is aimed at advertisers deciding where to spend their budgets. It shows that *Time* magazine attracts the lion's share of advertising spending. Or does it? The numbers above the pens show that advertising spending in *Time* is 1.64 times as great as in *Newsweek*. Why does the graph suggest that *Time* is much farther ahead?

To magnify a picture, the artist must increase *both* height and width to avoid distortion. If both the height and width of *Time*'s pen are 1.64 times as large as *Newsweek*'s, the area is 1.64 × 1.64, or 2.7 times as large. Our eyes, responding to the area of the pens, see *Time* as the big winner.

Here is a challenge for the artist: make the graph both attractive and honest. Figure 4-5 shows that it can be done. This is another *Time* ad making a point similar to that of Figure 4-4. The "bars" have equal widths, so our eyes receive an honest impression.

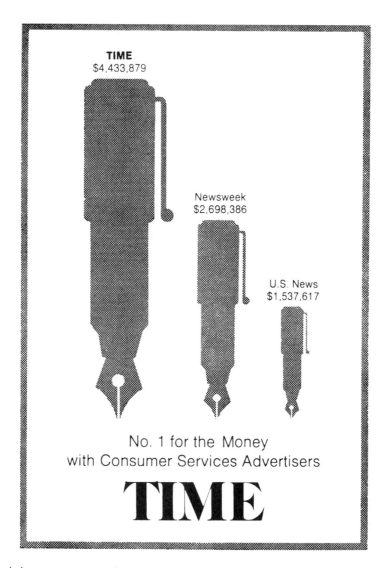

Figure 4-4 A pictogram. This variation of a bar graph is attractive but misleading. [Copyright © 1971 by Time, Inc. Reproduced by permission.]

LINE GRAPHS

Line graphs show the behavior of a variable over time. Mark the time scale on the horizontal axis, and put the scale for the variable being plotted on the vertical axis.

Figure 4-5 An attractive and accurate variation of a bar graph. [Copyright © 1972 by Time, Inc. Reproduced by permission.]

EXAMPLE 5. What has happened to the price of oranges? The Bureau of Labor Statistics records the price of fresh oranges each month at a probability sample of food outlets as part of its effort to measure consumer prices. Figure 4-6 is a line graph of the average retail price of fresh oranges for the years 1985 to 1995.

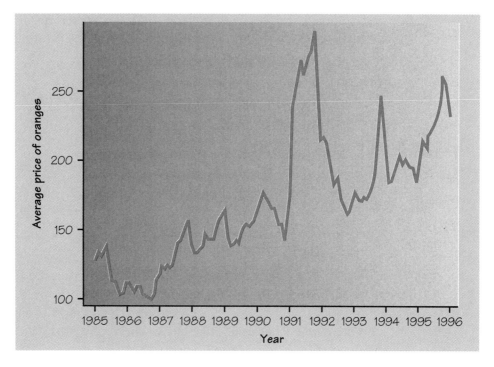

Figure 4-6 A line graph. The graph shows the average retail price of fresh oranges each month from 1985 to 1995. The price is given as a percent of the average price during an earlier base period.

The BLS gives each month's price not in dollars but as a percent of the average price during an earlier "base period." For example, the price of oranges in January 1985 was 127.6% of the base period price, so the first point on the plot is at 127.6. The February 1985 point is 133.8, and so on. Connecting the points for all months produces the line graph.

It would be difficult to see patterns in the long table of monthly prices. Figure 4-6 makes the patterns clear. We see at once that:

▶ The price of oranges was generally rising in the 1985 to 1995 period. This is a long-term **trend.**

▶ In most years, the price is highest in late summer and early fall, and lowest early in the year. (Each year's tick on the "Year" scale marks the beginning of the year.) This is **seasonal variation** caused by the fact that Florida's orange crop is harvested in the winter.

▶ Prices rose dramatically in 1991. The reason? A freeze in Florida destroyed much of the orange crop.

WATCH THOSE SCALES!

Just because graphs speak so strongly, they can mislead the unwary. The wary reader of a line graph looks closely at the *scales* marked off on the axes.

EXAMPLE 6. **The rise in divorce.** The number of divorced people in the United States has grown steadily over time. To see how divorced people have become more common, let's graph the number of divorced women per 1000 married women. For 1994, we can do the calculation from the information in Table 4-1:

$$\frac{\text{divorced women}}{\text{married women}} \times 1000 = \frac{10,120}{58,113} \times 1000 = 174$$

Figure 4-7 presents two line graphs of the data from 1960 to 1994. The graph on the left shows a rather gradual increase. The graph on the right shows that divorce is thundering upward.

How can two graphs of the same data appear so different? Look at the scales on the axes of the graphs. You can transform the left-hand graph into the right-hand graph by stretching the vertical scale, squeezing the horizontal scale, and cutting off the vertical scale just above and below the values to be plotted. Now you know how to either exaggerate or play down a trend in a line graph.

Which of these graphs is correct? Both are accurate graphs of the data, but both have scales chosen to create a specific effect. Because there is no one "right" scale for a line graph, correct graphs can give different impressions by their choices of scale. Watch those scales!

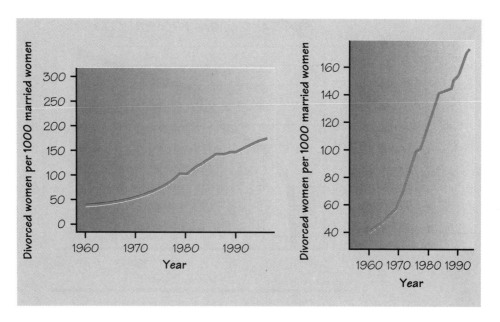

Figure 4-7 The effect of changing the scales in a line graph. Both graphs display the same data, but the right-hand graph makes the increase appear much more rapid.

Once in a while you may encounter a more barbarous line graph. Figure 4-8 is the least civilized I have seen. Note first that time is on the vertical axis. So when the graph goes straight up (1940 to 1946), taxes were not increasing at all. And when the graph is quite flat (1965 to 1972), taxes were rising rapidly. The graph gives an impression exactly the reverse of the truth. That's why time always belongs on the horizontal axis of a line graph. Second, the time scale does not have equal units. Equal lengths on the vertical axis represent first two years, then four years, then two years again, then whatever the interval between 1962 and 1964–1965 is, and so on. The graph is stretched and squeezed by the changing time scale, so we can't compare the rise in taxes in different time periods by looking at the steepness of the graph. These barbarisms were perpetrated by the Associated Press (not by the Census Bureau, which was only the source of the data). The newspaper then provided the caption below the chart, which no doubt refers to the "12 years" from 1940 to 1972.

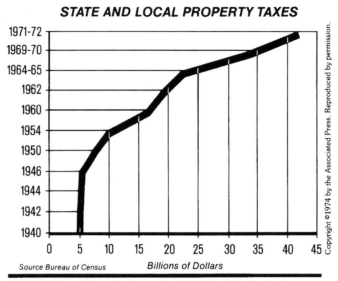

Figure 4-8 The world's worst line graph.

MAKING GOOD GRAPHS

Graphs are the most effective way to communicate using data. A good graph frequently reveals facts about the data that would be difficult or impossible to detect from a table. What is more, the immediate visual impression of a graph is much stronger than the impression made by data in numerical form—so strong that we must guard against false impressions.

Like tables, graphs should be clearly labeled to reveal the variables plotted, their units, the source of the data, and so on. Here is another important principle: *make the data stand out.* Be sure that the actual data, not labels, scale markings, or grids, catch the viewer's attention.[1]

EXAMPLE 7. **The rise in college education.** Compare the two graphs in Figure 4-9. Both are line graphs of the same data on the same scales. They show the increase over time in the percent of adult Americans who have finished four years of college. The graph

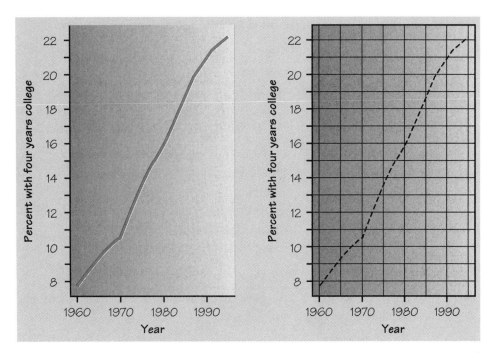

Figure 4-9 Two line graphs of the same data using the same scales. Both graphs display the percent of people over 25 years of age who have completed at least four years of college. Unnecessary clutter makes the graph on the right hard to read.

on the right, however, is cluttered with unneeded grid lines that hide a weak plot of the data. The grid lines serve no purpose—if your audience must know the actual numbers, add a table rather than grid lines on the graph. A good graph uses no more ink than is needed to present the data clearly. Always ask if your graph is more complicated than is necessary.

SECTION 1 EXERCISES

4.1   Answer the following questions using the data in Table 4-1.

(a)   What is the sum of the percents in the last column? Why do the percents not add to 100.0%?

(b)   How many unmarried adult women are there?

(c)   What percent of adult women are either widowed or divorced?

4.2   **Female doctorates.** Here are data on the percent of females among people earning doctorates in 1992 in several fields of study (from the 1995 *Statistical Abstract of the United States*).

Computer science	13.3%
Life sciences	38.3%
Education	59.5%
Engineering	9.6%
Physical sciences	21.9%
Psychology	59.7%

(a)   Present these data in a well-labeled bar graph.

(b)   Would it also be correct to use a pie chart to display these data? Explain your answer.

4.3   **Accidental deaths.** In 1992 there were 86,777 deaths from accidents in the United States. Among these were 40,982 deaths from motor vehicle accidents, 12,646 from falls, 3524 from drowning, 3958 from fires, and 7082 from poisoning. (Data from the 1995 *Statistical Abstract of the United States.*)

(a)   Find the percent of accidental deaths from each of these causes, rounded to the nearest percent. What percent of accidental deaths were due to other causes?

(b)   Make a well-labeled bar graph of the distribution of causes of accidental deaths. Be sure to include an "other causes" bar.

(c)   Would it also be correct to use a pie chart to display these data? Explain your answer.

4.4   **Murder weapons.** The 1995 *Statistical Abstract of the United States* reports FBI data on murders for 1993. In that year, 56.9% of all murders were committed with handguns, 12.7% with other firearms, 12.7% with knives, 5.0% with a part of the body (usually the hands or feet), and 4.4% with blunt objects. Make a graph to display these data. Do you need an "other methods" category?

4.5 **Civil unrest.** The years around 1970 brought unrest to many U.S. cities. Here are government data on the number of civil disturbances in each 3-month period during the years 1968 to 1972.

Period	Count	Period	Count
1968, Jan.–Mar.	6	1970, July–Sept.	20
Apr.–June	46	Oct.–Dec.	6
July–Sept.	25	1971, Jan.–Mar.	12
Oct.–Dec.	3	Apr.–June	21
1969, Jan.–Mar.	5	July–Sept.	5
Apr.–June	27	Oct.–Dec.	1
July–Sept.	19	1972, Jan.–Mar.	3
Oct.–Dec.	6	Apr.–June	8
1970, Jan.–Mar.	26	July–Sept.	5
Apr.–June	24	Oct.–Dec.	5

(a) Make a line graph of these data.

(b) The data show both a longer-term trend and seasonal variation within years. Describe the nature of both patterns. Can you suggest an explanation for the seasonal variation in civil disorders?

4.6 **Deaths from cancer.** Here are data on the rate of deaths from cancer (deaths per 100,000 population) in the United States from 1935 to 1990. The data are from the *Statistical Abstract* and *Historical Statistics of the United States, Colonial Times to 1970.* (The *Historical Statistics* volumes supplement the *Statistical Abstract* by providing data from earlier years on many of the same subjects.)

Year	1935	1940	1945	1950	1955	1960
Death rate	108.2	120.3	134.0	139.8	146.5	149.2

Year	1965	1970	1975	1980	1985	1990
Death rate	153.5	162.8	169.7	183.9	193.3	203.2

(a) Draw a line graph of these data designed to emphasize the rise in the cancer death rates. (Imagine that you are trying to persuade Congress to appropriate more money to fight cancer.)

Figure 4-10 A newspaper graph, for Exercise 4.8.

(b)   Draw another line graph of the same data designed to show only a moderate increase in the death rate.

(c)   Describe in words the overall pattern of these data.

4.7   Use Figure 4-8 to make an approximate table of the total amount of state and local property taxes in the years 1940, 1942, 1944, 1946, 1950, 1954, 1960, 1962, 1964, 1969, and 1971. Then draw a line graph from your table to see what a correct version of Figure 4-8 would look like.

4.8   Figure 4-10 shows a graph that appeared in the Lexington, Kentucky, *Herald-Leader* of October 5, 1975. Discuss the correctness of this graph.

4.9   **Trucks versus cars.** Plotting two line graphs on the same axes is a useful way to compare the change in two quantities over time. Here (again from the *Statistical Abstract*) are the number (in thousands) of new cars made in the United States and the number of new trucks made. (The government's definition of "truck" includes sports utility vehicles and minivans.) Make a plot and describe the trend that you see. Now extend this trend (this is risky) to estimate when truck sales will pass car sales.

Year	1981	1983	1985	1987	1989	1991	1993
Cars	6255	6739	8002	7085	6807	5407	5960
Trucks	1701	2414	3357	3821	4062	3375	4895

4.10 Figure 4-12 on the next page is a full-page advertisement for *Fortune* magazine. It contains eight separate graphic presentations of data. Comment briefly on the correctness of each one.

4.11 There are 517 colleges in the United States that enroll fewer than 2500 students. Another 715 colleges enroll between 2500 and 9999 students, and 370 colleges enroll 10,000 or more students. The two graphs in Figure 4-11 both display these counts. Comment on the accuracy of each graph.

4.12 **Fuel used by motor vehicles.** The table on page 219 shows the amount of fuel (in billions of gallons) consumed by motor vehicles in the United States for years between 1965 and 1993. The data are taken from the *Statistical Abstract*.

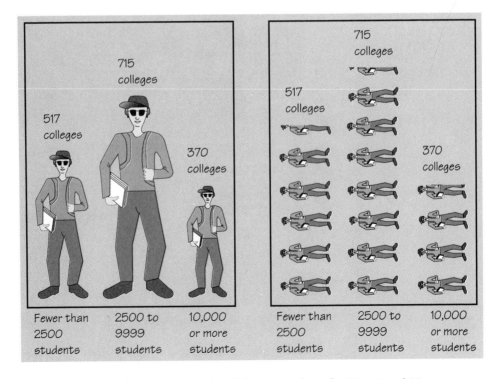

Figure 4-11 Two graphs of the same data, for Exercise 4.11.

We asked America's top businessmen about business magazines. This is what they said.

Which one contains the best writing?

Fortune 56%   Forbes 23%   Business Week 19%

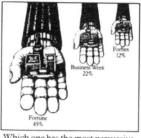

Which one has the most persuasive advertising?

Forbes 12%   Business Week 22%   Fortune 45%

Which one is easiest to read?

Business Week 45%   Forbes 41%   Fortune 11%

Which one best keeps its readers up to date on business events?

Business Week 82%   Forbes 15%   Fortune 2%

Which one carries the most interesting advertising?

Fortune 59%   Business Week 20%   Forbes 10%

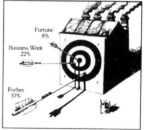

Which one is least accurate?

Fortune 8%   Business Week 22%   Forbes 37%

In which one would you like to see a major story on your company?

Fortune 59%   Business Week 26%   Forbes 14%

Erdos and Morgan recently asked officers of the top one thousand companies—chairmen, presidents, vice presidents, treasurers, secretaries and controllers—for their opinions of Business Week, Forbes and Fortune. 999 executives responded.

You can see the results for yourself. In nearly every instance, Fortune was the winner. Not just by a hair—but overwhelmingly.

Most authoritative? Best writing? Where would they most like to see their company story? Of course they named Fortune. You'd expect them to.

But why did they see the advertising in Fortune as more persuasive and more interesting—when the same advertising often runs in all three magazines?

Obviously, the Fortune climate makes something happen to advertising that doesn't happen anyplace else. It's a valuable edge.

Business leaders get more involved with Fortune, so they get more involved with the advertising. They respond to Fortune, so they respond to the advertising. The survey proves it.

The conclusion is clear and simple: dollar for dollar, your advertising investment gets more impact in Fortune.

You get more than mere advertising exposure in Fortune. You get real communication with the people who can *act* on your business or consumer message. Isn't that what advertising is all about?

FORTUNE
Nobody takes you to the top like Fortune.

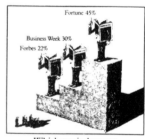

Which one is the most authoritative?

Fortune 45%   Business Week 30%   Forbes 22%

Figure 4-12 Graphs for Exercise 4.10.

Year	1965	1968	1970	1972	1974	1976	1978	1980
Fuel	71.1	82.9	92.3	105.1	106.3	115.7	125.1	115.0

Year	1982	1984	1986	1988	1990	1993
Fuel	113.4	118.7	125.2	130.1	130.8	137.2

(a)   Make a graph to show how fuel consumption has changed over time.

(b)   Beginning in 1973, the price of oil rose sharply, speed limits were reduced, and fuel consumption standards were imposed on vehicles. In the late 1980s and 1990s, fuel was relatively cheap and powerful cars and sports utility vehicles became popular. Do the data show that these events influenced the growth of fuel consumption?

4.13   College students. In 1992, 14,491,000 students were enrolled in U.S. colleges. According to the U.S. Department of Education, there were 119,000 American Indian students, 697,000 Asian, 1,394,000 non-Hispanic black, 954,000 Hispanic, and 10,870,000 non-Hispanic white students. In addition, 458,000 foreign students were enrolled in U.S. colleges.

(a)   Check these data for internal consistency. Does the fact that each number is rounded to the nearest thousand affect your result?

(b)   Present the data in a graph.

4.14   How much can a dollar buy? The buying power of a dollar changes over time. The Bureau of Labor Statistics measures the cost of a "market basket" of goods and services to compile its Consumer Price Index (CPI). If the CPI is 120, goods and services that cost $100 in the base period now cost $120. Here are the average values of the CPI for years between 1970 and 1994. The base period is the years 1982 to 1984.

Year	CPI	Year	CPI	Year	CPI
1970	38.8	1980	82.4	1990	130.7
1972	41.8	1982	96.5	1992	140.3
1974	49.3	1984	103.9	1994	148.2
1976	56.9	1986	109.6		
1978	65.2	1988	118.3		

(a)   Make a graph that shows how the CPI has changed over time.

(b)   What was the overall trend in prices during this period? Were there any years in which this trend was reversed?

(c)   In which years were prices rising fastest? In what period were they rising slowest?

▶ 2 DISPLAYING DISTRIBUTIONS

Variables that just record group membership, such as the marital status of a woman or the race of a college student, are measured in a nominal scale. We can use a pie chart or bar graph to display the distribution of such variables because they have relatively few values. What about variables measured in ordinal or interval/ratio scales, such as the SAT scores of students admitted to a college or the income of families? These variables take so many values that a graph of the distribution is clearer if nearby values are grouped together. The most common graph of the distribution of a variable is a *histogram*.

DRAWING HISTOGRAMS

Table 4-2 presents the percent of residents age 65 years and over in each of the 50 states. To make a histogram of this distribution, proceed as follows.

EXAMPLE 8. How to make a histogram.

Step 1. Divide the range of the data into classes of equal width.
The data in Table 4-2 range from 4.6 to 18.4, so we choose as our classes:

$$4.0 < \text{ percent over } 65 \ \leq 5.0$$
$$5.0 < \text{ percent over } 65 \ \leq 6.0$$
$$\vdots$$
$$18.0 < \text{ percent over } 65 \ \leq 19.0$$

TABLE 4-2 Percent of population 65 years old and over, by state (1994)

State	Percent	State	Percent	State	Percent
Alabama	12.9	Louisiana	11.4	Ohio	13.4
Alaska	4.6	Maine	13.9	Oklahoma	13.6
Arizona	13.4	Maryland	11.2	Oregon	13.7
Arkansas	14.8	Massachusetts	14.1	Pennsylvania	15.9
California	10.6	Michigan	12.4	Rhode Island	15.6
Colorado	10.1	Minnesota	12.5	South Carolina	11.9
Connecticut	14.2	Mississippi	12.5	South Dakota	14.7
Delaware	12.7	Missouri	14.1	Tennessee	12.7
Florida	18.4	Montana	13.3	Texas	10.2
Georgia	10.1	Nebraska	14.1	Utah	8.8
Hawaii	12.1	Nevada	11.3	Vermont	12.1
Idaho	11.6	New Hampshire	11.9	Virginia	11.1
Illinois	12.6	New Jersey	13.2	Washington	11.6
Indiana	12.8	New Mexico	11.0	West Virginia	15.4
Iowa	15.4	New York	13.2	Wisconsin	13.4
Kansas	13.9	North Carolina	12.5	Wyoming	11.1
Kentucky	12.7	North Dakota	14.7		

SOURCE: *Statistical Abstract of the United States*, 1995.

Be sure to specify the classes precisely so that each observation falls into exactly one class. A state with 5.0% of its residents age 65 or older falls into the first class, but 5.1% falls into the second.

Step 2. Count the number of observations in each class. Here are the counts.

Class	Count	Class	Count	Class	Count
4.1 to 5.0	1	9.1 to 10.0	0	14.1 to 15.0	7
5.1 to 6.0	0	10.1 to 11.0	5	15.1 to 16.0	4
6.1 to 7.0	0	11.1 to 12.0	9	16.1 to 17.0	0
7.1 to 8.0	0	12.1 to 13.0	12	17.1 to 18.0	0
8.1 to 9.0	1	13.1 to 14.0	10	18.1 to 19.0	1

Step 3. Draw the histogram. First mark the scale for the variable whose distribution you are displaying on the horizontal axis. That's "percent of residents over 65" in this example. The scale runs from 4 to 20 because that spans the classes we chose. Then mark the scale of counts on the vertical axis. Each bar represents a class. The base of the bar covers the class, and the bar height is the count for the class. Draw the graph with no horizontal space between the bars (unless a class is empty, so that its bar has height zero). Figure 4-13 is our histogram.

Histograms look like bar graphs, but they differ from bar graphs in several respects:

▶ The base scale of a histogram is marked off in equal units. There is no base scale in a bar graph.

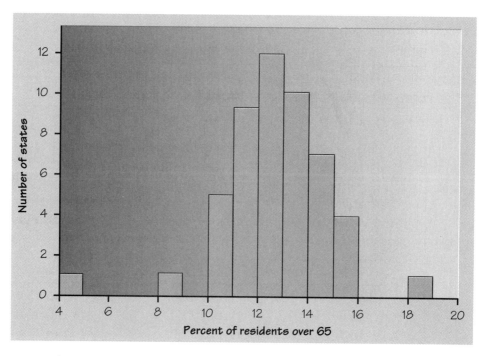

Figure 4-13 A histogram of the distribution of the percent of residents over 65 years of age in the 50 states.

▶ The width of the bars in a histogram has meaning—the base of each bar covers a class of values of the variable. The height of the bar is the count for that class. The width of the bars in a bar graph has no meaning.

▶ The bars in a histogram touch each other (unless a class has count zero), because their bases must cover the entire range of observed values of the variable with no gaps.

Just as with bar graphs, our eyes respond to the area of the bars in a histogram. When you choose the classes for a histogram, be sure they have equal widths. There is no one right choice of the classes. Too few classes will give a "skyscraper" histogram, with all values in a few classes with tall bars. Too many classes will produce a "pancake" graph, with most classes having one or no observations. Neither choice will give a good picture of the shape of the distribution. You must use your judgment in choosing classes to display the shape. As a rough starting point, use about as many classes as the square root of the count of observations—about 5 classes for 25 observations, or 10 classes for 100 observations.

INTERPRETING HISTOGRAMS

Making a statistical graph is not an end in itself. The purpose of the graph is to help us understand the data. After you (or your computer) make a graph, always ask, "What do I see?" Here is a general strategy for looking at graphs:

> **Pattern and deviations**
>
> In any graph of data, look for an **overall pattern** and also for striking **deviations** from that pattern.

The line graph of orange prices in Figure 4-6, for example, shows a clear overall pattern—an upward trend in prices with seasonal variation within each year. The upward leap of prices in 1991 is a deviation from this pattern.

In the case of Figure 4-13, it is easiest to begin with deviations from the overall pattern of the histogram. Three states stand out. You can find them in the table once the histogram has called attention to them. Florida has 18.4% of its residents over age 65, and Alaska has only 4.6%. These states are clear *outliers*.

Outliers

An **outlier** in any graph of data is an individual observation that falls outside the overall pattern of the graph.

We might also call Utah, with 8.8% over 65, an outlier, though it is not as far from the overall pattern as Florida and Alaska. Whether an observation is an outlier is to some extent a matter of judgment. It is much easier to spot outliers in the histogram than in the data table. Once you have spotted outliers, look for an explanation. Many outliers are due to mistakes, such as typing 4.0 as 40. Other outliers point to the special nature of some observations. Explaining outliers usually requires some background information. It is not surprising that Florida, with its many retired people, has many residents over 65 and that Alaska, the northern frontier, has few.

To see the *overall pattern* of a histogram, ignore any outliers. Here is a simple way to organize your thinking.

Overall pattern of a distribution

To describe the overall pattern of a distribution:

▶   Give the **center** and the **spread.**

▶   See if the distribution has a simple **shape** that you can describe in a few words.

Later sections of this chapter tell in detail how to measure center and spread. For now, describe the center by finding a value that divides the observations so that about half have larger values and about half have smaller values. The center in Figure 4-13 is about 13%. That is, about 13% of the residents of a typical state are over 65 years old. You can describe the spread by the extent of the data from smallest to largest value. The spread in Figure 4-13 is about 10% to 16% if we ignore the outliers. Again ignoring the outliers, the histogram in Figure 4-13 is roughly *symmetric* in shape.

> **Symmetric distributions**
> A distribution is **symmetric** if the right and left sides of the histogram are approximately mirror images of each other.

In mathematics, symmetry means that the two sides of a figure such as a histogram are exact mirror images of each other. Data are almost never exactly symmetric, so we are willing to call histograms like that in Figure 4-13 approximately symmetric as an overall description. Here is another example of a symmetric distribution.

> **EXAMPLE 9. Sampling distributions again.** The sampling distribution of a statistic shows the pattern of values that the statistic would take in many samples from the same population. The histogram in Figure 4-14 displays a sampling distribution that we met

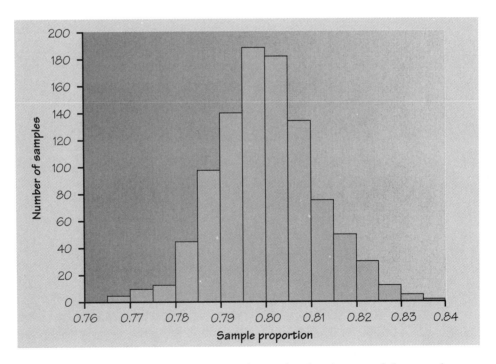

Figure 4-14 A sampling distribution. This is the distribution of the sample proportion $\hat{p}$ in 1000 samples of size 1373 from a population having population proportion $p = 0.8$.

in Chapter 1. Take 1000 simple random samples of 1373 parents. Ask each parent, "If you had it to do over again, would you have children?" Suppose that in fact 80% of all parents would say "Yes." The proportion of the sample who say "Yes" varies from sample to sample and has the distribution shown in Figure 4-14.

The center of this sampling distribution is at 0.80, reflecting the lack of bias of the statistic. The spread is from 0.765 to 0.84. The shape is roughly symmetric.

Of course, not all distributions are symmetric. Here is an example of another common shape.

EXAMPLE 10. **Shakespeare's words.** Figure 4-15 shows the distribution of lengths of words used in Shakespeare's plays.[2] This distribution is *skewed to the right*. That is, there are many short words (3 and 4 letters) and few very long words (10, 11, or 12

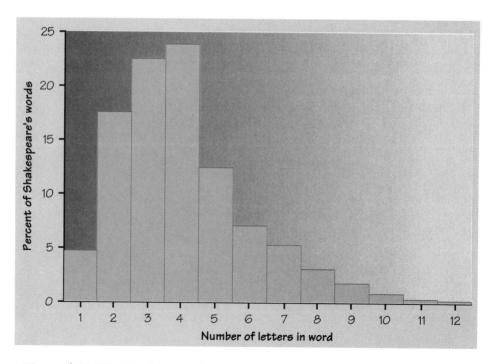

Figure 4-15 The distribution of lengths of words used in Shakespeare's plays.

letters), so that the right tail of the histogram extends out much farther than the left tail. The center of the distribution is about 4. That is, about half of Shakespeare's words have 4 or fewer letters. The spread is from 1 letter to 12 letters.

Notice that the vertical scale in Figure 4-15 is not the *count* of words but the *percent* of all of Shakespeare's words that have each length. A histogram of percents rather than counts is convenient when the counts are very large or when we want to compare several distributions.

Skewed distributions

A distribution is **skewed to the right** if the right side of the histogram (containing the upper half of the observations) extends much farther out than the left side (containing the lower half of the observations). It is **skewed to the left** if the left side of the histogram extends much farther out than the right side.

The distributions in Figures 4-14 and 4-15 both have a *single peak* marking the most common value in the data. Single or multiple peaks is another aspect of the shape of a distribution. Some distributions, however, have irregular shapes that are not easy to describe. Don't force a distribution into a category such as "symmetric" or "skewed" unless the label really fits.

STEMPLOTS

Histograms are not the only graphical display of distributions. For small data sets, a **stemplot** is quicker to make and presents more detailed information.

EXAMPLE 11. How long do presidents live? Table 4-3 lists the ages at death of American presidents. Figure 4-16 shows the steps in making a stemplot of these data.

Step 1. Use all but the final digit of the observations as *stems*. In this example, the first digit of age is the stem. Write the stems vertically from smallest to largest, with a vertical line to their right.

TABLE 4-3   Age at death of U.S. presidents

Washington	67	Fillmore	74	Roosevelt	60
Adams	90	Pierce	64	Taft	72
Jefferson	83	Buchanan	77	Wilson	67
Madison	85	Lincoln	56	Harding	57
Monroe	73	Johnson	66	Coolidge	60
Adams	80	Grant	63	Hoover	90
Jackson	78	Hayes	70	Roosevelt	63
Van Buren	79	Garfield	49	Truman	88
Harrison	68	Arthur	56	Eisenhower	78
Tyler	71	Cleveland	71	Kennedy	46
Polk	53	Harrison	67	Johnson	64
Taylor	65	McKinley	58	Nixon	81

```
4 |                 4 | 9 6                    4 | 9 6
5 |                 5 | 3 6 6 8 7              5 | 3 6 6 7 8
6 |                 6 | 7 8 5 4 6 3 7 0 7 0 3 4  6 | 0 0 3 3 4 4 5 6 7 7 7 8
7 |                 7 | 3 8 9 1 4 7 0 1 2 8    7 | 0 1 1 2 3 4 7 8 8 9
8 |                 8 | 3 5 0 8 1              8 | 0 1 3 5 8
9 |                 9 | 0 0                    9 | 0 0

Step 1. The stems   Step 2. The leaves       Step 3. Order the leaves
```

Figure 4-16   The steps in making a stemplot. The data appear in Table 4-3.

Step 2. Proceed through Table 4-3, writing the second digit of each age as a *leaf* to the right of the proper stem. George Washington's death at age 67 appears as a 7 following the stem 6. Stems can have any number of digits, but leaves must be just one digit.

Step 3. Finally, arrange the leaves following each stem in increasing order from left to right.

A stemplot looks like a histogram turned on its side. We can see from Figure 4-16 that the distribution of the age at which presidents die is quite symmetric and is centered at about age 67.

The chief advantage of a stemplot is that it retains the actual values of the observations. We can see, as we could not from a histogram, that the earliest death of a president occurred at age 46 and that two are tied for the latest death at age 90. Stemplots are also faster to draw than histograms. A stemplot requires that we use the one or more initial digits as stems. This amounts to an automatic choice of classes and can give a poor picture of the distribution. It is possible to split the stems when twice as many classes give a better picture. See Exercise 4.22 for an example. Stemplots do not work well with large data sets, since the natural stems then have too many leaves.

Displaying distributions

To display the distribution of a variable measured in an ordinal or interval/ratio scale, use a histogram or a stemplot.

SECTION 2 EXERCISES

4.15 **The statistics of writing style.** Numerical data can distinguish different types of writing, and sometimes even individual authors. Here are data on the percentages of words of 1 to 15 letters used in articles in *Popular Science* magazine.[3]

Length	1	2	3	4	5	6	7
Percent	3.6	14.8	18.7	16.0	12.5	8.2	8.1

Length	8	9	10	11	12	13	14	15
Percent	5.9	4.4	3.6	2.1	0.9	0.6	0.4	0.2

(a) Make a histogram of this distribution. Then describe its overall shape, center, and spread.

(b) How does the distribution of lengths of words used in *Popular Science* compare with the similar distribution in Figure 4-15 for Shakespeare's plays? Look in particular at short words (2, 3, and 4 letters) and very long words (more than 10 letters).

4.16 **The changing age distribution of the U.S.** The distribution of the ages of a nation's residents has a strong influence on economic and social conditions. The following table shows the age distribution of U.S. residents in 1950 and 2075, in millions of persons. The 1950 data come from that year's census. The 2075 data are projections made by the Census Bureau.

Age group	1950	2075
Under 10 years	29.3	34.9
10 to 19 years	21.8	35.7
20 to 29 years	24.0	36.8
30 to 39 years	22.8	38.1
40 to 49 years	19.3	37.8
50 to 59 years	15.5	37.5
60 to 69 years	11.0	34.5
70 to 79 years	5.5	27.2
80 to 89 years	1.6	18.8
90 to 99 years	0.1	7.7
100 to 109 years	—	1.7
TOTAL	151.1	310.6

(a) Because the total population in 2075 is much larger than the 1950 population, comparing percents in each age group is clearer than comparing counts. Make a table of the percent of the total population in each age group for both 1950 and 2075.

(b) Make a histogram of the 1950 age distribution (in percents). Describe the main features of the distribution. In particular, look at the percent of children relative to the rest of the population.

(c) Make a histogram of the projected age distribution for the year 2075. Use the same scales as in (b) for easy comparison. What are the most important changes in the age distribution projected for the 125-year period between 1950 and 2075?

4.17 A study of bacterial contamination in milk counted the number of coliform organisms (fecal bacteria) per milliliter in 100 specimens of milk purchased in East Coast groceries. The U.S. Public Health Service recommends no more than 10 coliform bacteria per milliliter. Here are the data:

5 8 6 7 8   3 2 4 7 8   6 4 4 8 8   8 6 10 6 5
6 6 6 6 4   3 7 7 5 7   4 5 6 7 4   4 4 3 5 7
7 5 8 3 9   7 3 4 6 6   8 7 4 8 5   7 9 4 4 7
8 8 7 5 4   10 7 6 6 7   8 6 6 6 0   4 5 10 4 5
7 9 8 9 5   6 3 6 3 7   1 6 9 6 8   5 2 8 5 3

(a) Make a table of the counts of each of the values 0 to 10 in this set of 100 observations.

(b) Draw a histogram of these data.

(c) Describe the shape of the distribution. Are the data symmetric? If so, approximately where is the center about which they are symmetric? Are the data strongly skewed? If so, in which direction? Are there any outliers?

4.18   Returns on common stocks. The total return on a stock is the change in its market price plus any dividend payments made. Total return is usually expressed as a percent of the beginning price. Figure 4-17 is a histogram of the distribution of total returns for all 1528 stocks listed on the New York Stock Exchange in one year.[4]

(a) Describe the overall shape of the distribution of total returns.

(b) What is the approximate center of this distribution? (For now, take the center to be a value with roughly half the stocks having lower returns and half having higher returns.)

(c) Approximately what were the smallest and largest total returns? (This describes the spread of the distribution.)

(d) A return less than zero means that owners of the stock lost money. About what percent of all stocks lost money?

4.19   Make a stemplot of the data in Table 4-2. Take the whole numbers (the first two digits of each table entry) as stems and the tenths (the final digit) as leaves. Do you think the stemplot gives as clear a picture of the data as the histogram in Figure 4-13?

4.20   How many calories does a hot dog have? *Consumer Reports* magazine (June 1986) presented the following data on the

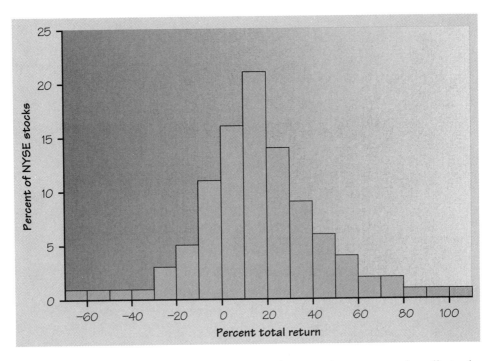

Figure 4-17 The distribution of percent total return in one year for all stocks listed on the New York Stock Exchange, for Exercise 4.18.

number of calories in a hot dog for each of 17 brands of meat hot dogs.

173   191   182   190   172   147   146   139   175
136   179   153   107   195   135   140   138

Make a stemplot of the distribution of calories in meat hot dogs and briefly describe the shape of the distribution. Most brands of meat hot dogs contain a mixture of beef and pork, with up to 15% poultry allowed by government regulations. The only brand with a different makeup was *Eat Slim Veal Hot Dogs*. Which point on your stemplot do you think represents this brand?

4.21   A marketing consultant observes 50 consecutive shoppers at a grocery store, recording how much each shopper spends in the store. Here are the data (in dollars), arranged in increasing order for convenience.

3.11	8.88	9.26	10.81	12.69	13.78	15.23	15.62
17.00	17.39	18.36	18.43	19.27	19.50	19.54	20.16
20.59	22.22	23.04	24.47	24.58	25.13	26.24	26.26
27.65	28.06	28.08	28.38	32.03	34.98	36.37	38.64
39.16	41.02	42.97	44.08	44.67	45.40	46.69	48.65
50.39	52.75	54.80	59.07	61.22	70.32	82.70	85.76
86.37	93.34						

Make a histogram of these data. Then describe the shape, center, and spread of the distribution. Are there any clear outliers?

4.22   **Variations on stemplots.** To make a stemplot of the data in Exercise 4.21, first *truncate* the data by discarding the cents, leaving only the dollar amounts spent. (You could also *round* the data to the nearest dollar, but truncating is faster.) Make a stemplot with tens of dollars as the stems and dollars as the leaves.

You can make a stemplot with twice as many stems by *splitting the stems.* Write two 0s, two 1s, and so on, as stems to the left of your vertical bar. The first of each pair of stems gets leaves 0 through 4, and the second gets leaves 5 through 9. Spreading out the stemplot in this manner is helpful if many observations fall on only a few stems. Make a stemplot of the truncated grocery spending data with split stems. Which of your two stemplots do you prefer?

4.23   Climatologists interested in flooding gather statistics on the daily rainfall in various cities. The following data set gives the *maximum* daily rainfall (in inches) for each of the years 1941 to 1970 in South Bend, Indiana. (Successive years follow each other across the rows in the table.)

1.88	2.23	2.58	2.07	2.94	2.29	3.14	2.15	1.95	2.51
2.86	1.48	1.12	2.76	3.10	2.05	2.23	1.70	1.57	2.81
1.24	3.29	1.87	1.50	2.99	3.48	2.12	4.69	2.29	2.12

Make a stemplot for these data, truncating and splitting stems as described in Exercise 4.22 if necessary. Describe the general shape of the distribution and any prominent deviations from the overall pattern.

4.24   **Back-to-back stemplots.** Corn is an important animal food. Normal corn lacks certain amino acids, which are building blocks for protein. Plant scientists have developed new corn varieties that have more of these amino acids. To test a new corn as an animal food, a group of 20 one-day-old male chicks was fed a ration containing the new corn. A control group of another 20 chicks was fed a ration that was identical except that it contained normal corn. Here are the weight gains (in grams) after 21 days.[5]

Normal corn				New corn			
380	321	366	356	361	447	401	375
283	349	402	462	434	403	393	426
356	410	329	399	406	318	467	407
350	384	316	272	427	420	477	392
345	455	360	431	430	339	410	326

(a)   This experiment was designed using the principles of Chapter 2. Briefly outline the proper design of this experiment.

(b)   To compare the two distributions of weight gains, make a *back-to-back stemplot*: draw a single column of stems with vertical lines on both sides of it. Now add leaves to these stems on both sides, the experimental group on the right and the control group on the left. Finally, order the leaves so that they increase as you move away from the stem.

(c)   What does your plot show about the effect of the improved corn variety on weight gain?

4.25   Although the back-to-back stemplot of Exercise 4.24 is effective for comparing the two weight-gain distributions, the stems divide the data into too many classes to show clearly the shape of the distributions.

(a)   Make separate tables of counts for the weight gains in the experimental group and the control group. Use the classes 270–299, 300–329, 330–359, and so on.

(b) Draw separate histograms for the two groups. For easy comparison, draw them one above the other, with the scales on the horizontal axes aligned. What does this plot show about the effect of the improved corn on weight gain?

(c) Are the distributions of weight gains symmetric or skewed?

4.26 **Babe Ruth versus Roger Maris.** Here are the number of home runs that Babe Ruth hit in each of his 15 years with the New York Yankees, from 1920 to 1934:[6]

54 59 35 41 46 25 47 60 54 46 49 46 41 34 22

Ruth's record of 60 home runs in a season was broken by another Yankee, Roger Maris, who hit 61 home runs in 1961. Here are Maris's home run totals for his ten years in the American League:

14 28 16 39 61 33 23 26 8 13

Make a back-to-back stemplot to compare Ruth and Maris. (See Exercise 4.24 for back-to-back stemplots.) What are the most important features of the comparison?

4.27 Sketch a histogram for a distribution that is skewed to the left.

4.28 **The distribution of random digits.** The entries in the table of random digits have the property that each value 0, 1, 2, 3, 4, 5, 6, 7, 8, and 9 occurs equally often in the long run.

(a) Make a table of the counts of these values in the first three rows of Table A (120 digits in all). Draw a histogram of the distribution.

(b) Is this distribution approximately symmetric? What is its center?

(c) Draw a histogram for the percents of each value that you expect to find in a very long table of random digits. Does your result for 120 random digits resemble this histogram?

4.29 **Automobile gas mileage.** Environmental Protection Agency regulations require auto makers to give the city and highway gas mileages for each model of car. Here are the highway mileages (miles per gallon) for 22 midsize 1996 car models:[7]

Model	MPG	Model	MPG
Buick Century	31	Lexus GS300	24
Buick Regal	29	Lexus LS400	26
Cadillac Eldorado	26	Lincoln Mark VIII	26
Chrysler Cirrus	29	Mazda 626	31
Dodge Stratus	36	Nissan Maxima	28
Ford Taurus	29	Oldsmobile Aurora	26
Ford Thunderbird	26	Oldsmobile 88	30
Hyundai Sonata	29	Rolls-Royce Silver Spur	17
Infiniti I30	28	Toyota Camry	27
Infiniti Q45	22	Volkswagen Passat	27
Jaguar XJ12	16	Volvo 850	26

(a)   Make a graph that displays the distribution. Describe its main features (overall pattern and any outliers) in words.

(b)   The government imposes a "gas guzzler" tax on cars with low gas mileage. Which of these cars do you think are subject to the gas guzzler tax?

▶ 3 MEASURING CENTER OR AVERAGE

How long do U.S. presidents live? Table 4-3 on page 228 gives the age at death of all presidents who are no longer alive. The stemplot in Figure 4-18 displays the distribution. But how long *on the average* do

```
4 | 96
5 | 36678
6 | 003344567778
7 | 0112347889
8 | 01358
9 | 00
```

Figure 4-18 Stemplot of the ages at which presidents have died.

presidents live? That calls for a single number that summarizes one aspect of these data, their center or average.

We often read or hear statements such as "The median income of households headed by a college graduate is $56,116, but the median for households headed by someone who stopped at high school is only $28,700." The median is one of the numerical measures that summarize specific features of a data set. Numerical summaries of data are as important in your statistical vocabulary as random samples and histograms. We have already met *counts* and *rates or proportions.* Now we will meet a famous trio: the mean, the median, and the mode.

Mean, median, and mode

▶ The **mean** of a set of observations is their arithmetic *average.* It is the sum of the observations divided by the number of observations.

▶ The **median** is the *middle value.* It is the midpoint of the observations when they are arranged in increasing order.

▶ The **mode** is the *most frequent value.* It is any value that occurs most often among the observations.

Note that we have defined mean, median, and mode for any set of observations. The mean, the median, and the mode are statistics if our observations are a sample; they are parameters if the observations form an entire population.

MEASURING CENTER: THE MEAN

The most common measure of center is the ordinary arithmetic average, or *mean.* Means are so common that a compact notation is useful.

The mean $\overline{x}$

To find the **mean** of a set of observations, add their values and divide by the number of observations. If our n observations are $x_1, x_2, \ldots, x_n$, their mean is

$$\overline{x} = \frac{x_1 + x_2 + \cdots + x_n}{n}$$

The notation in this definition is common in statistics. We call our data $x_1, x_2, \ldots, x_n$ to show that they are n observations on the same variable, like the ages at death of n presidents. The bar over the x indicates the mean of all the x-values. Pronounce the mean $\overline{x}$ as "x-bar."

EXAMPLE 12. To find the mean age at death of the 36 presidents in Table 4-3, add the 36 ages and divide by 36.

$$\overline{x} = \frac{x_1 + x_2 + \cdots + x_n}{n}$$

$$= \frac{67 + 90 + 83 + \cdots + 81}{36}$$

$$= \frac{2489}{36} = 69.14 \text{ years}$$

In practice, you can key the data into your calculator and hit the $\overline{x}$ key. You don't actually have to add and divide. You should, however, know that this is what the calculator is doing.

MEASURING CENTER: THE MEDIAN

The mean is not the only way to describe the center of a distribution. Another natural idea is to use the "middle value" in a histogram or stemplot. That is, find the number such that half the observations are smaller and the other half are larger. This is the *median* of the distribution. We will call the median M for short. Although the idea of the median as the midpoint of a distribution is simple, we need a precise rule for putting the idea into practice.

EXAMPLE 13. To find the median of the numbers

8 4 9 1 3

arrange them in increasing order as

1 3 **4** 8 9

The bold 4 is the center observation, because there are 2 observations to its left and 2 to its right. When the number of observations n

is odd, there is always one observation in the center of the ordered list. This is the median, $M = 4$.

If n is even, there is no one middle observation. But there is a middle pair, and we take the median to be the mean of this middle pair, the point halfway between them. So the median of

$$8\ 4\ 1\ 9\ 1\ 5$$

is found by arranging these numbers in increasing order,

$$1\ 1\ \mathbf{4}\ \mathbf{5}\ 8\ 9$$

and averaging the middle pair,

$$M = \frac{4 + 5}{2} = 4.5$$

Here is a fast way to locate the median in the ordered list: count up $(n + 1)/2$ places from the beginning of the list. Try it. In the first example, $n = 5$ and $(n + 1)/2 = 3$, so the median is the third entry in the ordered list. When $n = 6$, we get $(n + 1)/2 = 3.5$. This means "halfway between the third and fourth" entries, so M is the average of these two entries. Be sure to note that $(n + 1)/2$ does *not* give the median M, just its position in the ordered list of observations. Here is our rule for finding medians.

The median M

To find the **median** of a distribution:

1. Arrange all observations in order of size, from smallest to largest.

2. If the number of observations n is odd, the median M is the center observation in the ordered list. Find the location of the median by counting $(n + 1)/2$ observations up from the start of the list.

3. If the number of observations n is even, the median M is the mean of the two center observations in the ordered list. The location of the median is again $(n + 1)/2$ from the start of the list.

EXAMPLE 14. To find the median age at death of the 36 presidents in Table 4-3, first arrange the data in increasing order, from 46 (John F. Kennedy) to 90 (John Adams and Herbert Hoover). The stemplot in Figure 4-18 does this. Because

$$\frac{n+1}{2} = \frac{36+1}{2} = 18.5$$

count up 18 places from the beginning of the stemplot and average the 18th and 19th ages. The median is

$$M = \frac{67+68}{2} = 67.5 \text{ years}$$

THE MODE

The *mode* is any value occurring most frequently in the set of observations. It is convenient to arrange observations in increasing order as an aid to seeing how often each value occurs.

EXAMPLE 15. From the stemplot in Figure 4-18, we see that three presidents died at age 67 and that no more than two died at any other age. The mode is therefore 67.

If the next president to die lives to age 90, there will be *two* modes because 67 and 90 will each occur three times.

The mode is not really a measure of center or average. It records only the most frequent value, and this may be far from the center of the distribution of values. We have also seen that there may be more than one mode. The chief advantage of the mode is that, of our three measures, it alone makes sense for variables measured in a nominal scale. It is nonsense to speak of the median sex or mean race of American ambassadors, but the most frequent (modal) sex is male and the modal race is white.

COMPARING THE MEAN AND THE MEDIAN

"Midpoint" and "arithmetic average" are both reasonable ideas for describing the center of a set of data, but they are different ideas with different uses. The most important distinction is that the mean (the average) is strongly influenced by a few extreme observations and the median (the midpoint) is not.

> **EXAMPLE 16.** A professional basketball team has 5 players earning $200,000 per year, 6 who earn $500,000, and a star who earns

"Should we scare the opposition by announcing our mean height or lull them by announcing our median height?"

$3,200,000. The mean salary for the team's 12 players is therefore

$$\bar{x} = \frac{x_1 + x_2 + \cdots + x_{12}}{12}$$

$$= \frac{200,000 + 200,000 + \cdots + 3,200,000}{12}$$

$$= \frac{7,200,000}{12} = \$600,000$$

The star's very high salary pulls the mean above the amount paid to any other player. The median, however, is $500,000 and does not change even if the star earns $10 million. His salary is just one number falling above the midpoint.

The mean and median of a symmetric distribution are close to each other. In fact, $\bar{x}$ and M are exactly equal if the distribution is exactly symmetric. In skewed distributions, however, the mean and median may lie far apart. Many distributions of monetary values—incomes, house prices, wealth—are strongly skewed to the right. There are many moderately priced houses, for example, a fair number of very expensive houses, and a few immensely costly mansions that make the right tail of the distribution of house prices very long. The long right tail pulls the mean up but does not affect the median. The mean price of new houses sold in mid-1994 was $152,000, but the median price for these same houses was only $127,000.

Because monetary data often have a few extremely high observations, descriptions of these distributions usually employ the median—"half of all houses sold for more than $127,000, and half sold for less." The median gives the price of a typical house. This is not to say that the median always should be used for data containing extreme observations. A real estate broker is interested in the total selling price of the houses she sells, because her commission is a percentage of the total. The median price tells her nothing. The mean is directly related to the total. Always ask which of "midpoint" or "arithmetic average" best represents the data for your intended use.

Now you are an expert on mean, median, and mode. I hope they have lived up to your expectations. Be warned, however, that averages of all kinds can play tricks if you are not alert. Some examples of these tricks appear in the exercises. Arithmetic is never a substitute for understanding.

SECTION 3 EXERCISES

4.30   You read that the median income of U.S. households in 1994 was $32,264. Explain in plain language what "the median income" is.

4.31   Last year a small accounting firm paid each of its five clerks $25,000, two junior accountants $60,000 each, and the firm's owner $255,000. What is the mean salary paid at this firm? How many of the employees earn less than the mean? What is the median salary? What is the mode of the salaries?

4.32   **Athletes' salaries.** The mean and median salaries paid to major league baseball players in 1995 were $275,000 and $1,089,000. Which of these numbers is the mean, and which is the median? Explain your answer.

4.33   **Rich magazine subscribers.** The business magazine *Forbes* estimates (November 6, 1995) that the "average" household wealth of its readers is either about $800,000 or about $2.2 million, depending on the choice of "average" it reports. Which of these numbers is the mean wealth and which is the median wealth? Explain your answer.

4.34   **Ruth versus Maris again.** Here are the number of home runs Babe Ruth hit in each of his 15 years with the New York Yankees:

> 54 59 35 41 46 25 47 60 54 46 49 46 41 34 22

Roger Maris, who broke Ruth's single-year record, had these home run counts in his 10 years in the American League:

> 14 28 16 39 61 33 23 26 8 13

You made a back-to-back stemplot of these data in Exercise 4.26.

(a)   Find the mean and median number of home runs hit in a season by each player.

(b)   Now omit each player's record year (60 for Ruth, 61 for Maris). Find the mean and the median for their remaining years.

(c)   Why does removing Maris's highest count reduce his mean by more than it reduces his median?

(d)   Why does removing each player's highest count reduce Maris's mean and median more than it reduces Ruth's?

4.35   From the data in Exercise 4.23, find the mean and median of the maximum one-day rainfall amounts. Explain in terms of the shape of the distribution why the mean is larger than the median.

4.36   Here is a sample of 100 reaction times of a subject to a stimulus, in milliseconds:

```
10  14  11  15   7   7  20  10  14   9   8   6  12  12  10  14  11  13   9  12
13  11  12  10   8   9  14  18  12  10  10  11   7  17  12   9   9  11   7  10
14  12  12  10   9   7  11   9  18   6  12  12  10   8  14  15  12  11   9   9
11   8  11  10  13   8  11  11  13  20   6  13  13   8   9  16  15  11  10  11
20   8  17  12  19  14  17  12  18  16  15  16  10  20  11  19  20  13  11  20
```

(a)   Compute the mean and the median of these data.

(b)   Find the counts of each of the outcomes 6, 7, 8, ..., 20. Draw a histogram for these data.

(c)   Explain in terms of the shape of the distribution why these measures of center fall as they do (close together or apart).

4.37   Return to the sample of 100 counts of coliform bacteria in milk given in Exercise 4.17. You drew a histogram for the data as part of that exercise.

(a)   Compute the mean and the median of the data.

(b)   Explain in terms of the shape of the distribution why these measures of center fall as they do (close together or apart).

4.38   Identify which measure of center (mean, median, or mode) is the appropriate "average" in each of the following situations:

(a)   Someone declares, "The average American is a white female."

(b)   Middletown is considering imposing an income tax on citizens. The city government wants to know the average income of citizens so that it can estimate the total tax base.

(c) In an attempt to study the standard of living of typical families in Middletown, a sociologist estimates the average family income in that city.

4.39   **Statistical exaggeration.** As part of its twenty-fifth reunion celebration, the class of '72 of Central New Jersey University mails a questionnaire to its members. One of the questions asks the respondent to give his or her total income last year. Of the 820 members of the class of '72, the university alumni office has addresses for 583. Of these, 421 return the questionnaire. The reunion committee computes the mean income given in the responses and announces, "The members of '72 have enjoyed resounding success. The average income of class members is $134,000."

This report exaggerates the income of the members of the class of '72 for (at least) three reasons. What are these reasons?

4.40   The mean age of 5 people in a room is 30 years. A 36-year-old person walks in. What is now the mean age of the people in the room?

4.41   Consider this small set of data: 8, 4, 1, 9, 1, 5. Find the mean, median, and mode of these data. Now add 3 to each observation and again find the mean, median, and mode. How did these measures of center change? How would they change if you subtracted 1 from each observation?

4.42   Make up a list of numbers of which only 10% are "above the average" (that is, above the mean). What percent of the numbers in your list fall above the median?

4.43   Which of the mean, median, and mode of a list of numbers must always appear as one of the numbers in the list?

4.44   In computing the median income of any group, federal agencies omit all members of the group who had zero income. Give an example to show how the median income of a group (as reported by the federal government) can go *down* when the group becomes better off economically.

4.45   The mean age of members of the class of '45 at their fiftieth reunion was 71.9 years. At their fifty-first reunion the next year, the mean age was 71.5 years. How can the mean age decrease when all the class members are a year older?

4.46   You drive 5 miles at 30 miles per hour, then 5 more miles at 50 miles per hour. Have you driven at an average speed of 40 miles per hour? (Your average speed is total miles driven divided by the time you took to drive it.)

4.47   The following paragraph contains three major statistical blunders. Describe them in one sentence each.

In response to protests over the firing of coach Rockne, the university administration released the results of a questionnaire mailed to all 90,000 living alumni. Of the 8000 responses, 65% favored firing the coach, showing that the alumni want Rockne to go. And these alumni are supporters of the athletic program—the mean contribution of those responding was over $200 last year, including James Barkaddidy's handsome gift of $1 million for a new scoreboard. What is more, 25% of the faculty want Rockne fired, so that when faculty and alumni are considered together, an overwhelming 90% want a new coach.

▶ 4 MEASURING SPREAD OR VARIABILITY

The mean and median provide two different measures of the center of a distribution. Useful as the measures of center are, they are incomplete and often misleading without some accompanying indication of how spread out or variable the data are. The median household income of $31,241 in 1993 masks the fact that 14.2% of all households received less than $10,000 and that the top 5.8% had incomes over $100,000. The distribution of household incomes is skewed to the right and very spread out. Knowing the median alone gives us an inadequate description of the distribution of incomes for American households. It lumps the Rockefellers in with the welfare cases. The simplest adequate summary of the distribution of a single variable requires both a measure of center and a measure of spread.

MEASURING SPREAD: THE QUARTILES

One way to describe the spread of a distribution is to give the smallest and largest values. These two numbers show the full range of the data.

> **EXAMPLE 17. Highway gas mileage.** Table 4-4 gives the government's fuel economy ratings for two types of motor vehicle. What

TABLE 4-4   City and highway gas mileage for 1996 model vehicles

Mid-size cars			Four-wheel drive		
Model	City	Highway	Model	City	Highway
Buick Century	24	31	Chevrolet Blazer	16	21
Buick Regal	20	29	Chevrolet Suburban	13	17
Cadillac Eldorado	17	26	Chrysler Town	16	22
Chrysler Cirrus	20	29	& Country		
Dodge Stratus	25	36	Ford Bronco	13	17
Ford Taurus	20	29	Ford Explorer	15	20
Ford Thunderbird	19	26	Geo Tracker	23	24
Hyundai Sonata	21	29	Isuzu Trooper	14	18
Infiniti I30	21	28	Jeep Cherokee	19	22
Infiniti Q45	17	22	Jeep Grand Cherokee	15	20
Jaguar XJ12	12	16	Mazda MPV	15	19
Lexus GS300	18	24	Mitsubishi Montero	16	19
Lexus LS400	19	26	Nissan Pathfinder	15	19
Lincoln Mark VIII	18	26	Suzuki Sidekick	21	24
Mazda 626	23	31	Toyota Landcruiser	13	15
Nissan Maxima	21	28	Toyota RAV4	22	27
Oldsmobile Aurora	17	26			
Oldsmobile 88	19	30			
Rolls Royce Silver Spur	12	17			
Toyota Camry	21	27			
Volkswagen Passat	20	27			
Volvo 850	19	26			

SOURCE: Environmental Protection Agency *1996 Fuel Economy Guide.*

kind of highway gas mileage do midsize cars get? The median is 27 miles per gallon. The smallest and largest values are 16 miles per gallon for the Jaguar XJ12 and 36 miles per gallon for the Dodge Stratus.

Both the Jaguar and the Stratus are rather unusual midsize cars. As the histogram in Figure 4-19 shows, both their highway mileages are outliers. That's the weakness of giving just the smallest and largest observations to measure spread—they may be outliers that don't reflect the spread of most of the data.

We can improve our description of spread by also looking at the spread of the middle half of the data. The *quartiles* span the middle half.

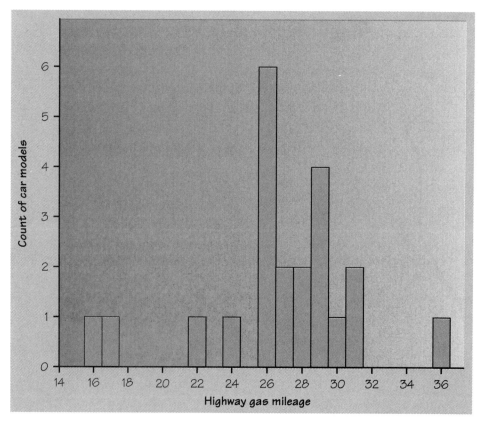

Figure 4-19 Histogram of the highway gas mileages of 1996 model midsize cars.

Count up the ordered list of observations, starting from the smallest. The *first quartile* lies one-quarter of the way up the list. The *third quartile* lies three-quarters of the way up the list. In other words, the first quartile is larger than 25% of the observations, and the third quartile is larger than 75% of the observations. The second quartile is the median, which is larger than 50% of the observations. That is the idea of quartiles. We need a rule to make the idea exact. The rule for calculating the quartiles uses the rule for the median.

The quartiles Q_1 and Q_3

To calculate the **quartiles**:

1. Arrange the observations in increasing order and locate the median M in the ordered list of observations.

2. The **first quartile** Q_1 is the median of the observations whose position in the ordered list is to the left of the location of the overall median.

3. The **third quartile** Q_3 is the median of the observations whose position in the ordered list is to the right of the location of the overall median.

EXAMPLE 18. Calculating the quartiles. Here are the highway mileages for the 22 midsize cars in Table 4-4 arranged in increasing order:

16 17 22 24 26 26 26 26 26 26 27 | 27 28 28 29 29 29 29 30 31 31 36

Because $(n + 1)/2 = 11.5$, the median M is located halfway between the 11th and 12th observations in the list. That location is marked by a vertical line in the list. When n is even, the location of the median is always between two entries in the ordered list of observations. That is true even if, as here, the two entries have the same value.

The first quartile is the median of the 11 observations to the left of the line that marks the location of the median. That is, $Q_1 = 26$. The third quartile is the median of the 11 observations to the right of this line, $Q_3 = 29$.

Table 4-4 also gives mileage data for 15 four-wheel drive sports utility vehicles. Here are the highway mileages for these vehicles arranged in increasing order:

15 17 17 18 19 19 19 **20** 20 21 22 22 24 24 27

Because $n = 15$ is odd, the median is a member of the list, $M = 20$. To find the quartiles, ignore this central observation. The first quartile is the median of the 7 observations to the left of the median, $Q_1 = 18$; the third quartile is the median of the 7 observations to its right in the list, $Q_3 = 22$.

THE FIVE-NUMBER SUMMARY AND BOXPLOTS

After years of improving fuel economy, the average gas mileage of motor vehicles sold in the United States stopped improving in the mid-1990s. One reason was the popularity of four-wheel drive sports utility vehicles, most of which get poor gas mileage. Let's compare the highway mileage of these vehicles with that of the midsize cars they often replace in our driveways.

EXAMPLE 19. **Comparing gas mileage.** We can quickly describe both the center and spread of the distribution of highway gas mileage by giving the median (center), the two quartiles, and also the smallest and largest values. Here are these five numbers for midsize cars,

16 26 27 29 36

and for sports utility vehicles,

15 18 20 22 27

Except for the lowest value (the Jaguar again), each point in the distribution of highway mileage for midsize cars is 7 to 9 miles per gallon higher than the matching point for sports utility vehicles. The data make clear the cost to fuel economy of the four-wheel drive craze. We have given *five-number summaries* of the two gas mileage distributions.

> **The five-number summary**
>
> The **five-number summary** of a data set consists of the smallest observation, the first quartile, the median, the third quartile, and the largest observation, written in order from smallest to largest. In symbols, the five-number summary is
>
> $$\text{Minimum} \quad Q_1 \quad M \quad Q_3 \quad \text{Maximum}$$

The five-number summary of a distribution leads to a new graph, the **boxplot.** Figure 4-20 shows boxplots for our two gas mileage distributions. The central box in a boxplot has its ends at the quartiles and therefore spans the middle half of the data. The line within the box marks the median. The "whiskers" at either end extend to the

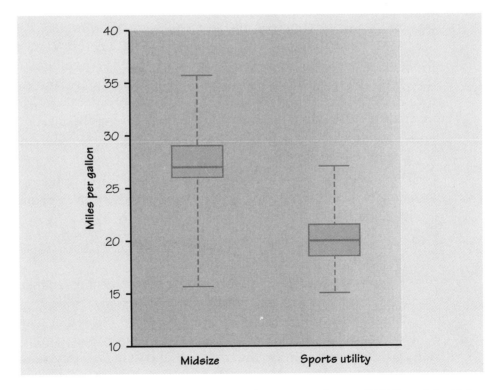

Figure 4-20 Boxplots comparing the highway gas mileage of 1996 model midsize cars and four-wheel drive sports utility vehicles.

smallest and largest observations. We can see at once by comparing the boxes that typical midsize cars get much better mileage than sports utility vehicles. The whiskers show the greater spread of the mileages for midsize cars, caused by the outliers.

Boxplots are effective for side-by-side comparisons, as in Figure 4-20. They are generally not good choices for displaying a single distribution because they lose so much detail. The histogram in Figure 4-19 shows the irregular shape and the outliers in the mileages for midsize cars, details that the boxplot in Figure 4-20 cannot show.

MEASURING SPREAD: THE STANDARD DEVIATION

Though the five-number summary is the most generally useful numerical description of a distribution, it is not the most common. That distinction belongs to the combination of the mean with the *standard deviation.* The mean, like the median, is a measure of center. The standard deviation, like the quartiles and extremes in the five-number summary, measures spread. The standard deviation and its square, the *variance,* measure spread by looking at how far the observations are from their mean. They should be used only in the company of the mean.

The standard deviation *s*

The **variance *s²*** of a set of observations is an average of the squares of the deviations of the observations from their mean. In symbols, the variance of n observations $x_1, x_2, \ldots, x_n$ is

$$s^2 = \frac{(x_1 - \overline{x})^2 + (x_2 - \overline{x})^2 + \cdots + (x_n - \overline{x})^2}{n - 1}$$

The **standard deviation *s*** is the square root of the variance s^2.

Many calculators have a standard deviation button that allows you to obtain s from keyed-in data. Doing a few step-by-step examples will help you understand how the variance and standard deviation work, however. Here is such an example.

EXAMPLE 20. Calculating standard deviation. A person's metabolic rate is the rate at which the body consumes energy.

Metabolic rate is important in studies of weight gain, dieting, and exercise. Here are the metabolic rates of 7 men who took part in a study of dieting. (The units are calories per 24 hours. These are the same calories used to describe the energy content of foods.)

1792 1666 1362 1614 1460 1867 1439

To find the standard deviation, first calculate the mean:

$$\overline{x} = \frac{1792 + 1666 + 1362 + 1614 + 1460 + 1867 + 1439}{7}$$

$$= \frac{11,200}{7} = 1600 \text{ calories}$$

The variance and standard deviation measure spread by using the deviations of the observations from the mean. Here is a table of those deviations:

Observations x	Deviations $x - \overline{x}$	Squared deviations $(x - \overline{x})^2$
1792	$1792 - 1600 = 192$	$(192)^2 = 36,864$
1666	$1666 - 1600 = 66$	$(66)^2 = 4,356$
1362	$1362 - 1600 = -238$	$(-238)^2 = 56,644$
1614	$1614 - 1600 = 14$	$(14)^2 = 196$
1460	$1460 - 1600 = -140$	$(-140)^2 = 19,600$
1867	$1867 - 1600 = 267$	$(267)^2 = 71,289$
1439	$1439 - 1600 = -161$	$(-161)^2 = 25,921$
		sum $= 214,870$

The variance is the sum of the squared deviations divided by one less than the number of observations:

$$s^2 = \frac{214,870}{6} = 35,811.67$$

The standard deviation is the square root of the variance:

$$s = \sqrt{35,811.67} = 189.24 \text{ calories}$$

Figure 4-21 displays the data of the example as points above the number line, with their mean marked by an asterisk (*). The arrows

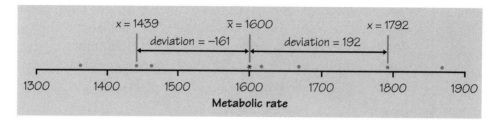

Figure 4-21 Metabolic rates for seven men, with their mean (*) and the deviations of two observations from the mean.

indicate two of the deviations from the mean. These deviations show how spread out the data are about their mean. Some of the deviations will be positive and some negative because some of the observations fall on each side of the mean. Squaring the deviations makes them all positive, so that observations far from the mean in either direction will have large positive squared deviations. The variance s^2 is the average squared deviation. The variance is large if the observations are widely spread about their mean; it is small if the observations are all close to the mean.

Because the variance involves squaring the deviations, it does not have the same unit of measurement as the original observations. If we measure energy consumed in calories, the variance is measured in calories squared. Taking the square root remedies this. The standard deviation s measures spread about the mean in the original scale.

It is usual—for good but somewhat technical reasons—to average the squared deviations by dividing their total by $n - 1$ rather than by n. Many calculators have two standard deviation buttons, giving you a choice between dividing by n and dividing by $n - 1$. Be sure to choose $n - 1$.

USING THE STANDARD DEVIATION

More important than the details of the calculation are the properties that show how the standard deviation measures spread.

Properties of the standard deviation

▶ s measures spread about the mean $\overline{x}$. Use s to describe the spread of a distribution only when you use $\overline{x}$ to describe the center.

▶ $s = 0$ only when there is *no spread.* This happens only when all observations have the same value. So standard deviation zero means no spread at all. Otherwise $s > 0$. As the observations become more spread out about their mean, s gets larger.

▶ s, like the mean $\overline{x}$, is strongly influenced by extreme observations. A few outliers can make s very large.

A skewed distribution with a few observations in the single long tail will have a large standard deviation. The number s does not give much helpful information in such a case. Because the two sides of a strongly skewed distribution have different spreads, no single number describes the spread well. The five-number summary, with its two quartiles and two extremes, does a better job.

Choosing numerical descriptions

The five-number summary is usually better than the mean and standard deviation for describing a skewed distribution. Use $\overline{x}$ and s only for reasonably symmetric distributions.

You may rightly feel that the importance of the standard deviation is not yet clear. It is a bit complicated and is not a good description for skewed distributions. We will see in the next section that the standard deviation is the natural measure of spread for an important class of symmetric distributions, the normal distributions.

Do remember that a graph gives the best overall picture of a distribution. Numerical measures of center and spread report specific facts about a distribution, but they do not describe its entire shape. Numerical summaries do not disclose the presence of multiple peaks or gaps, for example. *Always start with a graph of your data.*

SECTION 4 EXERCISES

4.48   **Comparing gas mileages.** Find the five-number summaries
for the city gas mileages of midsize cars and sports utility
vehicles, given in Table 4-4. Make side-by-side boxplots of these
distributions along with the highway mileages from Example 19,
using a common miles-per-gallon scale. Comment carefully on
what your plot shows about the comparison between city and
highway gas mileage and between midsize cars and sports utility
vehicles.

4.49   **Ruth versus Maris again.** Exercise 4.34 (page 243) gives the
number of home runs hit each year by Babe Ruth and Roger
Maris.

(a)   Find the five-number summary for each player.

(b)   Make side-by-side boxplots to compare Ruth and Maris as
home run hitters. What do you conclude?

4.50   **How long do presidents live?** Return to the ages at death of
the presidents, given in Table 4-3 (page 228) or in the stemplot
of Figure 4-18 (page 236). We have seen that the mean age at
death is 69.14 years (Example 12) and that the median is 67.5
years (Example 14).

(a)   Find the five-number summary. What range contains the
middle half of the ages at which presidents have died?

(b)   Because the distribution is quite symmetric, we might use
$\bar{x}$ and s to describe its center and spread. Calculate s (use a
calculator). In what units is s measured?

4.51   Calculate five-number summaries for each of the following sets
of data, and make a boxplot of each distribution. How do the
boxplots reflect the fact that one distribution is quite symmetric
and the other is strongly skewed?

(a)   The 100 coliform bacteria counts in Exercise 4.17

(b)   The 100 reaction times in Exercise 4.36

4.52   **SAT scores for the states.** The states vary greatly in the median
Scholastic Assessment Test (SAT) scores earned by their high

school seniors. I entered the SAT math and verbal scores for the 50 states and the District of Columbia into statistical software and asked for basic numerical descriptions. Here is what the computer reported:

SATVERBAL

N	MEAN	MEDIAN	STDEV	MIN	MAX	Q1	Q3
51	448.16	443.00	30.82	397.00	511.00	422.00	476.00

SATMATH

N	MEAN	MEDIAN	STDEV	MIN	MAX	Q1	Q3
51	497.39	490.00	34.57	437.00	577.00	470.00	523.00

Use this output to make side-by-side boxplots of SAT math and verbal scores for the states. Briefly compare the two distributions in words.

4.53 Compute five-number summaries for the weight gains of the experimental and control groups in the chicken nutrition experiment of Exercise 4.24 on page 234. Draw boxplots for the two groups with a common scale, as in Figure 4-20. Now discuss what the data show:

(a) The researchers hoped to show that chicks fed the new corn variety grow faster. Do the data confirm that? By how much does the weight gain of a typical chick fed the new corn exceed that of a typical chick fed normal corn?

(b) Can we conclude that the change in corn type actually *caused* the faster weight gain? Why?

(c) What other features do your plots suggest: Are the distributions roughly symmetric or clearly skewed? Do they differ in variability?

4.54 Calculate the mean, the variance, and the standard deviation of each of the following sets of numbers:

(a) 4 0 1 4 3 6

(b) 5 3 1 3 4 2

Which of the sets is more spread out? Draw a histogram of set (a) and one of set (b) to see how the set with the larger variance is more spread out.

4.55 The researchers in the chicken nutrition experiment of Exercise 4.24 reported the mean $\bar{x}$ and the standard deviation s as summaries of their data. Calculate these measures for both groups of chicks. (Use a calculator.) How do the groups differ in center? In spread?

4.56 Add 2 to each of the numbers in set (a) in Exercise 4.54. The data are now

$$6\ 2\ 3\ 6\ 5\ 8$$

(a) Find the mean and the standard deviation of this set of numbers.

(b) Compare your answers with those for set (a) in Exercise 4.54. How did adding 2 to each number change the mean? How did it change the standard deviation?

(c) Can you guess, without doing the arithmetic, what will happen to the mean and standard deviation of set (a) in Exercise 4.54 if we add 10 to each number in that set?

This exercise should help you see that the standard deviation only measures spread about the mean and ignores changes in where the data are centered.

4.57 **A standard deviation contest.** For (a) and (b), choose numbers from 0, 1, 2, 3, 4, 5, 6, 7, 8, and 9. Repeats are allowed.

(a) Give a list of 4 numbers with the largest standard deviation such a list can possibly have.

(b) Give a list of 4 numbers with the smallest standard deviation such a list can possibly have.

(c) Does either part (a) or part (b) have more than one correct answer?

4.58 Scores of adults on the Stanford-Binet IQ test have mean 100 and standard deviation 15. What is the variance of scores on this test?

4.59 If two distributions have exactly the same mean and standard deviation, must their histograms have the same shape? If they have the same five-number summary, must their histograms have the same shape? Explain.

4.60   A school system employs teachers at salaries between $25,000 and $55,000. The teachers' union and the school board are negotiating the form of next year's increase in the salary schedule.

(a)   If every teacher is given a flat $1000 raise, what will this do to the mean salary? To the median salary? To the extremes and quartiles of the salary distribution?

(b)   What will a flat $1000 raise do to the standard deviation of teachers' salaries? (Do Exercise 4.56 if you need help.)

(c)   If, instead, each teacher receives a 5% raise, the amount of the raise will vary from $1250 to $2750, depending on the present salary. What will this do to the mean salary? To the median salary?

(d)   A flat raise would not increase the spread of the salary distribution. What about a 5% raise? Specifically, will a 5% raise increase the distance of the quartiles from the median? Will it increase the standard deviation?

4.61   Exercise 4.23 (page 233) gives the maximum daily rainfall each year for a 30-year period at South Bend, Indiana. Based on the shape of this distribution, do you prefer the five-number summary or $\overline{x}$ and s as a brief numerical description? Why? Compute the summary you chose.

4.62   Exercise 4.21 (page 232) presents data on spending by shoppers at a grocery store. Based on the shape of this distribution, do you prefer the five-number summary or $\overline{x}$ and s as a brief numerical description? Why? Compute the summary you chose.

▶ 5 THE NORMAL DISTRIBUTIONS

We now have a kit of graphical and numerical tools for describing distributions. What is more, we have a clear strategy for exploring the distribution of a numerical variable:

▶ First make a stemplot or histogram.

▶ Look for the overall pattern (shape, center, and spread) and for striking deviations such as outliers.

▶ Choose either the five-number summary or $\bar{x}$ and s to briefly describe center and spread.

Here is one more step to add to this strategy:

▶ Sometimes the overall pattern of a large number of observations is so regular that we can describe it by a smooth curve.

DENSITY CURVES

Look at Figure 4-22. The histogram is familiar—it is the distribution arising from taking 1000 simple random samples of 1373 parents and

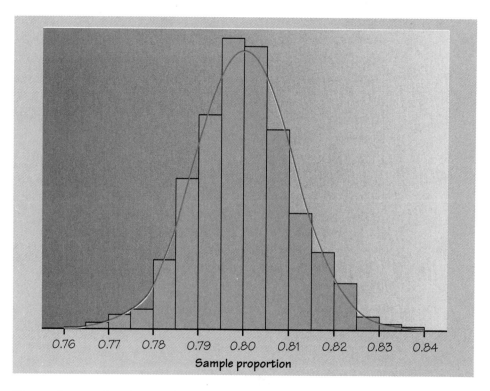

Figure 4-22 A distribution described by a histogram and by a smooth curve that summarizes the histogram.

asking each parent "If you had it to do over again, would you have children?" The distribution shows how the sample proportion who say "Yes" would vary if the truth about the population is that 80% would say yes.

The distribution is symmetric and quite smooth. There are no outliers or other deviations from the smooth shape. We can describe the overall shape by drawing a curve through the tops of the bars in the histogram. The curve is faster to draw and easier to work with than the histogram, because it does not depend on the choice of classes.

Curves that describe the overall shape of a distribution are called **density curves.** Of course, no set of real data is exactly described by a density curve. Density curves are ideal patterns that are accurate enough for practical purposes. For example, the density curve in Figure 4-22 is exactly symmetric, but the actual data are only approximately symmetric.

THE CENTER AND SPREAD OF A DENSITY CURVE

Density curves can help us better understand our measures of center and spread. Because a density curve replaces a histogram, areas under the curve, like the areas of the bars in a histogram, represent counts or percents of observations. We can see that:

▶ The *mode* of a distribution is the *peak point* of a density curve, the value where the curve is highest.

▶ The *median* is the *equal-areas point,* the value such that half the area under the curve lies to the left and half to the right.

Figure 4-23 locates the mode and the median on two density curves, one symmetric and one skewed to the right. You can locate the quartiles (at least roughly) by dividing the area under the curve into four equal parts. It is not so clear where to place the mean on a density curve. Here are the facts:

▶ The *mean* of a distribution is the *balance point,* the point about which the density curve would balance if cut out of solid material.

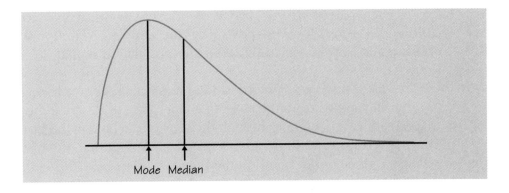

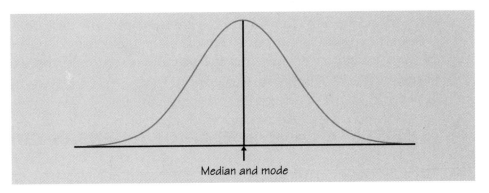

Figure 4-23 The mode of a density curve is its peak point. The median is the equal-areas point.

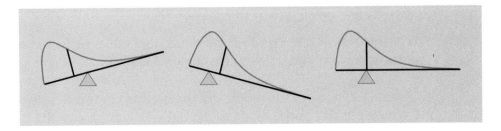

Figure 4-24 The mean of a density curve is the balance point.

Figure 4-24 illustrates the mean as balance point. If a density curve is symmetric, the balance point and the equal-areas point are the same. That is, the mean and median are the same for symmetric distributions.

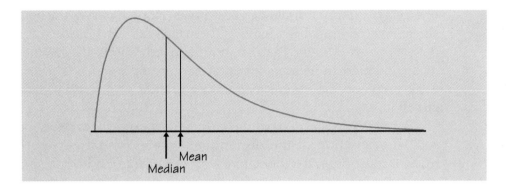

Figure 4-25 The mean and median of a right-skewed density curve. The mean is pulled toward the long tail.

Skewed distributions have density curves with a long tail in the direction of the skew. A little weight far out in one direction moves the balance point quite a bit in that direction, so the mean moves farther toward the long tail than does the median. Figure 4-25 gives a pictorial view of this familiar fact.

NORMAL DISTRIBUTIONS

The density curve that summarizes the distribution in Figure 4-22 belongs to a particularly important family: the normal curves. Figure 4-26

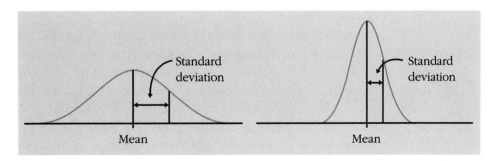

Figure 4-26 Two normal density curves, showing the means and standard deviations.

presents two more normal density curves. Normal curves are symmetric, single-peaked, and bell-shaped. Their tails fall off quickly, so that we do not expect outliers. Because normal distributions are symmetric, the mean and median lie together at the center of the curve. This is also the peak point, so the mean, median, and mode of a normal distribution are all identical.

Normal curves also have the special property that we can locate the standard deviation of the distribution by eye on the curve. This isn't true for most other density curves. Near its center, a normal curve falls ever more steeply as we move away from the center, like this:

But in either tail, the curve falls ever less steeply as we move away from the center, like this:

The points at which the curvature changes from the first type illustrated to the second are located one standard deviation on either side of the mean. The standard deviations are marked on the two curves in Figure 4-26. With a bit of practice, you can learn to find the change-of-curvature points by running a pencil along the curve and feeling where the curvature changes.

Normal curves have the special property that giving the mean and the standard deviation completely specifies the curve. The mean fixes the center of the curve, and the standard deviation determines its shape. Changing the mean of a normal distribution does not change its shape, only its location on the axis. Changing the standard deviation does change the shape of a normal curve, as Figure 4-26 illustrates. The distribution with the smaller standard deviation is less spread out and more sharply peaked. Here is a summary of basic facts about normal curves:

Normal density curves

The normal curves are symmetric, bell-shaped curves that have these properties:

▶ A specific normal curve is completely described by giving its mean and its standard deviation.

▶ The mean determines the center of the distribution. It is located at the center of symmetry of the curve.

▶ The standard deviation determines the shape of the curve. It is the distance from the mean to the change-of-curvature points on either side.

WHY ARE NORMAL DISTRIBUTIONS IMPORTANT?

Our first normal curve appeared in Figure 4-22 as an idealized description of the sampling distribution of a sample proportion. The sampling distribution shows how a statistic would vary if we took many random samples from the same population. In fact, the sampling distributions of sample proportions and sample means are close to normal and become more nearly normal as the size of the sample increases. Normal curves are important in statistics in part because sampling distributions are important.

It can be proved mathematically that any variable that is the sum or average of many small independent effects will have a distribution of values that is close to normal. This general fact helps explain why sample means (averages of many independent observations) are nearly normal. It also suggests that scores on long multiple-choice examinations such as the SATs will come close to following a normal curve.

Many types of data have distributions whose shape is approximately described by a normal curve. Normal curves were first applied to data by the great mathematician Carl Friedrich Gauss (1777–1855), who used them to describe the small errors made by astronomers or surveyors in repeated careful measurements of the same quantity. You will sometimes see normal distributions labeled "Gaussian" in honor of

Gauss. For much of the nineteenth century normal curves were called "error curves" because they were first used to describe the distribution of measurement errors. As it became clear that the distributions of some biological and psychological variables were at least roughly normal, the "error curve" terminology was dropped. The curves were first called "normal" by the American logician Charles S. Peirce in 1873.

One kind of data that are often normal is physical measurements of many members of a biological population. As an example, Figure 4-27 presents measurements on 2000 Hungarian skulls. However, even though many sets of data follow a normal distribution, many do not.

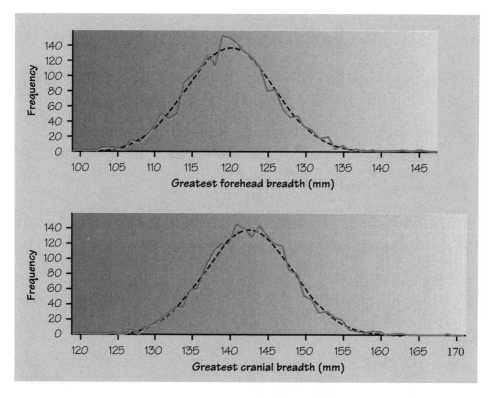

Figure 4-27 Measurements on 2000 Hungarian skulls. The solid curve connects the tops of the bars in a histogram, and the dashed curve is the normal density curve that summarizes the distribution. [From Karl Pearson, "Craniological notes," *Biometrika,* June 1903, p. 344. Reproduced by permission of the Biometrika Trustees.]

Most income distributions are skewed to the right and hence are not normal. Nonnormal data, like nonnormal people, meet us quite often and are sometimes more interesting than their normal counterparts.

THE 68–95–99.7 RULE

There are many normal curves, each described by its mean and standard deviation. All normal curves share many properties. In particular, all normal distributions obey the following rule:

The 68–95–99.7 rule

In any normal distribution, approximately:

▶ **68%** of the observations fall within one standard deviation of the mean.

▶ **95%** of the observations fall within two standard deviations of the mean.

▶ **99.7%** of the observations fall within three standard deviations of the mean.

Figure 4-28 illustrates the 68–95–99.7 rule. By remembering these three numbers, you can think about normal distributions without constantly making detailed calculations. Remember also, though, that no set of data is exactly described by a normal curve. The 68–95–99.7 rule will be only approximately true for SAT scores or the forehead breadth of 2000 Hungarian skulls.

> **EXAMPLE 21. Heights of young women.** The distribution of heights of women aged 18 to 24 is approximately normal with mean 65 inches and standard deviation 2.5 inches. Figure 4-29 shows the application of the 68–95–99.7 rule in this example.
>
> Half of the observations in any normal distribution lie above the mean, so half of all young women are taller than 65 inches.
>
> The central 68% of any normal distribution lies within one standard deviation of the mean. Half of this central 68%, or 34%, lie above the mean. So 34% of young women are between 65 inches and 67.5 inches tall.

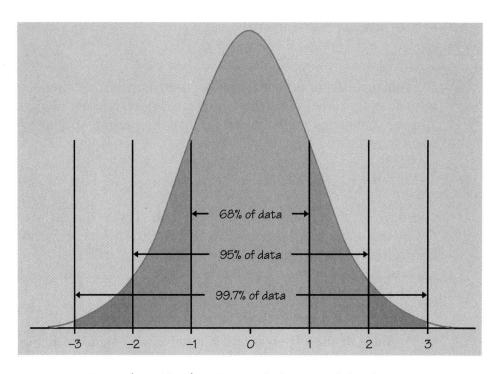

Figure 4-28 The 68–95–99.7 rule for normal distributions.

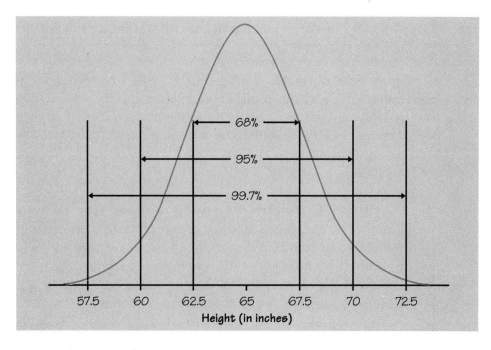

Figure 4-29 The 68–95–99.7 rule applied to the distribution of heights of young women.

The central 95% of any normal distribution lies within two standard deviations of the mean. Two standard deviations is 5 inches here, so the middle 95% of young women's heights are between 60 inches (that's 65 − 5) and 70 inches (that's 65 + 5).

The other 5% of young women have heights outside the range from 60 to 70 inches. Because the normal distributions are symmetric, half of these women are on the tall side. The tallest 2.5% of young women are taller than 70 inches.

Almost all (99.7%) of the observations in any normal distribution lie within three standard deviations of the mean. Almost all young women are between 57.5 and 72.5 inches tall.

STANDARD SCORES

The standard deviation is the natural unit of measurement for normal distributions. In fact, there is only one normal distribution if we measure in standard deviation units around the mean as the zero point. Observations expressed in standard deviation units about the mean as zero point are called *standard scores*.

Standard scores

The **standard score** for any observation is

$$\text{standard score} = \frac{\text{observation} - \text{mean}}{\text{standard deviation}}$$

A standard score of 1 says that the observation in question lies one standard deviation above the mean. An observation with standard score −2 is two standard deviations below the mean. Standard scores can be used to compare values in different distributions. Of course, you should not use standard scores unless you are willing to use the standard deviation to describe the spread of the distributions. That requires that the distributions be at least roughly symmetric.

EXAMPLE 22. ACT versus SAT scores. Eleanor scores 680 on the mathematics part of the SAT. Scores on the SAT are scaled to follow a normal distribution with mean 500 and standard deviation 100.

Gerald takes the American College Testing (ACT) mathematics test and scores 27. ACT scores are normally distributed with mean 18 and standard deviation 6. Assuming that both tests measure the same kind of ability, who has the higher score?

Eleanor's standard score is

$$\frac{680 - 500}{100} = \frac{180}{100} = 1.8$$

Compare this with Gerald's standard score, which is

$$\frac{27 - 18}{6} = \frac{9}{6} = 1.5$$

Because Eleanor's score is 1.8 standard deviations above the mean and Gerald's is only 1.5 standard deviations above the mean, Eleanor's performance is better.

PERCENTILES OF NORMAL DISTRIBUTIONS*

For normal distributions, standard scores allow a quite exact comparison because they translate directly into *percentiles*.

> **Percentiles**
>
> The **cth percentile** of a distribution is a value such that c percent of the observations lie below it and the rest lie above.

The median of any distribution is the 50th percentile, and the quartiles are the 25th and 75th percentiles. In any normal distribution, the point one standard deviation above the mean (standard score 1) is the 84th percentile. Figure 4-30 shows why. Half of the observations are less than the mean. The 68 part of the 68–95–99.7 rule says that 68% of the data lie within one standard deviation of the mean. Half of these, or 34%, lie above the mean. Adding 50% and 34% gives 84%.

Every standard score for a normal distribution translates into a specific percentile, which is the same no matter what the mean and

---

*This section is not needed to read the rest of the book.

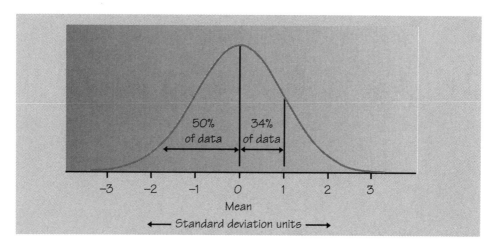

Figure 4-30 Standard score 1 is the 84th percentile for any normal distribution.

standard deviation of the original normal distribution are. Table B at the end of this book gives the percentiles corresponding to various standard scores. This table enables us to do calculations in greater detail than the 68–95–99.7 rule.

> **EXAMPLE 23. Percentiles for college entrance exams.** Eleanor's score of 680 on the SAT translates into a standard score of 1.8, as we saw in Example 22. Table B shows that this is the 96.41 percentile of a normal distribution. Eleanor did better than 96% of the people who took the exam. Gerald's 27 on the ACT is a standard score of 1.5, which is the 93.32 percentile. Gerald did well, but not as well as Eleanor. The percentile is easier to understand than either the raw score or the standard score. That's why reports of exams such as the SAT usually give both the score and the percentile.

SECTION 5 EXERCISES

4.63 (a) Sketch a density curve that is symmetric but has a shape different from that of the normal curves.

(b) Sketch a density curve that is strongly skewed to the left.

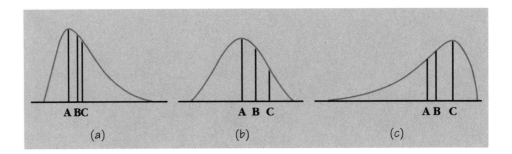

Figure 4-31 Three density curves, for Exercise 4.64.

4.64   **Mean, median, and mode.** Figure 4-31 presents three density curves, each with several points marked. At which of these points on each curve do the mean, the median, and the mode fall? (More than one measure may fall at the same point.)

4.65   Figure 4-32 is a normal density curve. What are the mean and the standard deviation of this distribution?

4.66   **Hungarian skulls.** Figure 4-27 records two distributions for measurements on 2000 Hungarian skulls. Both are approximately normal. Estimate the mean and the standard deviation of each set of data.

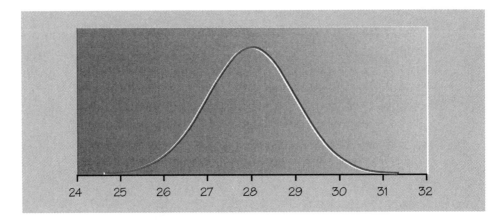

Figure 4-32 A normal density curve, for Exercise 4.65.

4.67 **Length of pregnancies.** The length of human pregnancies from conception to birth varies according to a distribution that is approximately normal with mean 266 days and standard deviation 16 days. Sketch a normal curve for this distribution. (Hint: First draw the curve, then locate the mean and the change-of-curvature points and use these to mark a scale in days on the axis.)

4.68 **Is this distribution normal?** Return once more to the coliform counts of Exercise 4.17 (page 230). The histogram that you drew in that exercise suggests that the distribution may be roughly normal. You can calculate (don't do it now) that the mean of the 100 counts is 5.88 and the standard deviation is 2.026. What percent of the 100 observations fall within one, two, and three standard deviations of the mean? Do you feel that the distribution of coliform counts is approximately normal?

4.69 **Length of pregnancies.** The length of human pregnancies from conception to birth varies according to a distribution that is approximately normal with mean 266 days and standard deviation 16 days. Use the 68–95–99.7 rule to answer the following questions. (Whenever you apply the 68–95–99.7 rule, start by sketching a picture like Figure 4-29 that shows how the rule applies to your specific setting.)

(a)   Between what values do the lengths of the middle 95% of all pregnancies fall?

(b)   How short are the shortest 2.5% of all pregnancies?

4.70 The army reports that the distribution of head circumference among male soldiers is approximately normal with mean 22.8 inches and standard deviation 1.1 inches. Use the 68–95–99.7 rule to answer the following questions. (Whenever you apply the 68–95–99.7 rule, start by sketching a picture like Figure 4-29 that shows how the rule applies to your specific setting.)

(a)   What percent of soldiers have head circumference greater than 23.9 inches?

(b)   The army plans to make helmets in advance to fit the middle 95% of head circumferences. Soldiers who fall outside this range will get custom-fitted helmets. What head circumferences are small enough or big enough to require custom fitting?

4.71 **IQ tests.** Scores on the Wechsler Adult Intelligence Scale for the 20–34 age group are approximately normally distributed with mean 110 and standard deviation 25. Use the 68–95–99.7 rule to learn what percent of people in this age group have scores:

(a)   above 110

(b)   above 160

(c)   below 85

(Start by sketching a normal curve like that in Figure 4-29 that applies the 68–95–99.7 rule to this setting.)

4.72 **Scoring 800 on the SAT.** SAT scores are approximately normal with mean 500 and standard deviation 100. Scores of 800 or higher are reported as 800, so a perfect paper is not required to score 800 on the SAT. What percent of students who take the SAT score 800?

4.73 **ACT versus SAT.** Jim scores 700 on the mathematics part of the SAT. Scores on the SAT follow the normal distribution with mean 500 and standard deviation 100. Julie takes the ACT test of mathematical ability, which has mean 18 and standard deviation 6. She scores 24. If both tests measure the same kind of ability, who has the higher score?

4.74 **Comparing IQ scores.** Scores on the Wechsler Adult Intelligence Scale for the 60–64 age group are approximately normally distributed with mean 90 and standard deviation 25.

(a)   Sarah, who is 30, scores 135 on this test. Use the information of Exercise 4.71 to restate Sarah's result as a standard score.

(b)   Sarah's mother, who is 60, also takes the test and scores 120. Express this as a standard score using the information given in this exercise.

(c)   Who scored higher relative to her age group, Sarah or her mother? Who has the higher absolute level of the variable measured by the test?

4.75 **Three great hitters.** Three landmarks of baseball achievement are Ty Cobb's batting average of .420 in 1911, Ted Williams' .406 in 1941, and George Brett's .390 in 1980. These batting averages cannot be compared directly, because the distribution of major league batting averages has changed over the decades. The distributions are quite symmetric and (except for outliers such as Cobb, Williams, and Brett) reasonably normal. Although the mean batting average has remained roughly constant, the standard deviation has decreased over time. Here are the facts.[8]

Decade	Mean	Std Dev
1910's	.266	.0371
1940's	.267	.0326
1970's	.261	.0317

Compute the standard scores for the batting averages of Cobb, Williams, and Brett. Compare how far each stood above his peers.

4.76 The mean height of men ages 18 to 24 is about 69.5 inches. Women that age have a mean height of about 65 inches. Do you think that the distribution of heights for all Americans ages 18 to 24 is approximately normal? Explain your answer.

4.77 **Sampling distributions.** Suppose that the proportion of all adult Americans who are afraid to go out at night because of crime is $p = 0.45$. The distribution of the sample proportion $\hat{p}$ in Gallup poll samples of size 1500 is then approximately normal with mean 0.45 and standard deviation 0.015. Use this fact and the 68–95–99.7 rule to answer these questions.

(a) In many samples, what percent of the values of $\hat{p}$ fall above 0.465? Below 0.42?

(b) In a large number of samples, what range contains the central 95% of values of $\hat{p}$?

The following exercises concern the optional material at the end of Section 5.

4.78 Use the 68–95–99.7 rule and a drawing like Figure 4-30 to explain why the point one standard deviation below the mean in a normal distribution is roughly the 16th percentile. Explain why the point two standard deviations above the mean is roughly the 97.5th percentile.

4.79 **Locating the quartiles.** Use Table B to answer these questions.

(a) The quartiles of any normal distribution are equally distant from the mean on either side. About how many standard deviations from the mean are the quartiles?

(b) About how many standard deviations from the mean in each direction must we go in order to span the central 90% of any normal distribution?

4.80 **IQ tests.** The Wechsler Intelligence Scale for Children is used (in several languages) in the United States and Europe. Scores in each case are approximately normally distributed with mean 100 and standard deviation 15. When the test was standardized in Japan, the mean was 111. To what percentile of the American-European distribution of scores does the Japanese mean correspond?

4.81 **IQ tests.** The IQ scores on the Wechsler Adult Intelligence Scale for the 20–34 age group are approximately normal with mean 110 and standard deviation 25.

(a) What percent of people ages 20 to 34 have IQs below 100? What percent have IQs 100 or above?

(b) What percent of people ages 20 to 34 have IQs above 150?

4.82 **IQ tests.** If only 1% of people ages 20 to 34 have IQs higher than April's, what is April's IQ? (Use the distribution of IQ scores given in the previous exercise and the entry in Table B that comes closest to the point with 1% of the observations above it.)

4.83 **Sampling distributions.** Suppose that the proportion of all adult Americans who are afraid to go out at night because of crime is $p = 0.45$. The distribution of the sample proportion $\hat{p}$ in Gallup poll samples of size 1500 is then approximately normal with mean 0.45 and standard deviation 0.015. Use this fact and Table B to answer these questions:

(a)   What percent of all Gallup poll samples would give a $\hat{p}$ of 0.43 or lower? Of 0.47 or higher?

(b)   What percent of all Gallup poll samples would give a $\hat{p}$ that misses the true $p = 0.45$ by two percentage points (± 0.02) or more?

4.84   **Sampling distributions.** Consider again the Gallup poll described in Exercise 4.83. Use Table B and the results of Exercise 4.79 to answer these questions:

(a)   In what range centered at the true $p = 0.45$ do half of all values of $\hat{p}$ fall in many Gallup poll samples?

(b)   In what range do the middle 90% of values of $\hat{p}$ fall in many Gallup poll samples?

NOTES

1. Two excellent books on graphing are: W. S. Cleveland, *The Elements of Graphing Data,* Wadsworth, 1985 and E. R. Tufte, *The Visual Display of Quantitative Information,* Graphics Press, 1983. These books elaborate the principles of graphing with abundant examples both good and bad.
2. The Shakespeare data appear in C. B. Williams, *Style and Vocabulary: Numerical Studies,* Griffin, 1970.
3. These data were collected by students as a class project.
4. Data from J. K. Ford, "Diversification: how many stocks will suffice?" *American Association of Individual Investors Journal,* January 1990, pp. 14–16.
5. Based on data summaries in G. L. Cromwell et al., "A comparison of the nutritive value of *opaque-2, floury-2* and normal corn for the chick," *Poultry Science,* 57 (1968), pp. 840–847.
6. Data from *The Baseball Encyclopedia,* 3rd ed., Macmillan, 1976. Maris's home run data are from the same source.
7. The gas mileage data are from the Environmental Protection Agency's *1996 Fuel Economy Guide.* This annual publication is available online at http://www.epa.com. In the table, the data are for the basic engine/transmission combination for each model, and models that are essentially identical (such as the Ford Taurus and Mercury Sable) appear only once.

8. Data from Stephen Jay Gould, "Entropic homogeneity isn't why no one hits .400 any more," *Discover,* August 1986, pp. 60–66. Gould does not standardize, giving instead a speculative discussion.

9. *Consumer Reports,* June 1986, pp. 366–367. A more recent study of hot dogs appears in *Consumer Reports,* July 1993, pp. 415–419. The newer data cover few brands of poultry hot dogs and take calorie counts mainly from the package labels, resulting in suspiciously round numbers.

REVIEW EXERCISES

4.85 **The class of '80.** A study of 10,500 students who were high school seniors in 1980 asked what the educational attainment of these students was by 1986. The results for men and women are given below. Present these data clearly in a graph. What are the most important facts about the education achieved by these students in the six years after their senior year of high school?

Sex	No HS diploma	HS only	License	Associate degree	Bachelor's degree	Professional or graduate
Men	1.0%	64.0%	10.5%	5.9%	17.6%	0.9%
Women	0.8%	59.6%	13.3%	7.0%	18.8%	0.6%

4.86 **The Border Patrol.** The U.S. Immigration and Naturalization Service reports each year the number of deportable aliens caught by the Border Patrol. Here are the counts, in thousands of persons, for 1970 through 1993. Display these data in a graph. What are the most important facts that the data show?

Year	1970	1975	1980	1981	1982	1983	1984	1985
Count	231.1	596.8	759.4	825.3	819.9	1105.7	1138.6	1262.4

Year	1986	1987	1988	1989	1990	1991	1992	1993
Count	1692.5	1159.0	971.1	893.0	1103.4	1132.9	1199.6	1263.5

4.87 **Making colleges look good.** Colleges announce an "average" SAT score for their entering freshmen. Usually the college would like this "average" to be as high as possible. A *New York Times* article (May 31, 1989) noted that "Private colleges that buy lots of top students with merit scholarships prefer the mean, while open-enrollment public institutions like medians." Use what you know about the behavior of means and medians to explain these preferences.

4.88 **Medical malpractice awards.** In the early 1990s, the median award to people who won medical malpractice lawsuits was $350,000. The mean award in these same lawsuits was about $1.7 million. Explain how this great difference between two measures of center can occur.

4.89 **Presidential elections.** Here are the percentages of the popular vote won by the successful candidate in each of the presidential elections from 1948 to 1992:

Year	1948	1952	1956	1960	1964	1968	1972	1976	1980	1984	1988	1992
%	49.6	55.1	57.4	49.7	61.1	43.4	60.7	50.1	50.7	58.8	53.9	43.2

(a) Round the data to the nearest percent; then make a stemplot. Are there any outliers?

(b) What is the median percent of the vote won by the successful candidate in presidential elections? (Use the rounded percentages.)

(c) Call an election a landslide if the winner's percent falls at or above the third quartile. Which elections were landslides?

4.90 **A hot stock.** The rate of return on a stock is its change in price plus any dividends paid. Rate of return is usually measured in percent of the starting value. We have data on the monthly rates of return for the stock of Wal-Mart stores for the years 1973 to 1991, the first 19 years Wal-Mart was listed on the New York Stock Exchange. There are 228 observations.

Below is output from a computer statistical package that describes the distribution of these data. The stems in the stemplot are the tens digits of the percent returns. The leaves are the ones digits. The stemplot uses split stems (Exercise 4.22, page 233) to give a better display. The software gives high and low outliers separately from the stemplot rather than spreading out the stemplot to include them.

```
Mean    =    3.064
Standard deviation   =   11.49

N = 228    Median = 3.469
Quartiles = -2.950, 8.451

Low outliers:  -34 -31 -27 -27

  -1 : 985
  -1 : 444443322222110000
  -0 : 999988777666666665555
  -0 : 444444443333333322222222222221111111100
   0 : 00000111111111111222223333333344444444
   0 : 5555555555555555555556666666666677777778888888888899999
   1 : 000000000011111111122233334444
   1 : 55566667889
   2 : 011334

High outliers: 32 42 42 58 59
```

(a) Give the five-number summary for monthly returns on Wal-Mart stock.

(b) Describe in words the main features of the distribution.

(c) If you had $1000 worth of Wal-Mart stock at the beginning of the best month during these 19 years, how much would your stock be worth at the end of the month? If you had $1000 worth of stock at the beginning of the worst month, how much would your stock be worth at the end of the month? (Message: Hot stocks move both down and up a lot.) If you had $1000 worth of stock at the beginning of a typical month (described by the median return), how much would your stock be worth at the end of the month?

4.91 Leading the league in home runs. Table 4-5 gives the number of home runs hit by the American League leader each year between 1972 and 1995. Describe these data with both a graphical display and a numerical summary of your choice. Are there any outliers? Describe the overall pattern, ignoring any outliers.

TABLE 4-5 American League home run leaders, 1972–1995

Year	Player	Home runs	Year	Player	Home runs
1972	Dick Allen	37	1984	Tony Armas	43
1973	Reggie Jackson	32	1985	Darrell Evans	40
1974	Dick Allen	32	1986	Jesse Barfield	40
1975	Scott and Jackson	36	1987	Mark McGwire	49
1976	Graig Nettles	32	1988	Jose Canseco	42
1977	Jim Rice	39	1989	Fred McGriff	36
1978	Jim Rice	46	1990	Cecil Fielder	51
1979	Gorman Thomas	45	1991	Canseco and Fielder	44
1980	Reggie Jackson	41	1992	Julio Gonzalez	43
1981	Four players	22	1993	Julio Gonzalez	46
1982	Thomas and Jackson	39	1994	Ken Griffey, Jr.	40
1983	Jim Rice	39	1995	Albert Belle	50

4.92 **How many calories in a hot dog?** Government regulations recognize three types of hot dogs: beef, meat, and poultry. Do these types differ in the number of calories they contain? *Consumer Reports* magazine, in a story on hot dogs, measured the calories in 20 brands of beef hot dogs, 17 brands of meat hot dogs, and 17 brands of poultry hot dogs.[9] Here is computer output describing the beef hot dogs:

```
Mean = 156.8  Standard deviation = 22.64  Min = 111  Max = 190
N = 20  Median = 152.5  Quartiles = 140, 178.5
```

the meat hot dogs:

```
Mean = 158.7  Standard deviation = 25.24  Min = 107  Max = 195
N = 17  Median = 153  Quartiles = 139, 179
```

and the poultry hot dogs:

```
Mean = 122.5  Standard deviation = 25.48  Min = 87  Max = 170
N = 17  Median = 129  Quartiles = 102, 143
```

Use this information to make side-by-side boxplots of the calorie counts for the three types of hot dogs. Write a brief comparison of the distributions. Will eating poultry hot dogs usually lower your calorie consumption compared with eating beef or meat hot dogs?

4.93 **Mexican Americans.** The Acculturation Rating Scale for Mexican Americans (ARSMA) is a psychological test that evaluates the degree to which Mexican Americans are adapted to Mexican/Spanish versus Anglo/English culture. The distribution of ARSMA scores in a population used to develop the test is approximately normal with mean 3.0 and standard deviation 0.8. The range of possible scores is 1.0 to 5.0, with higher scores showing more Anglo/English acculturation. Between what values do the ARSMA scores of the central 95% of Mexican Americans lie? How high must an ARSMA score be in order to fall in the top 2.5% of the population?

4.94 The ARSMA test is described in the previous exercise. A researcher believes that Mexicans will have an average score near 1.7, and that first generation Mexican Americans will average about 2.1 on the ARSMA scale. What proportion of the original population has scores below 1.7? Between 1.7 and 2.1? (Use Table B.)

WRITING PROJECTS

4.1 **Statistical graphics in the press.** Graphs good and bad fill the news media. Some publications, such as *USA Today,* make particularly heavy use of graphs to present data. Collect several graphs (at least five) from newspapers and magazines (not from advertisements). Use them as examples in a brief essay about the clarity, accuracy, and attractiveness of graphs in the press. You can find more information about what makes good graphs in the books by Tufte and Cleveland mentioned in Note 1.

4.2 **High school dropouts.** Locate in a current *Statistical Abstract* data on the percent of each state's population who are high school dropouts. (In the 1995 *Statistical Abstract,* Table 242, Educational Attainment—States, contains the information. You can expect a similar table in later editions.) Give a complete description of the data, using a graph, appropriate numerical measures, and a summary in words.

The data show that the states vary greatly in school dropout rates. What factors do you think might account for this? The high dropout rate for the District of Columbia, for example, suggests that a large urban population may contribute to a high state rate. Look in the *Statistical Abstract* for information you think is relevant. For example, the *Statistical Abstract* contains a table that gives the percent of each state's population that lives in metropolitan areas. ("Metropolitan areas" cover suburbs along with central cities and small cities as well as large.) Write a brief essay suggesting some explanations, and accompany it with data and graphs that explore at least one of your suggestions.

CHAPTER 5

UNDERSTANDING RELATIONSHIPS

A medical study finds that short women are more likely to have heart attacks than women of average height, while tall women have fewer heart attacks. An insurance group reports that heavier cars have fewer deaths per 10,000 vehicles registered than do lighter cars. These and many other statistical studies look at the relationship between two variables. To understand such a relationship, we must often examine other variables as well. To conclude that shorter women have higher risk from heart attacks, for example, the researchers had to eliminate the effect of other variables such as weight and exercise habits. Our topic in this chapter is relationships between variables. One of our main themes is that the relationship between two variables can be strongly influenced by other variables that are lurking in the background.

To study the relationship between two variables, we measure both variables on the same individuals. If we measure both the height and the weight of each of a large group of people, we know which height goes with each weight. These data allow us to study the connection between height and weight. A list of the heights and a separate list of the weights, two sets of single-variable data, do not show the connection between the two variables. In fact, taller people also tend to be heavier. And people who smoke more cigarettes per day tend not to live as long as those who smoke fewer. We say that pairs of variables such as height and weight or smoking and life expectancy are *associated*.

284

> **Association between variables**
>
> Two variables measured on the same individuals are **associated** if some values of one variable tend to occur more often with some values of the second variable than with other values of that variable.

Association between variables is most clearly visible in the physical sciences, where values of one variable are often connected to values of another by a "law." For example, one of the laws of motion states that if you drop a ball from a height, the downward speed of the ball is directly proportional to the time it has been falling. After four seconds it is moving twice as fast as after two seconds. This association can be graphed as a perfect straight-line relation between time and speed.

Statistics is less concerned with ironclad relationships of the kind expressed by "physical laws" than with relationships that hold "on the average." There is an association between the height and the weight of individuals because, on the average, tall people are heavier than shorter people. There is an association between the sex of workers and their pay because, on the average, women earn less than do men. Yet many individual women earn more than many individual men, just as many individual short people are heavier than many individual taller people. Our goal is to describe and understand association, keeping in mind its "on-the-average" character.

The statistical techniques used to study relations among variables are more complex than the one-variable methods in Chapter 4. Fortunately, analysis of several-variable data builds on the tools used for examining individual variables. The principles that guide examination of data are also the same:

1. Start with a graph.

2. Look for an overall pattern and deviations from the pattern.

3. Add numerical descriptions of specific aspects of the data.

4. Sometimes there is a way to describe the overall pattern very briefly.

▶ 1 TWO-WAY TABLES

When both variables are nominal (or their values are grouped into classes), a **two-way table** of counts displays the association between them.

> **EXAMPLE 1.** **Who earns academic degrees?** How do women and men compare in the pursuit of academic degrees? Let's examine the relationship between the level of a degree and the sex of the recipient. The information we want appears in Table 5-1, taken from the *Statistical Abstract*. The values of one variable (sex) label the rows, and the values of the other (degree earned) label the columns of this two-way table. Each entry in the table is the count of people who fall in one of the eight sex-by-degree categories. Women earned about 16,000 doctorate degrees in 1993, for example. (Did you notice that the entries are in thousands?)

DESCRIBING RELATIONSHIPS

The bar graph in Figure 5-1 displays the counts in Table 5-1. We see at once that more women than men earn bachelor's and master's degrees, but that men outnumber women among recipients of professional and doctorate degrees. That is a description of the relationship between sex and degree level. To describe the relationship more exactly, we calculate and compare appropriate percents.

> **EXAMPLE 2.** **Women among degree recipients.** To put the association between level of degree and sex of the recipient in numerical

TABLE 5-1   Earned degrees, 1993, by level and sex (thousands)

	Bachelor's	Master's	Professional	Doctorates
Female	616	194	30	16
Male	529	171	44	26
Total	1145	364	74	42

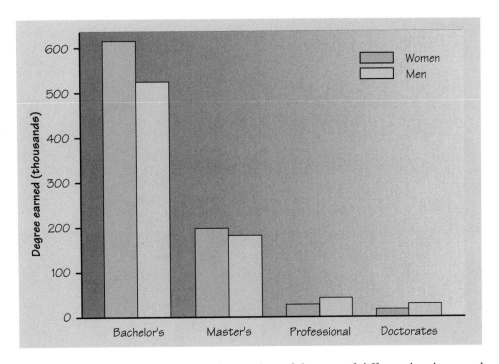

Figure 5-1 Bar graph comparing the number of degrees of different levels earned by men and women in 1993.

terms, compare the percent of each degree that is earned by women. For example, women earn

$$\frac{616}{1145} = 0.538 = 53.8\%$$

of all bachelor's degrees. Similarly, women earn 53.3% of all master's degrees (check this result from the table). However, women's share of professional degrees such as law and medicine is 40.1% and their share of academic doctorates is 38.1%. Women have achieved parity with men in the lower degrees but still earn markedly fewer than half of the most advanced degrees.

You must think carefully about which percents to calculate to answer the question you have in mind. You can find an entry as a percent of the table total, as a percent of the total for its row, or as a percent of the total for its column. These percents answer different questions.

Compare these questions:

1. What percent *of all doctorate degrees* are earned by women? To answer this, look at the "Doctorates" column in Table 5-1:

$$\frac{16}{42} = 0.381 = 38.1\%$$

2. What percent *of all degrees earned by women* are doctorates? Now look at the "Female" row in the table. The total for this row is $616 + 194 + 30 + 16 = 856$. So the answer to our question is

$$\frac{16}{856} = 0.019 = 1.9\%$$

3. What percent *of all degrees* are doctorates earned by women? The total number of degrees for 1993 is the total of all eight counts in the table, 1626 (thousands). The percent of these that are doctorates earned by women is

$$\frac{16}{1626} = 0.010 = 1.0\%$$

Answering these questions was mainly a matter of straight thinking. One way to think straight is to ask "What group is the total that I want a percent of?" The italics in the questions above identify that group in each case. The group total forms the denominator of the fraction needed to get the percent you want.

Although we made a bar graph of *counts* in Figure 5-1, bar graphs that compare *percents* are often clearer. This is particularly true when the total counts in the groups being compared are very different, so that the heights of the bars in a graph of counts would differ greatly. Because percents always have the same 0% to 100% scale, they are easier to compare. Exercise 5.7 leads you through a complete description of association using percents.

The column totals (bottom row) in Table 5-1 present the distribution of the level of degree. The row totals, which don't appear in the table, give the distribution of the sex of degree recipients. Both of these distributions can be obtained by adding counts from the two-way table. Exercise 5.3 demonstrates that we *cannot* recover the entries in the body of the table from the two sets of totals. A two-way table contains more information than the two individual distributions because it describes the relationship between the two variables.

SIMPSON'S PARADOX

The nature of the observed relationship between two variables can change radically when we take into account other variables that lie hidden in the situation. Let's look at a surprising example of this fact.

> **EXAMPLE 3. Discrimination in admissions?** A university offers only two degree programs, one in electrical engineering and one in English. Admission to these programs is competitive, and the women's caucus suspects discrimination against women in the admissions process. The caucus obtains the following data from the university, a two-way table of all applicants by sex and admission decision:
>
> | | Male | Female |
> |--------|------|--------|
> | Admit | 35 | 20 |
> | Deny | 45 | 40 |
> | Total | 80 | 60 |
>
> These data do show an association between the sex of applicants and their success in obtaining admission. To describe this association more precisely, we compute some percents from the data.
>
> $$\text{Percent of male applicants admitted} = \frac{35}{80} = 44\%$$
>
> $$\text{Percent of female applicants admitted} = \frac{20}{60} = 33\%$$
>
> Aha! Almost half the males but only one-third of the females who applied were admitted.

The university replies that although the observed association is correct, it is not due to discrimination. In its defense, the university produces a **three-way table** that classifies applicants by sex, admission decision, and the program to which they applied. We present a three-way table as several two-way tables side-by-side, one for each value of the third variable. In this case there are two two-way tables, one for each program.

	Engineering			English	
	Male	Female		Male	Female
Admit	30	10	Admit	5	10
Deny	30	10	Deny	15	30
Total	60	20	Total	20	40

Check that these entries add to the entries in the two-way table. The university has simply broken down that table by department. We now see that engineering admitted exactly half of all applicants, both male and female, and that English admitted one-fourth of both males and females. There is *no association* between sex and admission decision in either program.

How can no association in either program produce strong association when the two are combined? Look at the data: English is hard to get into, and mainly females applied to that program. Electrical engineering is easy to get into and attracted mainly male applicants. English had 40 female and 20 male applicants, while engineering had 60 male and only 20 female applicants. The original two-way table, which did not take account of the difference between programs, was misleading. This is an example of *Simpson's paradox*.

Simpson's paradox

The nature of an association can change, and even reverse direction, when data from several groups are combined to form a single group.

BEWARE THE LURKING VARIABLE

Although Example 3 was artificial for the sake of simplicity, it is based on a real controversy over admission to graduate programs.[1] The relationship that we observe between two variables is often influenced by other variables that we did not measure or even think about. Because these variables are lurking in the background, we call them *lurking variables*. The effect of lurking variables on an observed association is at the heart of most controversies over alleged discrimination in both public debate and court cases.

> **EXAMPLE 4. Discrimination in mortgage lending?** Studies of applications for home mortgage loans from banks show a strong racial pattern: banks reject a much higher percentage of black applicants than of white applicants. One lawsuit in the Washington, D.C., area contends that a bank rejected 17.5% of blacks, but only 3.3% of whites.[2]
>
> The bank replies that lurking variables explain the difference in rejection rates. Blacks have (on the average) lower incomes, poorer credit records, and less secure jobs than whites. Unlike race, these are legitimate reasons to turn down a mortgage application. It is because these lurking variables are confounded with race, the bank says, that it rejects a higher percentage of black applicants.
>
> Who is right? Both sides will hire statisticians to examine the effects of the lurking variables. The court will eventually decide.

> **EXAMPLE 5. Sex differences in pay.** Women who are employed full-time earn (on the average) about 76 percent as much as men. Does this difference reflect discrimination against women?
>
> It is easy to see that lurking variables might explain part of the earnings gap. Women workers (on the average) differ from male workers in age, years of schooling, labor force experience, and so on. In particular, many older women have spent time outside the labor force for family reasons. Because pay rises with experience, the average pay of men would exceed that of women even if there were no discrimination. The bar graph in Figure 5-2 shows that the male-female gap in earnings has narrowed over time, and that the gap is smaller for young workers. These changes reflect changes in the lurking variables as more women plan to have working careers.
>
> The differing characteristics of male and female workers appear to explain roughly half of the earnings gap. Outright discrimination—different pay for the same work—is illegal and therefore rare. Why does the earnings gap persist? Because some jobs (such as clerical work) are primarily held by women and others (such as maintenance work) are largely male. Jobs that are

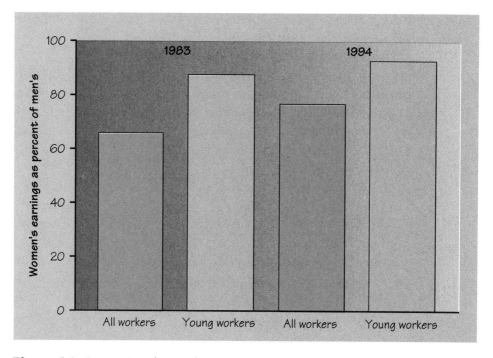

Figure 5-2 Comparing the median earnings of male and female full-time wage and salary workers. The graph shows women's earnings as a percent of the male median in 1983 and in 1994 for all full-time workers and for full-time workers under 25 years of age.

traditionally male pay more in general than jobs dominated by women.

Should public policy try to close the gap? We might try to assess the skill, training, and responsibility required by different jobs and require that jobs that are similar in these variables carry equal pay. A number of public employers, including the states of Iowa and Minnesota, have adopted this approach. The pay of clerical workers is increased to match the pay of "comparable" maintenance jobs.

The opposing view holds that we don't know or can't measure all the lurking variables that may influence how much a job pays. How much weight should we give to the fact that a job can be held by part-time workers? These jobs pay less, and are more often held by women. How much weight should we give to the risk of injury or death? Men hold most risky jobs—about 95% of all workers killed on the job are men.

The example of the male-female earnings gap is a reminder that in complex situations when many variables affect an outcome, you should be slow to jump to a conclusion suggested by a strong association alone.[3]

SECTION 1 EXERCISES

5.1 The U.S. Office of Civil Rights has among its jobs the investigation of possible sex discrimination in university hiring. The office wants to know if women are hired for the faculty in the same proportion as their availability in the pool of potential employees. If new university faculty in 1993 were recruited almost entirely from new doctorate degree recipients, which of the three questions asked on page 288 about Table 5-1 is useful to the office? Why?

5.2 In Table 5-1, the "Total" entry for Master's degrees differs slightly from the sum of the male and female entries. Why do you think this occurs?

5.3 Here are the row and column totals for a two-way table with two rows and two columns.

$$
\begin{array}{cc|c}
a & b & 50 \\
c & d & 50 \\
\hline
60 & 40 & 100
\end{array}
$$

Find *two different* sets of counts *a*, *b*, *c*, and *d* for the body of the table that give these same totals. This shows that the relationship between two variables cannot be obtained from the two individual distributions of the variables.

5.4 **High blood pressure.** High blood pressure is unhealthy. Here are the results of one of the studies that link high blood pressure to death from cardiovascular disease. The researchers classified a group of white males aged 35 to 64 as low blood pressure (systolic pressure less than 140 mm Hg) or high (140 mm Hg or higher), then followed the subjects for five years. The following two-way table gives the results of the study.[4]

Cardiovascular death?

Blood pressure	Yes	No	Total
Low	21	2655	2676
High	55	3283	3338

(a)   Calculate the mortality rate (deaths as a percent of the total) for each group of men.

(b)   Do these data support the idea that high blood pressure is associated with death from cardiovascular disease? Why?

(c)   Do these data give good evidence that high blood pressure *causes* death from cardiovascular disease? Why?

5.5   **Age and education.** Table 5-2 presents data from the *Statistical Abstract* on the educational attainment of the adult population. Many people under age 25 have not completed their education, so the table concerns only people at least 25 years old.

(a)   How many people are in the 25 to 34 age group? How many people have four or more years of college?

(b)   What percent of people 25 to 34 years of age have at least four years of college? What percent of people age 65 and over have at least four years of college?

TABLE 5-2   Years of school completed by age, 1994 (thousands of people)

Education	Age group		
	25–34	35–64	≥ 65
Did not complete high school	5,705	14,152	11,561
Completed high school	14,472	31,539	10,504
1–3 years of college	11,913	19,107	4,853
4 or more years of college	9,816	22,887	3,843

(c)   What percent of people with at least four years of college are age 25 to 34? What percent of people with at least four years of college are age 65 and over?

5.6   **Age and education, continued.** A two-way table such as Table 5-2 yields the distributions of both individual variables as its row totals and column totals.

(a)   Add the totals for each row to Table 5-2. These row totals are the distribution of education among the entire over-25 population.

(b)   What percent of the over-25 population falls into each of the four education categories? These percents are another form of the distribution of education.

(c)   Make a bar graph of the percents you found in (b). This graph displays the distribution of education in the over-25 population. Would it make sense to make a pie chart of these data?

5.7   **Age and education, continued.** To show the nature of the association between the variables in a two-way table, we calculate and compare percents. To show the association between age and education in Table 5-2, we can compare the percents in each age group who reach each level of education.

(a)   Consider first young people age 25 to 34 years. What percent of this age group did not complete high school? Completed high school but reached no higher level? Have one to three years of college? Have at least four years of college? Check your work by adding these four percents—what should the total be?

(b)   Repeat part (a) for the middle-age (35 to 64 years) population.

(c)   Repeat part (a) for the senior (65 and over) population.

(d)   In parts (a), (b), and (c), you found the distributions of education separately for the three age groups. Comparing these distributions shows the association between age and education. Make three bar graphs, one for each distribution. Use the same scale and place the graphs one above the other for easy comparison.

(e)   Based on your percents and the bar graphs, write a brief but clear description of how education changes with age in the United States.

5.8   **Degrees in foreign languages.** Here is a table of earned degrees in foreign languages for 1992, classified by level and by the sex of the recipient. Write a brief description of the relationship between sex and level. What are the most important differences between foreign languages degrees and the facts for all degrees given by Table 5-1?

	Bachelor's	Master's	Doctorates
Male	3,990	971	378
Female	9,913	1,955	472
Total	13,903	2,926	850

5.9   **Suicides.** Here is a pleasant little two-way table that describes all suicides committed in 1992.

	Male	Female
Firearms	15,802	2,367
Poison	3,262	2,233
Hanging	3,822	856
Other	1,571	571
Total	24,457	6,027

(a)   How many suicides were reported in 1992?

(b)   Give the distribution of the methods used in all suicides. What method was most commonly used, and what percent of all suicides were committed by this method?

(c)   What percent of all women who committed suicide used poison?

5.10   **Suicides, continued.** Using the data in the previous exercise, make a bar graph that compares the percents of male and female suicides who used each method. Write a brief description of the chief differences between men and women in their choice of suicide methods.

TABLE 5-3	People below poverty level, 1993 (thousands of people)		
	White	Black	Other
Under 18 years	9,752	5,125	850
18 to 24 years	3,274	1,264	316
25 to 64 years	10,261	3,785	881
65 years and over	2,939	702	114

(handwritten margin notes: 15,727; 4854; 14,927; 3,755)

5.11   **Poverty in the United States.** Table 5-3 gives data on poverty in the United States, from the *Statistical Abstract.*

(a)   To understand these data, you need an essential definition that is given in the source. What definition?

(b)   How many people below the poverty level were there in 1993?

(c)   What percent of people below the poverty level were 65 or older?

(d)   What percent of people below the poverty level were black?

(e)   Of all whites below the poverty level, what percent were 65 or older?

(f)   Of all children under 16 below the poverty level, what percent were black?

(g)   You want to know what percent of all people 65 and older were below the poverty level. Can you learn the answer from this table? Explain.

5.12   **Poverty in the United States, continued.** What is the nature of the association between age and race among people below the poverty level? To answer this question, proceed as follows.

(a)   Use the counts in Table 5-3 to find the percent of the white poverty-level population in each age group. Then do the same for the black poverty-level population.

(b)   Use bar graphs to compare these percents visually. You can make two graphs, one for each race, or you can make one graph like Figure 5-1 that places black and white bars together for each age group.

(c) Based on your percents and graph, describe the association between race and age. For example, are blacks who are below the poverty level more likely than whites to be children?

5.13 **Child restraints in autos.** Do seat belts and other restraints prevent child injuries in automobile accidents? Some evidence is provided by a study of reported crashes of 1967 and later-model cars in North Carolina in 1973–1974. There were 26,971 passengers under the age of 15 in these cars. Here are data on their conditions.[5]

	Restrained	Unrestrained
Injured	197	3,844
Uninjured	1,749	21,181

(a) What percent of these 26,971 young passengers were wearing seat belts or were otherwise restrained?

(b) Compute appropriate percents to show the association between wearing restraints and escaping uninjured.

5.14 **Child restraints in autos, continued.** The study in Exercise 5.13 also looked at where the passengers were seated. Here is a three-way table of passengers by seat location, restraint, and condition.

	Front seat		Back seat	
	Restrained	Not	Restrained	Not
Injured	121	2,125	76	1,719
Uninjured	981	9,679	768	11,502

Unlike Example 3, the third variable here does *not* greatly change the observed association between the first two. Compute appropriate percents to demonstrate this.

5.15 **Salaries of young scientists.** A study by the National Science Foundation[6] found that the median salary of newly graduated female engineers and scientists was only 73% of the median salary for males. When the new graduates were broken down by field, however, the picture changed. Women earned at least 84% as much as men in *every* field of engineering and science. The median salary for women was higher than that of men in many engineering disciplines. How can women do nearly as well as men in every field, yet fall far behind men when we look at all young engineers and scientists?

5.16 **Smoking by students and their parents.** A study of the effect of parents' smoking habits on the smoking habits of students in eight Arizona high schools produced the following counts.[7]

	Student smokes	Student does not smoke
Both parents smoke	400	1380
One parent smokes	416	1823
Neither parent smokes	188	1168

Describe in words the association between the smoking habits of parents and children. Compute appropriate percents and make a bar graph to back up your statements.

5.17 **Hospital death rates.** A news report (*New York Times*, January 12, 1990) said:

The results of a government study on death rates in nearly 6,000 hospitals were challenged today by researchers who said the federal analysis failed to account for variations in the severity of patients' illness when they were hospitalized.

As a result, they said, some hospitals were treated unfairly in the findings, which named hospitals with higher-than-expected death rates.

We will examine a simplified example to show how overlooking severity of illness can indeed create an incorrect impression. Your community is served by two hospitals, Hospital A and Hospital B. Here are data on the survival of patients after surgery in these two hospitals. All patients undergoing surgery in a recent time period are included. "Survived" means that the patient lived at least six weeks following surgery.

Good condition			Poor condition		
	Hospital A	Hospital B		Hospital A	Hospital B
Died	6	8	Died	57	8
Survived	594	592	Survived	1443	192
Total	600	600	Total	1500	200

(a)   Combine the data in this three-way table to form a two-way table of outcome (died or survived) by hospital (A or B). Use your two-way table to find the percent of patients who died at each hospital. Which hospital has the higher death rate?

(b)   Now return to the three-way table. Find the percent of patients in good condition who died at each hospital. Then find the percent of patients in poor condition who died at each hospital. Which hospital has the higher death rate for patients in good condition? For patients in poor condition?

(c)   This is an example of Simpson's paradox: Hospital A has a lower death rate for both classes of patients, but has a higher overall death rate than Hospital B. Explain carefully, from the data given, how this can happen.

5.18   **Airline flight delays.** Here are the numbers of flights on time and delayed for two airlines at five airports in June 1991. Overall on-time percentages for each airline are often reported in the news. We will see that lurking variables can make such reports misleading.[8]

	Alaska Airlines		America West	
	On time	Delayed	On time	Delayed
Los Angeles	497	62	694	117
Phoenix	221	12	4840	415
San Diego	212	20	383	65
San Francisco	503	102	320	129
Seattle	1841	305	201	61

(a)   What percent of all Alaska Airlines flights were delayed? What percent of all America West flights were delayed? These are the numbers usually reported.

(b)   Now find the percent of delayed flights for Alaska Airlines at each of the five airports. Do the same for America West.

(c)   America West does worse at *every one* of the five airports, yet does better overall. That sounds impossible. Explain carefully, referring to the data, how this can happen. (The weather in Phoenix and Seattle accounts for this example of Simpson's paradox.)

▶ 2 SCATTERPLOTS AND CORRELATION

When our data are nominal, we use counts and percents to describe them. Variables measured in an interval/ratio scale, such as height in inches or age in years, contain more detailed information and allow more elaborate statistical descriptions. We display data on a single variable by a stemplot or histogram and summarize them numerically by the five-number summary or by the mean and standard deviation. When we have data on two variables, we add graphs and numerical measures that describe the relationship between the variables.

SCATTERPLOTS

The most effective way to display the relation between two numerical variables is a *scatterplot*. Here is an example of a scatterplot.

> **EXAMPLE 6. Explaining state SAT scores.** Table 5-4 gives the median Scholastic Assessment Test (SAT) mathematics score for high school seniors in each state, as well as the percent of seniors who choose to take the SAT. Let us examine the relationship between these variables.
>
> We expect that states in which only students applying to selective colleges take the SAT will have higher median scores than states in which almost all college-bound seniors take the SAT. That is, we expect that "percent taking" will help explain "median score." Therefore, "percent taking" is the *explanatory variable* and "median score" is the *response variable*. We want to see how median score

TABLE 5-4   Median SAT math score and percent taking SAT, by state

State	SAT math	Percent taking	State	SAT math	Percent taking
AL	514	8	MT	523	20
AK	476	42	NE	546	10
AZ	497	25	NV	487	24
AR	511	6	NH	486	67
CA	484	45	NJ	473	69
CO	513	28	NM	527	12
CT	471	74	NY	470	70
DE	470	58	NC	440	55
FL	466	44	ND	564	6
GA	443	57	OH	499	22
HI	481	52	OK	523	9
ID	502	17	OR	484	49
IL	528	16	PA	463	64
IN	459	54	RI	461	62
IA	577	5	SC	437	54
KS	548	10	SD	555	5
KY	521	10	TN	525	12
LA	517	9	TX	461	42
ME	463	60	UT	539	5
MD	478	59	VT	466	62
MA	473	72	VA	470	58
MI	514	12	WA	486	44
MN	542	14	WV	490	15
MS	519	4	WI	543	11
MO	522	12	WY	519	13

changes when percent taking changes, so we put percent taking (the explanatory variable) on the horizontal axis.

Figure 5-3 is our scatterplot. Each point represents a state. In Alabama, for example, 8% take the SAT, and the median SAT math score is 514. Find 8 on the x (horizontal) axis and 514 on the y (vertical) axis. Alabama appears as the point (8, 514) above 8 and to the right of 514. Alaska appears as the point (42, 476), and so on.

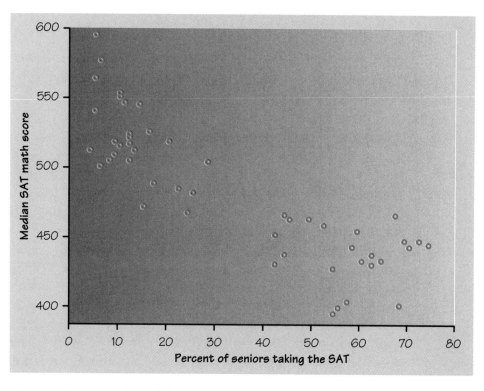

Figure 5-3 Scatterplot of the median SAT mathematics score for the 50 states versus the percent of high school seniors in each state who take the SAT.

Scatterplot

A **scatterplot** shows the relationship between two variables measured on the same individuals when both variables have ordinal or interval/ratio scales. The values of one variable appear on the horizontal axis, and the values of the other variable appear on the vertical axis. Each individual in the data appears as a point in the plot.

Always plot the explanatory variable, if there is one, on the horizontal axis (the x axis) of a scatterplot. As a reminder, we usually call the explanatory variable x and the response variable y. If there is no explanatory-response distinction, either variable can go on the horizontal axis.

INTERPRETING SCATTERPLOTS

To interpret a scatterplot, look first for an overall pattern. This pattern should reveal the *form*, *direction*, and *strength* of the relationship between the two variables.

> **EXAMPLE 7. Explaining state SAT scores.** The *form* of the relationship in Figure 5-3 strikes us at once: there are two distinct clusters of states. In one cluster, more than 40% of high school seniors take the SAT, and the state median score is low. Less than 30% of seniors in states in the other cluster take the SAT, and these states have higher median scores.
>
> What explains the clusters? There are two widely used college entrance exams, the SAT and the American College Testing (ACT) exam. Each state favors one or the other. The left cluster in Figure 5-3 contains the ACT states, and the SAT states make up the right cluster. In ACT states, most students who take the SAT are applying to a selective college that requires SAT scores. This group of students has a higher median score than the much larger group of students who take the SAT in SAT states.
>
> The *direction* of the overall pattern in Figure 5-3 is also clear. States where a higher percent of seniors take the SAT tend to have lower median scores. We say that percent taking and median score are *negatively associated.*

Positive association, negative association

Two variables are **positively associated** when larger values of one tend to go with larger values of the other.

Two variables are **negatively associated** when larger values of one tend to go with smaller values of the other.

Scatterplots not only display the pattern of the relationship between variables but also highlight individual observations that deviate from the overall pattern. These observations appear as *outliers*, data points that stand apart from the rest.

EXAMPLE 8. In Figure 5-3, the three states with the lowest SAT scores lie below the right-hand cluster. These states are Georgia, North Carolina, and South Carolina. We might consider them outliers, although they lie only slightly outside the overall pattern. As always, identifying outliers is in part a matter of judgment.

The *strength* of a relationship in a scatterplot is determined by how closely the points follow a clear form. Widely scattered points, as in Figure 5-3, show a weak relationship. It is easiest to describe strength when the form is simple. The simplest form for a scatterplot is a straight line.

EXAMPLE 9. **Classifying fossils.** *Archaeopteryx* is an extinct beast having feathers like a bird but teeth and a long bony tail like a reptile. Only six fossil specimens are known. Because these specimens differ greatly in size, some scientists think they are different species rather than individuals from the same species. We will examine data on the lengths in centimeters of the femur (a leg bone) and the humerus (a bone in the upper arm) for the five specimens that preserve both bones. Here are the data.[9]

| Femur | 38 | 56 | 59 | 64 | 74 |
| Humerus | 41 | 63 | 70 | 72 | 84 |

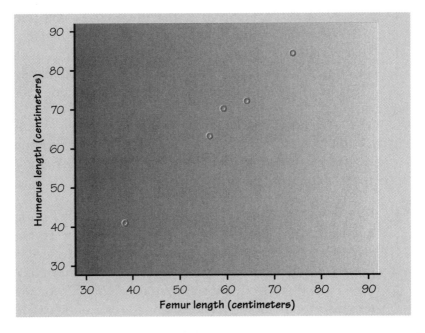

Figure 5-4 Scatterplot of the lengths of two bones in the five surviving fossil specimens of the extinct beast *Archaeopteryx*.

Because there is no explanatory/response distinction, we can put either measurement on the *x* axis of a scatterplot. The plot appears in Figure 5-4.

The plot shows a strong positive straight-line association. The association is strong because the points lie close to a line. It is positive because as the length of one bone increases, so does the length of the other bone. These data suggest that all five specimens belong to the same species and differ in size because some are younger than others. We expect that a different species would have a different relationship between the lengths of the two bones, so that it would appear as an outlier.

CORRELATION

A scatterplot displays the direction, form, and strength of the relationship between two variables. Straight-line relations are particularly important because a straight line is a simple pattern that is quite common. A straight-line relation is strong if the points lie close to a straight line, and weak if they are widely scattered about a line. A numerical measure can help us describe just how strong a straight-line association is.

Correlation

The **correlation** describes the direction and strength of a straight-line relationship between two variables. Correlation is usually written as r.

Correlation is a bit of a mess to calculate. In practice, we use a calculator or computer software that will do the arithmetic given only the data. Nonetheless, the recipe for r can help us understand how correlation works, so here it is. Suppose that we have data on variables x and y for n individuals. Think, for example, of measuring height and weight for n people. Then x_1 and y_1 are your height and your weight, x_2 and y_2 are my height and my weight, and so on. For the ith individual, height x_i goes with weight y_i.

Calculating the correlation

1. Find the mean $\overline{x}$ and standard deviation s_x of the values $x_1, x_2, \ldots, x_n$ of the first variable. Then find the standard score for each x-observation,

$$\frac{x_i - \overline{x}}{s_x}$$

2. Find the mean $\overline{y}$ and standard deviation s_y of the values $y_1, y_2, \ldots, y_n$ of the second variable. Then find the standard score for each y-observation,

$$\frac{y_i - \overline{y}}{s_y}$$

3. The correlation r is an average of the products of the standard scores for the n individuals,

$$r = \frac{\left(\frac{x_1-\overline{x}}{s_x}\right)\left(\frac{y_1-\overline{y}}{s_y}\right) + \left(\frac{x_2-\overline{x}}{s_x}\right)\left(\frac{y_2-\overline{y}}{s_y}\right) + \cdots + \left(\frac{x_n-\overline{x}}{s_x}\right)\left(\frac{y_n-\overline{y}}{s_y}\right)}{n - 1}$$

EXAMPLE 10. *Archaeopteryx* again. We will find the correlation between the lengths of two bones in the five surviving fossils of the extinct beast *Archaeopteryx*. We have $n = 5$ observations:

Individual	i	1	2	3	4	5
Femur length	x	38	56	59	64	74
Humerus length	y	41	63	70	72	84

First find the means and standard deviations. Use a calculator or follow the pattern of Example 20 in Chapter 4 (page 252).

$$\text{Femur lengths } x: \quad \overline{x} = 58.2 \text{ cm} \quad s_x = 13.20 \text{ cm}$$
$$\text{Humerus lengths } y: \quad \overline{y} = 66.0 \text{ cm} \quad s_y = 15.89 \text{ cm}$$

Then find the standard scores and their products.

i	x	y	$\dfrac{x - 58.2}{13.20}$	$\dfrac{y - 66.0}{15.89}$	$\left(\dfrac{x - 58.2}{13.20}\right)\left(\dfrac{y - 66.0}{15.89}\right)$
1	38	41	-1.530	-1.573	2.407
2	56	63	-0.167	-0.189	0.032
3	59	70	0.061	0.252	0.015
4	64	72	0.439	0.378	0.166
5	74	84	1.197	1.133	1.356
					3.976

The sum of the products of the standard scores is 3.976. The correlation is therefore

$$r = \frac{3.976}{4} = 0.994$$

Don't forget to divide by one fewer than the number of observations, just as in calculating a standard deviation.

UNDERSTANDING CORRELATION

More important than calculating r (a task for a machine) is understanding how correlation measures association. Here are the facts:

▶ *Positive r indicates positive association between the variables, and negative r indicates negative association.* The scatterplot in Figure 5-4 shows strong positive association, so the standard scores in Example 10 are either positive for both x and y or negative for both. The products are all positive, giving a positive r.

▶ *The correlation r always falls between −1 and 1.* Values of r near 0 indicate a very weak straight-line relationship. The strength of the relationship increases as r moves away from 0 toward either −1 or 1. Values of r close to −1 or 1 indicate that the points lie close to a straight line. The extreme values $r = -1$ and $r = 1$ occur only when the points in a scatterplot lie exactly along a straight line.

The result $r = 0.994$ in Example 10 reflects the strong positive straight-line pattern in Figure 5-4. The scatterplots in Figure 5-5 illustrate how r measures both the direction and strength of a straight-line relationship. Study them carefully. Note that the sign of r matches the direction of the slope in each plot, and that as r approaches −1 or 1 the pattern of the plot comes closer to a straight line.

▶ Because r uses standard scores, *the correlation between x and y does not change when we change the units of measurement of x, y, or both.* The bone lengths in Example 10 are given in centimeters. Changing to inches would not change r. The correlation r itself has no unit of measurement. It is just a number between −1 and 1.

▶ *Correlation ignores the distinction between explanatory and response variables.* If we reverse our choice of which variable to call x and which to call y, the correlation does not change.

▶ *Correlation measures the strength of only straight-line association between two variables.* Correlation does not describe curved relationships between variables, no matter how strong they are.

▶ Like the mean and standard deviation, *the correlation is strongly affected by a few outlying observations.* Use r with caution when outliers appear in the scatterplot.

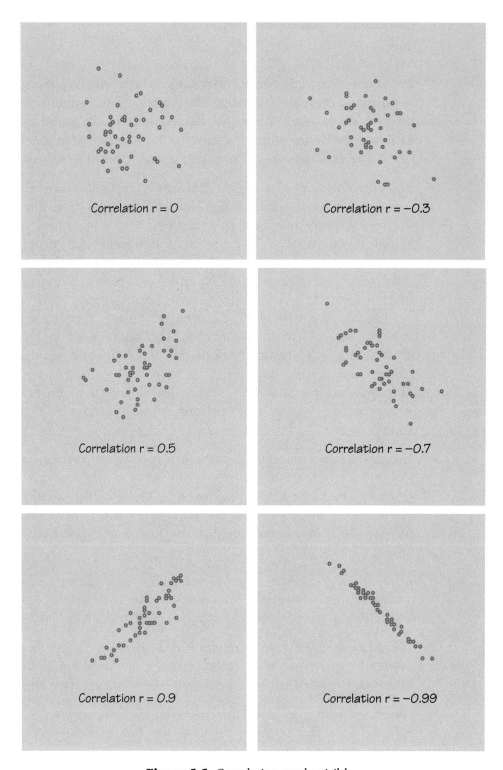

Figure 5-5 Correlation made visible.

There are many kinds of association, and many ways to measure them. Although correlation is very common, do remember its limitations. Correlation makes no sense for nominal variables—we can speak of the association between the sex of voters and the political party they prefer, but not of the correlation between these variables. Even for variables such as the length of bones that have an interval/ratio scale, correlation measures only straight-line association.

Remember also that correlation is not a complete description of two-variable data, even when there is a straight-line relationship between the variables. You should give the means and standard deviations of both x and y along with the correlation. Because the formula for correlation uses the means and standard deviations, these measures are the proper choice to accompany a correlation.

"He says we've ruined his positive association between height and weight."

SECTION 2 EXERCISES

5.19   **Mothers and fathers.** Figure 5-6 is a scatterplot of the heights of 53 pairs of mothers and fathers in a sample of parents.[10] Answer the following questions from this graph:

(a)   What is the smallest height of any mother in the sample? How many women have that particular height? What are the heights of the husbands of these women?

(b)   What is the greatest height of any father in the sample? How many men have that height, and what are the heights of their wives?

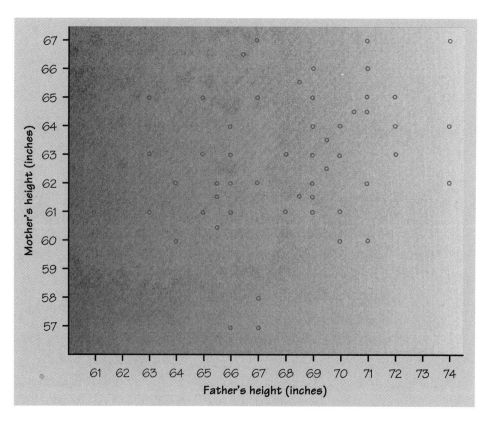

Figure 5-6 Heights of 53 pairs of parents.

(c)   Does the scatterplot show a positive or negative association between the heights of mothers and fathers? How strong is the association?

5.20   **Pollution from vehicle exhausts.** Auto manufacturers are required to test their vehicles for the amount of each of several pollutants in the exhaust. The amount of a pollutant varies even among identical vehicles, so that several vehicles must be tested. Figure 5-7 is a scatterplot of the amounts of two pollutants, carbon monoxide and oxides of nitrogen, for 46 vehicles of the same model. Both variables are measured in grams of the pollutant per mile driven.[11]

(a)   Describe the nature of the relationship. Is the association positive or negative? Is the relation close to a straight line or clearly curved? Are there any outliers?

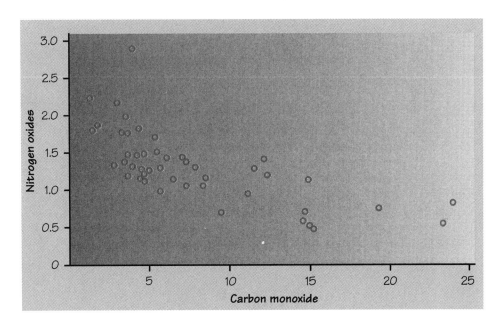

Figure 5-7 Amounts of two pollutants in the exhaust of 46 vehicles of the same model.

(b)   A car magazine says, "When a car's engine is properly tuned, it emits few pollutants. If the engine is out of tune, it emits more of all important pollutants. You can find out how badly a vehicle is polluting the air by measuring any one pollutant. If that value is acceptable, the other emissions will also be OK." Do the data in Figure 5-7 support this claim?

5.21   **Is wine good for your heart?** There is some evidence that drinking moderate amounts of wine helps prevent heart attacks. Table 5-5 gives data on yearly wine consumption (liters of alcohol from drinking wine, per person) and yearly deaths from heart disease (deaths per 100,000 people) in 19 developed nations.[12]

(a)   Make a scatterplot that shows how wine consumption affects heart disease.

(b)   Describe the form of the relationship. Is there a straight-line pattern? How strong is the relationship? (See also Exercise 5.37.)

TABLE 5-5   Wine consumption and heart attacks

Country	Alcohol from wine	Heart disease deaths	Country	Alcohol from wine	Heart disease deaths
Australia	2.5	211	Netherlands	1.8	167
Austria	3.9	167	New Zealand	1.9	266
Belgium	2.9	131	Norway	0.8	227
Canada	2.4	191	Spain	6.5	86
Denmark	2.9	220	Sweden	1.6	207
Finland	0.8	297	Switzerland	5.8	115
France	9.1	71	United Kingdom	1.3	285
Iceland	0.8	211	United States	1.2	199
Ireland	0.7	300	West Germany	2.7	172
Italy	7.9	107			

(c) Is the direction of the association positive or negative? Explain in simple language what this says about wine and heart disease. Does the relationship you observe give good evidence that drinking wine *causes* a reduction in heart disease deaths? Why?

5.22 Here are the golf scores of 12 members of a college women's golf team in two rounds of tournament play. (A golf score is the number of strokes required to complete the course, so that low scores are better.)

Player	1	2	3	4	5	6	7	8	9	10	11	12
Round 1	89	90	87	95	86	81	102	105	83	88	91	79
Round 2	94	85	89	89	81	76	107	89	87	91	88	80

(a) Make a scatterplot of the data, taking the first round score as the explanatory variable. (Because both variables have the same scale, you should use the same scale on both axes of your plot, as in Figure 5-4.)

(b) Is there an association between the two scores? If so, is it positive or negative? Explain why you would expect scores in two rounds of a tournament to have an association like that you observed.

(c) The plot shows one outlier. Circle it. The outlier may occur because a good golfer had an unusually bad round or because a weaker golfer had an unusually good round. Can you tell from the data given whether the outlier is from a good player or a poor player? Explain your answer.

5.23 **How many corn plants are too many?** How much corn per acre should a farmer plant to obtain the highest yield? To find the best planting rate, do an experiment: plant at different rates on several plots of ground and measure the harvest. (Be sure to treat all the plots the same except for the planting rate.) Here are data from such an experiment.[13]

Plants per acre	Yield (bushels per acre)			
12,000	150.1	113.0	118.4	142.6
16,000	166.9	120.7	135.2	149.8
20,000	165.3	130.1	139.6	149.9
24,000	134.7	138.4	156.1	
28,000	119.0	150.5		

(a)   Is yield or planting rate the explanatory variable?

(b)   Make a scatterplot of yield and planting rate.

(c)   Describe the overall pattern of the relationship. Is it a straight line? Is there a positive or negative association, or neither? Explain why increasing the number of plants per acre of ground has the effect that your graph shows.

(d)   Find the mean yield for each of the five planting rates. Plot each mean yield against its planting rate on your scatterplot and connect these five points with lines. This combination of numerical description and graphing makes the relationship clearer. What planting rate would you recommend to a farmer whose conditions were similar to those in the experiment?

5.24   For each of the following sets of data, draw a scatterplot. Then calculate the mean and standard deviation of each variable x and y separately. Finally, follow the table arrangement of Example 10 to obtain the correlation r. In each case, explain briefly how the value of r reflects the pattern of the scatterplot.

(a) x 4   4   −4   −4
    y −4   4   4   −4

(b) x 4   3   0   −3   −4
    y −4 −2   0   2   4

(c) x 4   2   −2   −4
    y 4   −2   2   −4

5.25   **Strong association, but no correlation.** The gas mileage of an automobile first increases and then decreases as the speed increases. Suppose that this relationship is very regular, as shown by the following data on speed (miles per hour) and mileage (miles per gallon):

$$\begin{array}{lccccc} \text{Speed} & 20 & 30 & 40 & 50 & 60 \\ \text{MPG} & 24 & 28 & 30 & 28 & 24 \end{array}$$

Make a scatterplot of mileage versus speed. Show that the correlation between speed and mileage is $r = 0$. Explain why the correlation is 0 even though there is a strong relationship between speed and mileage.

5.26 Make a scatterplot of the following data:

$$\begin{array}{lcccccc} x & 1 & 2 & 3 & 4 & 10 & 10 \\ y & 1 & 3 & 3 & 5 & 1 & 11 \end{array}$$

Show that the correlation is about 0.5. What feature of the data is responsible for reducing the correlation to this value despite a strong straight-line association between x and y in most of the observations?

5.27 Figure 5-8 displays five scatterplots. Match each to the r below that best describes it. (Some r's will be left over.)

$$\begin{array}{ll} r = -0.9 & r = 0.3 \\ r = -0.7 \qquad r = 0 & r = 0.7 \\ r = -0.3 & r = 0.9 \end{array}$$

5.28 **What are the units?** Your data consist of observations on the age of several subjects (measured in years) and the reaction times of these subjects (measured in seconds). In what units are each of the following descriptive statistics measured?

(a) The mean age of the subjects.

(b) The variance of the subjects' reaction times.

(c) The standard deviation of the subjects' reaction times.

(d) The correlation coefficient between age and reaction time.

(e) The median age of the subjects.

5.29 If the heights in Exercise 5.19 were given in centimeters rather than inches, how would the correlation between mother's height and father's height change? If alcohol consumption in Exercise 5.21 were given in quarts per person rather than liters per person, how would its correlation with death rates change?

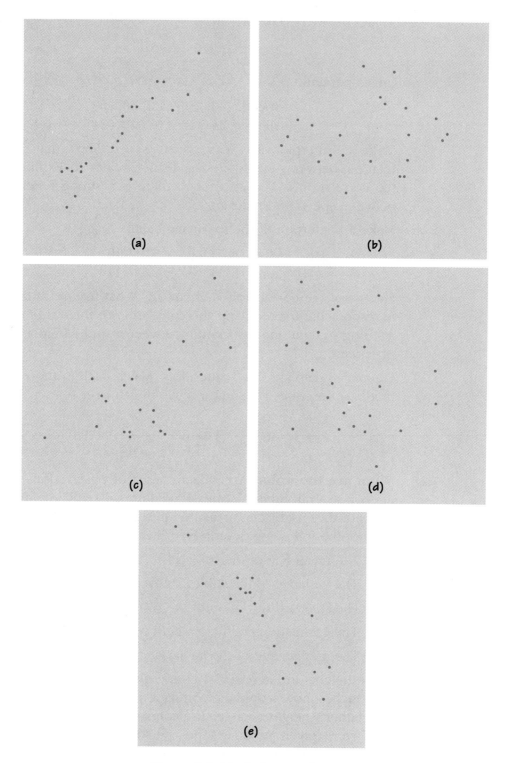

Figure 5-8 Match the correlations.

5.30 A psychologist speaking to a meeting of the American Association of University Professors says, "The evidence suggests that
there is nearly correlation zero between the teaching ability of
a faculty member and his or her research productivity." The
student newspaper reports this as: "Professor McDaniel said that
good teachers tend to be poor researchers and good researchers
tend to be poor teachers."

Explain what (if anything) is wrong with the newspaper's
report. If the report is not accurate, write your own plain-
language account of what the speaker meant.

5.31 Measurements in large samples show that the correlation

(a) between the heights of fathers and the heights of their adult
sons is about _____.

(b) between the heights of husbands and the heights of their
wives is about _____.

(c) between the heights of women at age 4 and their heights
at age 18 is about _____.

The answers (in scrambled order) are

$$r = 0.25 \qquad r = 0.5 \qquad r = 0.8$$

Match the answers to the statements and explain your choice.

5.32 If women always married men who were 2 years older than
themselves, what would be the correlation between the ages
of husband and wife? (*Hint:* Draw a scatterplot for the ages of
husband and wife when the wife is 20, 30, 40, and 50 years old.)

5.33 For each of the following pairs of variables, would you expect a
substantial negative correlation, a substantial positive correlation,
or a small correlation?

(a) The age of a second-hand car and its price.

(b) The weight of a new car and its gas mileage in miles per
gallon.

(c) The height and the weight of a person.

(d) The height of a person and the height of his or her mother.

(e) The height and the IQ of a person.

5.34   **Sloppy writing about correlation.** Each of the following statements contains a blunder. Explain in each case what is wrong.

(a)   There is a high correlation between the sex and income of American workers.

(b)   Student ratings of professors' teaching and colleagues' ratings of their research have correlation $r = 1.21$. This shows that better teachers also tend to be better researchers.

(c)   The correlation between pounds of nitrogen fertilizer applied to the field and the bushels per acre of corn harvested was $r = 0.63$ bushels. This shows that applying more fertilizer increases yields.

5.35   **Statistics for investing.** Financial experts use statistical measures to describe the performance of investments such as mutual funds. In the past, funds feared that their investors would not understand statistical descriptions, but mounting pressure to give better information is moving standard deviations and correlations into the public eye.

(a)   The T. Rowe Price mutual fund group reports the standard deviation of yearly percent returns for its funds. For the five years ending March 31, 1995, Equity Income Fund had standard deviation 9.94%, and Science & Technology Fund had standard deviation 23.77%. Explain to someone who knows no statistics how these standard deviations help investors compare the two funds.

(b)   Some mutual funds act much like the stock market as a whole, as measured by a market index such as the Standard & Poor's 500 stock index. Others are very different from the overall market. We can use correlation to describe the association. Monthly returns from Fidelity Magellan Fund, the largest mutual fund, have correlation $r = 0.83$ with the S&P 500 index. Fidelity Capital Appreciation Fund has correlation $r = 0.57$ with the S&P 500 index. (These correlations are for 36 months ending early in 1995.[14]) Explain to someone who knows no statistics how these correlations help investors compare the two funds.

The following exercises are best done using a calculator that will give the correlation from keyed-in data.

5.36 Find the correlation between golfers' scores in two rounds of a tournament (Exercise 5.22), both with and without the outlier. Explain from the scatterplot why dropping the outlier changes r in the way you observe.

5.37 Complete your description of the association between wine consumption and deaths from heart disease (Exercise 5.21) by calculating the correlation r.

5.38 **Tree seeds.** Ecologists collect data to study nature's patterns. Table 5-6 gives data on the mean number of seeds produced in a year by several common tree species and the mean weight of the seeds produced.[15] (Some species appear twice because their seeds were counted in two locations.) We might expect that trees with heavy seeds produce fewer of them, but what is the form of the relationship?

(a) Make a scatterplot showing how the weight of tree seeds helps explain how many seeds the tree produces. Describe the direction, form, and strength of the relationship.

TABLE 5-6   Count and weight of seeds produced by common tree species

Tree species	Seed count	Seed weight (mg)	Tree species	Seed count	Seed weight (mg)
Paper birch	27,239	0.6	American beech	463	247
Yellow birch	12,158	1.6			
White spruce	7,202	2.0	American beech	1,892	247
Engelmann spruce	3,671	3.3			
			Black oak	93	1,851
Red spruce	5,051	3.4	Scarlet oak	525	1,930
Tulip tree	13,509	9.1	Red oak	411	2,475
Ponderosa pine	2,667	37.7	Red oak	253	2,475
White fir	5,196	40.0	Pignut hickory	40	3,423
Sugar maple	1,751	48.0	White oak	184	3,669
Sugar pine	1,159	216.0	Chestnut oak	107	4,535

(b) When dealing with sizes and counts, the logarithms of the original data are often the "natural" variables. Use your calculator to obtain the logarithms of both the seed weights and seed counts in Table 5-6. Make a new scatterplot using these new variables. Now what are the direction, form, and strength of the relationship?

(c) Use a statistical calculator (or software) to find the correlation that accompanies your scatterplots in both (a) and (b). What does r tell you in each case?

▶ 3 ISSUES: THE QUESTION OF CAUSATION

There is a strong association between cigarette smoking and death rate from lung cancer. A study of British doctors found that smokers had 20 times the risk of nonsmokers, and a large study of American men ages 40 to 79 found 11 times higher death rates among smokers (see Figure 5-9).[16] Is this association good evidence that cigarette smoking causes lung cancer?

We are asking whether an observed association is due to changes in one variable (lung cancer) being *caused* by changes in another variable (smoking). Some associations are due to cause and effect. Others are not. Consider the following examples. In each case, there is a clearly observed association between an explanatory variable x and a response variable y. Moreover, the association is positive whenever the direction makes sense.

> **EXAMPLE 11. Examples of observed association.**
>
> **1.** $x =$ father's adult height, $y =$ son's adult height.
>
> **2.** $x =$ amount of the artificial sweetener saccharin in a rat's diet, $y =$ count of tumors in the rat's bladder.
>
> **3.** $x =$ median pay of public school teachers, $y =$ dollar value of yearly sales of alcoholic beverages.
>
> **4.** $x =$ a student's SAT score as a high school senior, $y =$ the student's grade index as a first-year college student.

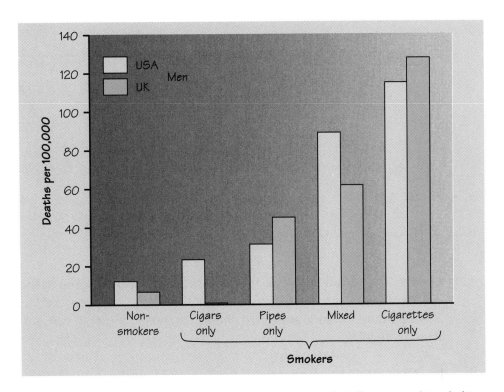

Figure 5-9 Deaths from lung cancer among men with different smoking habits. The bar graph displays data from two large studies, one of American men and one of British doctors.

5. $x =$ the anesthetic used in surgery, $y =$ whether the patient survives the surgery.

6. $x =$ the number of years of education a worker has, $y =$ the worker's income.

How shall we account for these examples of association? Let's consider each in turn.

EXPLAINING ASSOCIATION: CAUSATION

Figure 5-10 shows in outline form how a variety of underlying links between variables can explain association. The dashed line represents an

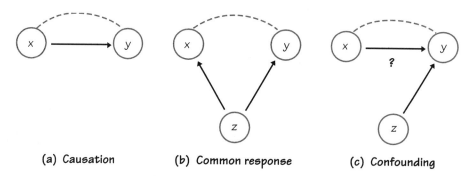

Figure 5-10 Some explanations for an observed association. The broken lines show an association. The arrows show a cause-and-effect link. We observe the variables x and y, but z is a lurking variable.

observed association between the variables x and y. Some associations are explained by a direct cause-and-effect link between the variables. The first diagram in Figure 5-10 shows "x causes y" by an arrow running from x to y.

> **EXAMPLE 12. Causation.** Items 1 and 2 in Example 11 are examples of direct causation. Thinking about these examples, however, shows that "causation" is not a simple idea.
>
> **1.** Height is in part determined by heredity. Sons inherit half their genes from their fathers. There is therefore a direct causal link between the height of fathers and the height of sons. Of course, the causal link is not perfect (the correlation r is not 1). Other factors such as nutrition influence the height of children. Even when there is a direct connection (such as heredity), "x causes y" is usually an incomplete explanation of the behavior of y.
>
> **2.** The best evidence for causation comes from randomized comparative experiments. Experiments show conclusively that large amounts of saccharin in the diet cause bladder tumors in rats. Should we avoid saccharin as a replacement for sugar in food? Rats are not people. Although we can't experiment with people, comparative observational studies of people who consume different amounts of saccharin show no association between saccharin and bladder tumors. Even well-established causal relations may not generalize to other settings.

EXPLAINING ASSOCIATION: COMMON RESPONSE

"Beware the lurking variable" is good advice when thinking about an association between two variables. The second diagram in Figure 5-10 illustrates *common response*. The observed association between the variables x and y is explained by a lurking variable z. Both x and y change in response to changes in z. This common response creates an association even though there may be no direct causal link between x and y.

> EXAMPLE 13. **Common response.** The third and fourth items in Example 11 illustrate how common response can create an association.
>
> **3.** If we look at median teachers' salaries (in dollars) and sales of alcoholic beverages (in dollars) each year, we find a strong positive association. Can we conclude that teachers are spending their extra pay on booze, thus driving up sales? Not at all. Both teachers' pay and alcohol sales have increased steadily over time due to inflation and prosperity. These same lurking variables create an association between almost any two variables measured in dollars if we record them each year for many years. Correlations like that between teachers' pay and alcohol sales are sometimes called "nonsense correlations." The correlation is real. What is nonsense is the "x causes y" interpretation.
>
> **4.** Students who are smart and who have learned a lot tend to have both high SAT scores and high college grades. The positive correlation is explained by this common response to students' ability and knowledge. Although SAT scores x don't cause college grades y, we can use x to predict y (only modestly well, because the correlation is not high). So an association can be useful even when it does not reflect a cause-and-effect link.

EXPLAINING ASSOCIATION: CONFOUNDING

When many variables interact with each other, *confounding* of several variables can prevent us from drawing conclusions about causation. The third diagram in Figure 5-10 illustrates confounding. Both the explanatory variable x and the lurking variable z may influence the response variable y. Because x is confounded with z, we cannot distinguish the influence of x from the influence of z. We cannot say how strong the direct effect of x on y is. In fact, it can be hard to say if x influences y at all.

> **EXAMPLE 14. Confounding.** The last two associations in Example 11 (items 5 and 6) are explained in part by confounding.
>
> **5.** Surgeons choose different anesthetics based in part on the seriousness of the operation and the condition of the patient. An anesthetic that is used mainly for difficult operations on gravely ill patients will have a higher death rate than one used mainly for minor surgery. This confounding of anesthetic x with the type of surgery z makes it hard to see the direct effect of x on patient deaths y.
>
> **6.** It is likely that more education does help cause higher income—many highly paid professions require advanced education. However, confounding is also present. People who are smart and come from prosperous homes are more likely to get many years of education than people who are less smart and poorer. Of course, people who start out smart and rich are more likely to have high earnings even without much education. It is hard to see how much of the higher income of well-educated people is actually caused by their education.

ASSOCIATION IS NOT CAUSATION

Many observed associations are at least partly explained by lurking variables. Both common response and confounding involve the influence of a lurking variable (or variables) z on the response variable y. The distinction between these two types of relationship is less important than the common element, the influence of lurking variables. In Chapter 2, we reached one firm conclusion about establishing causation: *to get firm evidence for causation, use randomized comparative experiments to defeat the influence of lurking variables.* Here is a second conclusion, less positive but equally important:

Association does not imply causation

In the absence of an experiment, even a very strong association between an explanatory variable x and a response variable y is not by itself good evidence that changes in x actually cause changes in y.

ESTABLISHING CAUSATION

Many of the sharpest disputes in which statistics plays a role involve questions of causation that cannot be settled by experiment. Does taking the drug Bendectin cause birth defects? Does living near power lines cause cancer? Has increased free trade helped to create the growing gap between the incomes of more educated and less educated American workers? All of these questions have become public issues. All concern associations among variables. And all have this in common: they try to pinpoint cause and effect in a setting involving complex relations among many interacting variables. Common response and confounding, along with the number of potential lurking variables, make observed associations misleading. Experiments are not possible for moral or practical reasons. We can't randomly assign people to live near power lines or compare two identical nations with and without free trade agreements.

> EXAMPLE 15. **Prescription drugs and birth defects.** Are children born to women who take a prescription drug during pregnancy more likely to have birth defects? This question has been asked about many drugs. It is not easy to answer. Consider Bendectin, once commonly prescribed for nausea during pregnancy, then accused of causing birth defects and studied intensively in and out of court.[17]
>
> In the absence of an experiment, we must rely on comparative observational studies. That is, we compare children whose mothers took Bendectin with similar children whose mothers did not. Who took Bendectin? Mothers of children with birth defects are more likely to remember taking a drug during pregnancy than those with

normal children. This recall bias will overstate the rate of birth defects in the Bendectin group. Perhaps we should instead look at prescription records. One study considered a child exposed to Bendectin early in pregnancy if a prescription was filled between 365 and 250 days before birth. There is no way of knowing whether the mothers actually took the drug, or how regularly they took it.

Constructing a group of "similar children" whose mothers did not take Bendectin is even more difficult than deciding who took the drug. What potential lurking variables should we consider? For example, users of Bendectin also took more other drugs than nonusers. Perhaps the alleged effects of Bendectin are due to confounding with some other drug, or to a combination of drugs. Or perhaps birth defects and use of Bendectin and other drugs are common responses to some other medical, economic, or psychological condition. It is hard to say what makes children "similar" to the Bendectin children.

Any comparative observational study must choose some specific rule for forming the two groups. Any specific rule can be criticized. In the case of Bendectin, more than 20 studies mostly found that the drug appeared safe. Most of the lawsuits charging that Bendectin caused birth defects failed in court. The legal expense nonetheless forced the manufacturer to withdraw the drug from the market. Bendectin probably carries little or no risk, but all we can say for sure is "not enough evidence."

EXAMPLE 16. Smoking and lung cancer. Despite the difficulties, it is sometimes possible to build a very strong case for causation in the absence of experiments. The evidence that smoking causes lung cancer is about as strong as nonexperimental evidence can be.

Doctors had long observed that most lung cancer patients were smokers. Comparative observational studies of smokers and "similar" nonsmokers, like those summarized in Figure 5-9, showed a very strong association between smoking and death from lung cancer. Could the association be due to common response? Might there be, for example, a genetic factor that predisposes people both to nicotine addiction and to lung cancer? Smoking and lung cancer would then be positively associated even if smoking had no direct effect on the lungs. Or perhaps confounding is to blame. It might be that smokers live unhealthy lives in other ways (diet, alcohol, lack of exercise) and that some other habit confounded with smoking is a cause of lung cancer. How were these objections overcome?

Let's answer this question in general terms: What are the criteria for establishing causation when we cannot do an experiment?

▶ *The association is strong.* The association between smoking and lung cancer is very strong. That between Bendectin and birth defects is weak.

▶ *The association is consistent.* Many studies of different kinds of people in many countries link smoking to lung cancer. That reduces the chance that a lurking variable specific to one group or one study explains the association.

▶ *Higher doses are associated with stronger responses.* People who smoke more cigarettes per day or who smoke over a longer period get lung cancer more often. People who stop smoking reduce their risk.

▶ *The alleged cause precedes the effect in time.* Lung cancer develops after years of smoking. The number of men dying of lung cancer rose as smoking became more common, with a lag of about 30 years. Lung cancer kills more men than any other form of cancer. Lung cancer was rare among women until women began to smoke. Lung cancer in women rose along with smoking, again with a lag of about 30 years, and has now passed breast cancer as the leading cause of cancer death among women.

▶ *The alleged cause is plausible.* Experiments with animals show that tars from cigarette smoke do cause cancer.

Medical authorities do not hesitate to say that smoking causes lung cancer. The U.S. Surgeon General states that cigarette smoking is "the largest avoidable cause of death and disability in the United States."[18] The evidence for causation is overwhelming—but it is not as strong as the evidence provided by randomized comparative experiments.

IS IT A COINCIDENCE?

We have seen the importance of searching for other variables that may be lurking in the background before drawing conclusions from an observed association. Here is another danger: casual observation can mislead us by suggesting a cause-and-effect relationship when only the play of chance is present. When we observe an unusual outcome, we tend to seek a cause just because the event was unusual. Yet it may be mere coincidence. After all, if we look around us long enough, we will see something unusual. It is indeed very unlikely that *this particular* unusual event would occur, but it is certain that *some* unusual event would eventually occur, simply by chance.

> **EXAMPLE 17. Winning the lottery twice.** In 1986, Evelyn Marie Adams won the New Jersey state lottery for the second time, adding $1.5 million to her previous $3.9 million jackpot. The *New York Times* (February 14, 1986) claimed that the odds of one person winning the big prize twice were about 1 in 17 trillion. Nonsense, said two Purdue statistics professors in a letter that appeared in the *Times* two weeks later. The chance that Evelyn Marie Adams would win twice in her lifetime is indeed tiny, but it is almost certain that *someone* among the millions of regular lottery players in the United States would win two jackpot prizes. The statisticians estimated even odds of another double winner within seven years. Sure enough, Robert Humphries won his second Pennsylvania lottery jackpot ($6.8 million total) in May 1988.

Example 17 makes clear the distinction between the chance of *some* unusual event and the chance of *this particular* unusual event in a setting where the chances can actually be calculated. The same distinction arises in more important and more complex settings, as the next example illustrates.

> **EXAMPLE 18. Cancer clusters.** In 1984, residents of a neighborhood in Randolph, Massachusetts, counted 67 cancer cases in their 250 residences. This cluster of cancer cases seemed unusual, and

the residents expressed concern that runoff from a nearby chemical plant was contaminating their water supply and causing cancer. In 1979, two of the eight town wells serving Woburn, Massachusetts, were found to be contaminated with organic chemicals. Alarmed citizens began counting cancer cases. Between 1964 and 1983, 20 cases of childhood leukemia were reported in Woburn. This is an unusual number of cases of this rather rare disease. The residents believed that the well water had caused the leukemia, and proceeded to sue two companies held responsible for the contamination.

Cancer is a common disease, accounting for more than 23% of all deaths in the United States. That cancer cases sometimes occur in clusters in the same neighborhood is not surprising; there are bound to be clusters *somewhere* simply by coincidence. But when a cancer cluster occurs in *our* neighborhood, we tend to suspect the worst and look for someone to blame.

Both of the Massachusetts cancer clusters were investigated by statisticians from the Harvard School of Public Health. The investigators tried to obtain complete data on everyone who had lived in the neighborhoods in the periods in question and to estimate their exposure to the suspect drinking water. They also tried to obtain data on other factors that might explain cancer, such as smoking and occupational exposure to toxic substances. Such data may establish a link between exposure to the water and cancer that is better evidence than the cluster alone. The verdict: chance is the likely explanation of the Randolph cluster, but there is evidence of an association between drinking water from the two Woburn wells and developing childhood leukemia.[19]

Analysis of cancer clusters is another example of the difficulties of trying to study cause and effect when experiments are not possible. My present point, however, is that many (not all) clusters are just chance happenings, like winning the lottery. Chance may explain other misfortunes as well. About 3% of all children are born with major birth defects. When some of the many children whose mothers took Bendectin are born with birth defects, you should at least consider coincidence as a possible explanation.

SECTION 3 EXERCISES

5.39   A study of grade-school children ages 6 to 11 years found a high positive correlation between reading ability y and shoe size x. Explain why common response to a lurking variable z accounts for this correlation. Present your suggestion in a diagram like one of those in Figure 5-10.

5.40   There is a negative correlation between the number of flu cases y reported each week through the year and the amount of ice cream x sold that week. It is unlikely that ice cream prevents flu. What is a more plausible explanation for this correlation? Draw a diagram like one of those in Figure 5-10 to illustrate the relationships among the variables.

5.41   **Hospital size and length of stay.** A study shows that there is a positive correlation between the size of a hospital (measured by its number of beds x) and the median number of days y that patients remain in the hospital. Does this mean that you can shorten a hospital stay by choosing a small hospital? Use a diagram like one of those in Figure 5-10 to explain the association.

5.42   **Miscarriages among workers.** A study showed that women who work in the production of computer chips have abnormally high numbers of miscarriages. The union claimed that exposure to chemicals used in production caused the miscarriages. Another possible explanation is that these workers spend most of their work time standing up. Illustrate these relationships in a diagram like one of those in Figure 5-10.

5.43   **Do artificial sweeteners cause weight gain?** People who use artificial sweeteners in place of sugar tend to be heavier than people who use sugar. Does this mean that artificial sweeteners cause weight gain? Give a more plausible explanation for this association.

5.44   **Education and income.** A study of the salaries paid to economists found a *negative* correlation between years of education and income. In particular, economists with only a bachelor's degree earned more on the average than economists with a doctorate.

Surely more education does not cause economists to earn less. In fact, there is a strong positive correlation between years of education and income for economists employed by business firms. There is also a strong positive correlation between years of education and income for economists employed by colleges and universities. These facts show that confounding explains the original negative correlation. Explain why.

5.45 **Is a little alcohol good for you?** A survey of 7000 California men found little correlation between alcohol consumption and chance of dying during the 5 1/2 years of the study. In fact, men who did not drink at all during these years had a slightly higher death rate than did light drinkers. This lack of correlation was somewhat surprising. Explain how common response might account for the higher death rate among men who did not drink at all over a short period.

5.46 **Beer drinking and cancer.** A study using data from 41 states found a positive correlation between per capita beer consumption and death rates from cancer of the large intestine and rectum. The states with the highest rectal cancer death rates were Rhode Island and New York. The beer consumption in those states was 80 quarts per capita. South Carolina, Alabama, and Arkansas drank only 26 quarts of beer per capita and had rectal cancer death rates less than one-third of those in Rhode Island and New York.

Suggest some lurking variables that may be confounded with a state's beer consumption. For a clue, look at the high- and low-consumption states given above.

5.47 A study of London double-decker bus drivers and conductors found that drivers had twice the death rate from heart disease as conductors. Because drivers sit while conductors climb up and down stairs all day, it was at first thought that this association reflected the effect of type of job on heart disease. Then a look at bus company records showed that drivers were issued consistently larger-size uniforms when hired than were conductors.

This fact suggested an alternative explanation of the observed association between job type and deaths. What is it?

5.48 Someone says, "There is a strong positive correlation between the number of firefighters at a fire and the amount of damage the fire does, so sending lots of firefighters just causes more damage." Explain why this reasoning is wrong.

5.49 Your friend Julie, during a month of travel in Europe, runs into Bob at a museum in London. Bob was a casual acquaintance from a history course the previous year. "Imagine meeting Bob in a place like that! It must have been fated!" says Julie. Try to convince her that there is nothing extraordinary about such a meeting.

5.50 You are getting to know your new roommate, assigned to you by the college. In the course of a long conversation, you find that both of you have sisters named Deborah. Should you be surprised? Explain your answer.

▶ 4 REGRESSION AND PREDICTION

A scatterplot displays the relationship between two variables. Correlation measures the strength and direction of a straight-line relationship. If a scatterplot shows a strong straight-line relationship, we would like to summarize this overall pattern by drawing a line through the scatterplot. A *regression line* summarizes the relationship between two variables, but only in a specific setting.

Regression line

A **regression line** is a straight line that describes how a response variable y changes as an explanatory variable x takes different values. We often use a regression line to predict the value of y for a given value of x. Regression, unlike correlation, requires that we have an explanatory variable and a response variable.

EXAMPLE 19. How much natural gas does a household use?
Joan is concerned about the amount of energy she uses to heat her home in the Midwest. She keeps a record of the natural gas she consumes each month over one year's heating season. Because the months are not all equally long, she divides each month's consumption by the number of days in the month to get the average number of cubic feet of gas used per day. Demand for heating is strongly influenced by the outside temperature. From local weather records, Joan obtains the average temperature for each month, in degrees Fahrenheit. Here are Joan's data.

Month	Oct.	Nov.	Dec.	Jan.	Feb.	Mar.	Apr.	May
Temperature x	49.4	38.2	27.2	28.6	29.5	46.4	49.7	57.1
Gas consumed y	520	610	870	850	880	490	450	250

A scatterplot of these data appears in Figure 5-11(a). Temperature is the explanatory variable x (plotted on the horizontal axis) because outside temperature helps explain gas consumption. The scatterplot shows a strong negative straight-line association: as temperature increases, gas consumption goes down because less gas is used for heating. The correlation is $r = -0.983$, so the points fall very close to a line. To draw a regression line that describes the relation, lay a transparent straightedge over the points. We want to predict y from x, so move the line about until the distances from the line to the points in the *vertical* (y) direction seem smallest. The resulting regression line appears in Figure 5-11(b).

Now for prediction. How much gas can Joan expect to use in a month that averages 30°F per day? First locate $x = 30$ on the horizontal axis. Draw a vertical line up to the fitted line and then a horizontal line over to the gas consumption scale. As Figure 5-11(b) shows, we predict that Joan will use slightly more than 800 cubic feet per day.

REGRESSION EQUATIONS

When a plot shows a straight-line relationship as strong as that in Figure 5-11, it is easy to draw a line close to the points by using a

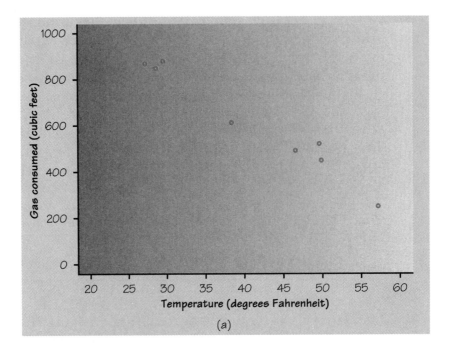

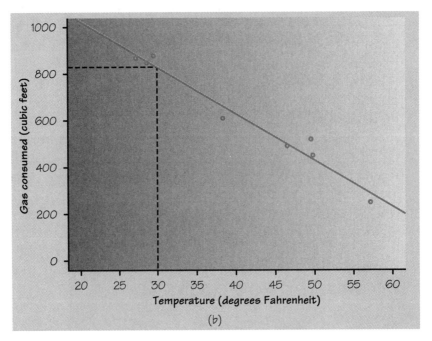

Figure 5-11 Home consumption of natural gas versus outdoor temperature. The regression line in (b) is used to predict gas consumption from temperature.

transparent straightedge. This gives us a line on the graph, but not an equation for the line. There is also no guarantee that the line we fit by eye is the "best" line. It is difficult to concentrate on just the vertical distances of the points from the line when fitting a line by eye. There are statistical techniques for finding from the data the equation of the best regression line (with various meanings of "best"). The most common of these techniques is called *least-squares regression*. Least-squares regression makes the vertical distances of the points in a scatterplot from the regression line as small as possible in a specific sense.

Least-squares regression line

The **least-squares regression line** of y on x is the line that makes the sum of the squares of the vertical distances of the data points from the line as small as possible.

The name "least-squares" reminds us that this line makes the squared distances small. The line in Figure 5-11(b) is in fact the least-squares regression line for predicting natural gas consumption from temperature. All statistical computer software packages and some calculators will calculate the least-squares line for you, so the equation of the line is often available with little work. You should therefore know how to use a regression equation even if you leave to a machine the details of finding the equation from the data.

In writing the equation of a line, x stands as usual for the explanatory variable and y for the response variable. The equation has the form

$$y = a + bx$$

The number b is the *slope* of the line, the amount by which y changes when x increases by one unit. The number a is the *intercept*, the value of y when $x = 0$. To use the equation for prediction, just substitute your x value into the equation and calculate the resulting y value.

> **EXAMPLE 20. Understanding a regression line.** A computer program tells us that the least-squares regression line computed

from Joan's data for predicting gas consumption y from temperature x is

$$y = 1425 - 19.9x$$

The slope of this line is $b = -19.9$. This means that gas consumption goes down by 19.9 cubic feet per day when the average outdoor temperature rises by one degree. The slope of a regression line is usually important for understanding the data. The slope is the rate of change, the amount of change in the predicted y when x increases by 1.

The intercept of the least-squares line is $a = 1425$. This is the value of the predicted y when $x = 0$. Although we need the intercept to draw the line, it is statistically meaningful only when x can actually take values close to zero. An average temperature of $0°$ for a month never happens in Joan's location.

To use the equation for prediction, just substitute the value of x and calculate y. For example, Joan's predicted gas consumption for a month with average temperature $30°$ is

$$y = 1425 - (19.9)(30)$$
$$= 828 \text{ cubic feet per day}$$

The prediction in Example 20 is based on fitting a regression line to the data for eight past months. Joan's house will almost certainly not use exactly 828 cubic feet of gas per day during the next month that has average temperature $30°$. Because the past data points lie so close to the line, however, we can be confident that gas consumption in such a month will be quite close to 828 cubic feet per day.

UNDERSTANDING PREDICTION

Computers make prediction easy and automatic, even from very large sets of data. Anything that can be done automatically is often done thoughtlessly. The computer will happily fit a straight line to a curved relationship, for example. Also, the computer cannot decide which is the explanatory variable and which is the response variable. This distinction is important, because the least-squares line for predicting gas consumption from temperature is not the same as the line for predicting temperature from gas consumption. And, as you might expect, the least-squares regression line can be heavily influenced by a few outliers in the data. Before you begin to calculate, choose your explanatory and response variables and plot your data to check that there is a straight-line overall pattern and that there are no striking outliers. Exercise 5.62 gives a graphic reminder of the perils of numerical descriptions not accompanied by plots.

The usefulness of an observed association for predicting a response y given the value of the explanatory variable x does not depend on a cause-and-effect relationship between x and y. An employer who uses an aptitude test to screen potential employees does not think that high test scores cause good job performance after the person is

"How did I get into this business? Well, I couldn't understand regression and correlation in college, so I settled for this instead."

hired. Rather, the test attempts to measure abilities that will usually result in good performance. It is a matter of common response. If there is a relationship between the test results and later performance on the job, then, in the language of Chapter 3, the test has predictive validity.

The usefulness of the fitted line for prediction does depend on the strength of the association. Not all straight-line associations are as strong as that between temperature and natural gas consumption. The correlation r is one measure of the strength of a straight-line relationship. There is a connection between correlation and least-squares regression that helps us understand both of these statistical ideas.

---

r^2 in regression

The **square of the correlation, r^2,** is the fraction of the variation in the values of y that is explained by the least-squares regression of y on x.

The idea is that when there is a straight-line relationship, some of the variation in y is accounted for by the fact that as x changes it pulls y along with it.

> **EXAMPLE 21. Correlation and regression.** Look again at Figure 5-11. There is a lot of variation in the observed y's, the gas consumption data. They range from a low of 250 to a high of 880. The scatterplot shows that almost all of this variation in y is accounted for by the fact that outdoor temperature was changing and pulled gas consumption along with it. There is only a little remaining variation in y, which appears in the scatter of points about the line. The strong correlation, $r = -0.983$, says that the line explains y very well.
>
> Now look at Figure 5-12. This scatterplot shows the relationship between the percent total return for American common stocks and the percent return for foreign stocks for the 25 years ending in 1995.[20] Investors look at this relationship to see if foreign stocks behave differently from U.S. stocks.
>
> Figure 5-12 shows a positive straight-line relationship—when returns on American stocks are high, foreign stocks also tend to

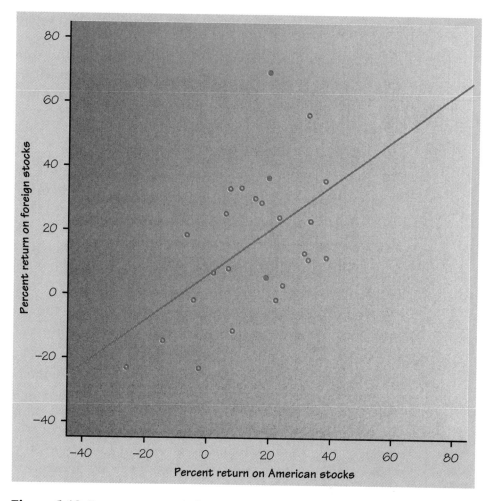

Figure 5-12 Percent return on foreign common stocks versus percent return on American stocks for the years 1971 to 1995. The line is the least-squares regression line for predicting foreign returns from American returns.

have high returns. The line drawn on the scatterplot is the least-squares regression line for predicting foreign return from the same year's American return. There is a great deal of scatter about the line. Consider the three solid points in Figure 5-12. In these three years, U.S. common stocks performed almost the same (the returns are 18.2%, 18.6%, and 18.9%). The returns on foreign stocks, however, varied greatly—they were 4.8%, 36.3% and a remarkable 69.4%.

The straight-line relationship explains some of the variation in returns on foreign stocks, but there is much variation left over. The smaller correlation, $r = 0.520$, tells us that the line does not explain y very well.

The idea that the straight-line tie between x and y explains part, but not all, of the variation in y can be expressed in algebra, though we won't do it. It is possible to break the total variation in the observed values of y, measured by their variance, into two parts. One part is the variation we expect as x moves and y moves with it along the regression line. The other measures the remaining distances of the data points from the line. The squared correlation r^2 is the first of these as a fraction of the whole:

$$r^2 = \frac{\text{variation in predicted } y \text{ as } x \text{ pulls it along the line}}{\text{total variation in observed values of } y}$$

EXAMPLE 22. Using r^2. Correlation $r = \pm 1$ says that the points lie exactly on a line. The straight-line relationship explains all ($r^2 = 1$) of the variation in y, and the regression line predicts y perfectly.

In Figure 5-11, $r = -0.983$ and $r^2 = 0.966$. About 97% of the variation in gas consumption is accounted for by the straight-line relationship with temperature. Predictions using the regression line will be quite accurate.

In Figure 5-12, $r = 0.520$ and $r^2 = 0.270$. The straight-line relationship between American and foreign stock returns explains only 27% of the variation in foreign returns. We can't predict foreign returns accurately from American returns.

In reporting a regression, it is usual to give r^2 as a measure of how successful the regression was in explaining the response. When you see a correlation, square it to get a better feel for the strength of the association. For example, correlation $r = \pm 0.7$ is about "half way" between $r = 0$ and $r = 1$, because $(0.7)^2 = 0.49$.

MULTIPLE REGRESSION AND BEYOND

How large an r^2, and how precise a prediction, you can expect depends on the variables you are interested in. In many fields outside the physical sciences, high correlations are rare.

> **EXAMPLE 23. Predicting law school grades.** American law schools use the Law School Admission Test (LSAT) to help predict performance and to aid in admissions decisions.[21] The correlation between LSAT scores and the first-year grades in law school of admitted applicants is only about $r = 0.36$. This is discouraging, for such a correlation means that LSAT scores explain only 13% of the variance in law school grades. Nonetheless, LSAT scores are a better predictor than the applicants' undergraduate grades, which have a correlation of only about $r = 0.25$ with later grades in law school.

When no single explanatory variable has a high correlation with a response variable, we can use several explanatory variables together to predict the response. This is called *multiple regression*. We can no longer fit a relationship among the variables by eye. A recipe that a computer can follow is essential. Fortunately, the computer will produce a *multiple correlation coefficient* (also called r) between the response variable and all the explanatory variables together. The square r^2 of the multiple correlation coefficient has the same interpretation as in the one-explanatory-variable case. For example, when both LSAT scores and undergraduate grades are used to predict law school grades, the multiple correlation coefficient is about $r = 0.45$. So straight-line dependence on both explanatory variables together accounts for $(0.45)^2$, or about 20%, of the variability in first-year law school grades.

I hope that our tour of association, causation, correlation, and regression has left you with that slightly winded feeling that follows a good workout. Overconfidence in interpreting correlation and regression is the root of many a statistical sin. Most users of statistics have learned that "correlation does not imply causation." The use of regression, especially multiple regression with many explanatory variables, to predict a response has many other pitfalls. It is hard to see outliers when your data have many variables. Lurking variables remain a threat to interpreting your regression results. There are often strong relationships among the explanatory variables themselves, relationships that complicate any attempt to say which variables are "most important" for explaining the response. Multiple regression often accompanies observational studies that try to explain a response. Skepticism is a healthy first reaction to any explanation not based on experiments.

SECTION 4 EXERCISES

5.51   Use the regression line in Figure 5-11 to predict the amount of natural gas Joan's house will use each day in a month with average temperature 45°F. Then substitute $x = 45$ into the regression equation given in Example 20 to get a more precise prediction.

5.52   Figure 5-13 displays data on the number of slices of pizza consumed by pledges at a fraternity party (the explanatory variable x) and the number of laps around the block the pledges could run immediately afterward (the response variable y). The line on the scatterplot is the least-squares regression line computed from these points for predicting y from x.

(a)   At the next party, a pledge eats 6 slices of pizza before running. How many laps do you predict he will complete?

(b)   Another pledge eats 9 pieces of pizza. Predict how many laps he will complete.

(c)   A third pledge shows off by eating 25 pieces of pizza. You should refuse to predict his performance from the scatterplot and regression line. Explain why.

5.53   **SAT verbal and math scores.** Data on the average Scholastic Assessment Test (SAT) verbal and mathematics scores for high

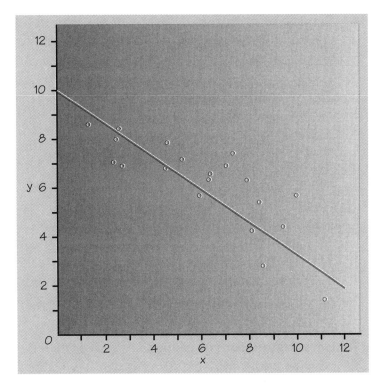

Figure 5-13 A fraternity pizza regression, for Exercise 5.52.

school seniors in the 50 states show a strong straight-line association. The least-squares regression line for predicting a state's median math score y from its median verbal score x is

$$y = 27 + 1.03x$$

(a)   Is the association between math and verbal scores positive or negative? (That is, do states with high average verbal score tend to have high or low average math score?) How does the regression equation tell you the answer to this question?

(b)   The median SAT verbal score in New York was 422. Use the regression line to predict New York's median math score. (New York's actual median SAT math score was 466.)

5.54   **Growth of children.** The growth of children follows a straight-line pattern between early childhood and adolescence. Here are data on Sarah's growth between the ages of 36 months and 60 months.

Age (months)	36	48	51	54	57	60
Height (cm)	86	90	91	93	94	95

(a)   The least-squares regression line for predicting height from age from these data is $y = 71.95 + 0.383x$. Sarah's growth rate is the number of centimeters she grows each month (on the average). What is her growth rate?

(b)   Make a scatterplot of the data and draw the least-squares line on the plot. (To draw the line, use the equation to find y for two values of x, such as $x = 40$ and $x = 60$. Plot the two (x, y) points on your graph and draw the line through them.)

(c)   Use your regression line to estimate Sarah's height at age 42 months.

5.55   **Is wine good for your heart?** Table 5-5 (page 314) presents data on wine consumption and heart disease deaths in 19 developed countries. Many people think that drinking wine in moderation helps prevent heart disease.

(a)   Make a scatterplot of the data. (Which is the explanatory variable?)

(b)   Draw a regression line by eye on your plot that allows you to use wine consumption to predict heart disease deaths.

(c)   Use your line to predict the heart disease death rate in another country in which adults average 4 liters of alcohol from wine each year.

(d)   The correlation for these variables is $r = -0.843$. Why is the correlation negative? About what percent of the variation among countries in heart disease death rates is explained by the straight-line relationship with wine consumption?

5.56   **The endangered manatee.** Manatees are large, gentle sea creatures that live along the Florida coast. Many manatees are killed or injured by power boats. Table 5-7 gives data on power boat

TABLE 5-7 Florida power boats and manatee deaths, 1977–1990

Year	Boats (thousands)	Manatees killed	Year	Boats (thousands)	Manatees killed
1977	447	13	1984	559	34
1978	460	21	1985	585	33
1979	481	24	1986	614	33
1980	498	16	1987	645	39
1981	513	24	1988	675	43
1982	512	20	1989	711	50
1983	526	15	1990	719	47

registrations (in thousands) and the number of manatees killed by boats in Florida in the years 1977 to 1990.

(a) We want to examine the relationship between number of power boats and number of manatees killed by boats. Which is the explanatory variable?

(b) Make a scatterplot of these data. Describe the form and direction of the relationship.

(c) Describe in words the strength of the relationship. The correlation is $r = 0.941$. What fraction of the variation in manatee deaths can be explained by the number of boats registered? Can the number of manatees killed be predicted accurately from power boat registrations?

(d) Predict the number of manatees that will be killed by boats in a year when 716,000 power boats are registered.

5.57 Wine and heart disease again. A computer software package tells us that the least-squares regression line for predicting heart disease death rate from wine consumption, calculated from the data in Table 5-5 (page 314), is

$$y = 260.56 - 22.969x$$

Use this equation to predict the heart disease death rate in another country where adults average 4 liters of alcohol from wine each year. Compare your result with the graphical prediction you found in Exercise 5.55.

5.58 **The endangered manatee again.** A computer software package tells us that the least-squares regression line for predicting manatees killed from power boat registrations, calculated from the data in Table 5-7, is

$$y = -41.4 + 0.125x$$

Use this equation to predict the number of manatees killed in a year when 716,000 power boats are registered. Compare this prediction with the graphical prediction you found in Exercise 5.56.

5.59 **Beware extrapolation.** Suppose that in some far future year 2 million power boats are registered in Florida. Use the regression line in Exercise 5.58 to predict manatees killed. Explain why this prediction is very unreliable. (Using a regression line to predict the response to an x value outside the range of the data used to fit the line is called *extrapolation*. Extrapolation often produces unreliable predictions.)

5.60 **More about the endangered manatee.** Here are four more years of manatee data, in the same form as in Table 5-7:

1991	716	53	1993	716	35
1992	716	38	1994	735	49

(a) Add these points to your scatterplot from Exercise 5.56. Florida took stronger measures to protect manatees during these years. Do you see any evidence that these measures succeeded?

(b) In Exercises 5.56 and 5.58, you predicted manatee deaths in a year with 716,000 power boat registrations. In fact, power boat registrations remained at 716,000 for the next three years. Compare the mean manatee deaths in these years with your prediction from either previous exercise. How accurate was your prediction?

5.61 **The strength of road pavement.** Concrete road pavement gains strength over time as it cures. Highway engineers use regression lines to predict the strength after 28 days (when curing is complete) from measurements made after 7 days. Let x be strength (in pounds per square inch) after 7 days and y the strength after 28 days. One set of data gave the following least-squares regression line:

$$y = 1389 + 0.96x$$

(a) Explain in words what the slope 0.96 tells us about the curing of concrete.

(b) A test of some new pavement after 7 days shows that its strength is 3300 pounds per square inch. Predict the strength of this pavement after 28 days.

5.62 **Always plot your data!** Table 5-8 presents three sets of data prepared by the statistician Frank Anscombe to illustrate the dangers of calculating without first plotting the data. *All three sets have the same correlation and the same least-squares regression line* to several decimal places. The regression equation is

$$y = 3 + 0.5x$$

TABLE 5-8 Three data sets with the same correlation and least-squares regression line

Data Set A

x	10	8	13	9	11	14	6	4	12	7	5
y	8.04	6.95	7.58	8.81	8.33	9.96	7.24	4.26	10.84	4.82	5.68

Data Set B

x	10	8	13	9	11	14	6	4	12	7	5
y	9.14	8.14	8.74	8.77	9.26	8.10	6.13	3.10	9.13	7.26	4.74

Data Set C

x	8	8	8	8	8	8	8	8	8	8	19
y	6.58	5.76	7.71	8.84	8.47	7.04	5.25	5.56	7.91	6.89	12.50

(a) Make a scatterplot for each of the three data sets and draw the regression line on each of the plots. (To draw the regression line, substitute two convenient values of x into the equation to find the corresponding values of y. Plot the two (x, y) points on your scatterplot and draw the line through them.)

(b) Use the regression equation to predict y given that $x = 14$. Mark the resulting point on all three scatterplots. (It is a point on the regression line.)

(c) In which of the three cases would you be willing to use the regression line to predict y given that $x = 14$? Explain your answer in each case.

5.63 A study of class attendance and grades among freshmen at a state university showed that in general students who attended a higher percent of their classes earned higher grades. Class attendance explained 16% of the variation in grade index among the freshmen studied. What is the numerical value of the correlation between percent of classes attended and grade index?

Notes

1. P. J. Bickel and J. W. O'Connell, "Is there a sex bias in graduate admissions?" *Science,* Volume 187 (1975), pp. 398–404.
2. The data are for Nationsbank, from S. A. Holmes, "All a matter of perspective," *New York Times,* October 11, 1995.
3. For a glimpse of the statistical complexities, see the studies collected in *Pay Equity: Empirical Inquiries,* National Academy Press, 1989.
4. From J. Stamler, "The mass treatment of hypertensive disease: defining the problem," *Mild Hypertension: To Treat or Not to Treat,* New York Academy of Sciences, 1978, pp. 333–358.
5. Adapted from data of Williams and Zador in *Accident Analysis and Prevention,* Volume 9 (1977), pp. 69–76.
6. National Science Board, *Science and Engineering Indicators, 1991,* U.S. Government Printing Office, 1991. The detailed data appear in Appendix Table 3-5, p. 274.
7. From S. V. Zagona (ed.), *Studies and Issues in Smoking Behavior,* University of Arizona Press, 1967, pp. 157–180.

8. These data, from reports submitted by airlines to the Department of Transportation, appear in A. Barnett, "How numbers can trick you," *Technology Review,* October 1994, pp. 38–45.

9. The data are from M. A. Houck et al., "Allometric scaling in the earliest fossil bird, *Archaeopteryx lithographica," Science,* Volume 247 (1990), pp. 195–198. The authors conclude from a variety of evidence that all specimens represent the same species.

10. The data are an SRS of 53 pairs of parents from a group of 1079 pairs whose heights are recorded in K. Pearson and A. Lee, "On the laws of inheritance in man," *Biometrika,* November 1903, p. 408.

11. Data from T. J. Lorenzen, "Determining statistical characteristics of a vehicle emissions audit procedure," *Technometrics,* Volume 22 (1980), pp. 483–493.

12. The data appear in the *New York Times,* December 28, 1994. The *Times* credits Dr. M. H. Criqui of the University of California, San Diego, School of Medicine.

13. The data are from W. L. Colville and D. P. McGill, "Effect of rate and method of planting on several plant characters and yield of irrigated corn," *Agronomy Journal,* Volume 54 (1962), pp. 235–238.

14. The correlations are reported by the *Fidelity Insight* newsletter, March 1995, p. 11.

15. Data from many studies compiled in D. F. Greene and E. A. Johnson, "Estimating the mean annual seed production of trees," *Ecology,* Volume 75 (1994), pp. 642–647.

16. From Royal College of Physicians, *Smoking and Health Now,* Pitman Medical Publishing, 1971, p. 5.

17. See L. Lasagna and S. R. Schulman, "Bendectin and the language of causation," in K. R. Foster, D. E. Bernstein, and P. W. Huber (eds.), *Phantom Risk: Scientific Inference and the Law,* The MIT Press, 1994, pp. 101–122. See also the first chapter in this book and the review by J. L. Gastwirth, *Journal of the American Statistical Association,* Volume 90 (1995), pp. 388–390.

18. From *The Health Consequences of Smoking: 1983,* U.S. Public Health Service.

19. For Woburn, see S. W. Lagakos, B. J. Wessen, and M. Zelen, "An analysis of contaminated well water and health effects in Woburn, Massachusetts," *Journal of the American Statistical Association,* Volume 81 (1986), pp. 583–596, and the following discussion. For Randolph, see R. Day, J. H. Ware, D. Wartenberg, and M. Zelen, "An investigation of a reported cancer cluster in Randolph, Ma.," Harvard School of Public Health technical report, June 27, 1988.

20. The variables plotted in Figure 5-12 are the annual percent total returns for two major indexes of the prices of many stocks, 1971 to 1995. The

Standard & Poor's 500 Stock Index represents American stocks. For foreign stocks, I used the Morgan Stanley EAFE (Europe, Australia, Far East) index.

21. The facts about the LSAT, along with much useful information about the alleged cultural bias of such tests, is found in D. Kaye, "Searching for the truth about testing," *The Yale Law Journal,* Volume 90 (1980), pp. 431–457.

22. From S. W. Hargarten et al., "Characteristics of firearms involved in fatalities," *Journal of the American Medical Association,* Volume 275 (1996), pp. 42–45.

23. From D. M. Barnes, "Breaking the cycle of addiction," *Science,* Volume 241 (1988), pp. 1029–1030.

24. From M. Radelet, "Racial characteristics and imposition of the death penalty," *American Sociological Review,* Volume 46 (1981), pp. 918–927.

25. From W. M. Lewis and M. C. Grant, "Acid precipitation in the western United States," *Science,* Volume 207 (1980), pp. 176–177.

26. From a Gannett News Service article appearing in the Lafayette, Indiana, *Journal and Courier,* April 23, 1994.

REVIEW EXERCISES

5.64   **Discrimination?** A study of the salaries of full professors at Upper Wabash Tech shows that the median salary for female professors is considerably less than the median male salary. Further investigation shows that the median salaries for male and female full professors are about the same in every department (English, physics, etc.) of the university. Explain how equal salaries in every department can still result in a higher overall median salary for men.

5.65   **Firearm deaths.** Firearms are second to motor vehicles as a cause of non-disease deaths in the United States. Here are counts from a study of all firearm-related deaths in Milwaukee, Wisconsin, between 1990 and 1994.[22] We want to compare the types of firearms used in homicides and in suicides. We suspect that long guns (shotguns and rifles) will more often be used in suicides because many people keep them at home for hunting. Make a careful comparison of homicides and suicides, with a bar graph. What do you find about long guns versus handguns?

	Homicides	Suicides
Handgun	468	124
Shotgun	28	22
Rifle	15	24
Not specified	13	5
Total	524	175

5.66 Treating cocaine addiction. Cocaine addiction is hard to break. Addicts need cocaine to feel any pleasure, so perhaps giving them an antidepressant drug will help. A 3-year study with 72 chronic cocaine users compared an antidepressant drug called desipramine with lithium and a placebo. (Lithium is a standard drug to treat cocaine addiction.) One-third of the subjects, chosen at random, received each drug. Here are the results.[23]

	Cocaine relapse?	
	Yes	No
Desipramine	10	14
Lithium	18	6
Placebo	20	4

(a) Compare the effectiveness of the three treatments in preventing relapse. Use percents and draw a bar graph.

(b) Do you think that this study gives good evidence that desipramine actually *causes* a reduction in relapses? Why?

5.67 Race and the death penalty. Whether a convicted murderer gets the death penalty seems to be influenced by the race of the victim. Here are data on 326 cases in which the defendant was convicted of murder.[24]

White defendant	Death penalty	
	Yes	No
White victim	19	132
Black victim	0	9

Black defendant	Death penalty	
	Yes	No
White victim	11	52
Black victim	6	97

(a)   Use these data to make a two-way table of defendant's race (white or black) versus death penalty (yes or no).

(b)   Show that Simpson's paradox holds: a higher percent of white defendants were sentenced to death overall, but for both black and white victims a higher percent of black defendants were sentenced to death.

(c)   Use the data to explain the paradox in language that a judge could understand.

5.68   **Obesity and health.** Recent studies have shown that earlier reports seriously underestimated the health risks associated with being overweight. The error was due to overlooking some important lurking variables. In particular, smoking tends both to reduce weight and to lead to earlier death. Illustrate Simpson's paradox by a simplified version of this situation. That is, make up a three-way table of overweight (yes or no) by early death (yes or no) by smoker (yes or no) such that:

▶   Overweight smokers and overweight nonsmokers both tend to die earlier than those who are not overweight.

▶   When smokers and nonsmokers are combined into a two-way table of overweight by early death, people who are not overweight tend to die earlier.

5.69   **Nematodes and tomatoes.** To demonstrate the effect of nematodes (microscopic worms) on plant growth, a botanist prepares 16 identical planting pots, then introduces different numbers of nematodes into the pots. A tomato seedling is transplanted into each plot. Here are data on the increase in height of the seedlings (in centimeters) during the 16 days after planting.

Nematodes	Seedling growth			
0	10.8	9.1	13.5	9.2
1000	11.1	11.1	8.2	11.3
5000	5.4	4.6	7.4	5.0
10,000	5.8	5.3	3.2	7.5

(a) Make a scatterplot of the response variable (growth) against the explanatory variable (nematode count).

(b) Briefly describe the conclusions about the effects of nematodes on plant growth that these data suggest.

(c) Do you recommend calculating the correlation coefficient r to describe the strength of the relationship between nematode count and growth? Explain your answer.

5.70 **Do heavy people burn more energy?** The following table gives data on the lean body mass (kilograms) and resting metabolic rate for 12 women and 7 men who are subjects in a study of obesity. The researchers suspect that lean body mass (that is, the subject's weight leaving out all fat) is an important influence on metabolic rate.

Subject	Sex	Mass	Rate	Subject	Sex	Mass	Rate
1	M	62.0	1792	11	F	40.3	1189
2	M	62.9	1666	12	F	33.1	913
3	F	36.1	995	13	M	51.9	1460
4	F	54.6	1425	14	F	42.4	1124
5	F	48.5	1396	15	F	34.5	1052
6	F	42.0	1418	16	F	51.1	1347
7	M	47.4	1362	17	F	41.2	1204
8	F	50.6	1502	18	M	51.9	1867
9	F	42.0	1256	19	M	46.9	1439
10	M	48.7	1614				

(a) Make a scatterplot of the data for the 12 female subjects. (Which is the explanatory variable?)

(b) Does metabolic rate increase or decrease as lean body mass increases? What is the overall shape of the relationship? Are there any outliers?

5.71 **Enhancing a scatterplot.** When observations on two variables fall into several categories, you can display more information in a scatterplot by plotting each category with a different symbol or a different color. Add the data for male subjects to your scatterplot in Exercise 5.70, using a different color or symbol than you used for females. Does the type of relationship you found for females hold for men also? How do the male subjects as a group differ from the female subjects as a group?

5.72 **Acid rain.** Researchers studying acid rain measured the acidity of precipitation in a Colorado wilderness area for 150 consecutive weeks. Acidity is measured by pH. Lower pH values show higher acidity. The acid rain researchers observed a straight-line pattern over time. They reported that the least-squares regression line

$$pH = 5.43 - (.0053 \times weeks)$$

fit the data well.[25]

(a) Draw a graph of this line. Is the association positive or negative? Explain in plain language what the direction of the association tells us.

(b) According to the regression line, what was the pH at the beginning of the study (weeks = 1)? At the end (weeks = 150)?

(c) What is the slope of the regression line? Explain clearly what this slope says about the change in the pH of the precipitation in this wilderness area.

5.73 **The declining farm population.** The number of people living on American farms has declined steadily during this century. Here are data on the farm population (millions of people) from 1935 to 1980:

Year	1935	1940	1945	1950	1955	1960	1965	1970	1975	1980
Pop.	32.1	30.5	24.4	23.0	19.1	15.6	12.4	9.7	8.9	7.2

(a) Make a scatterplot that shows the change in farm population over time. Is the association positive or negative? Is the pattern roughly a straight line?

(b) The equation of the least-squares regression line for predicting farm population y (in millions) from the year x is

$$y = 1166.93 - 0.5869x$$

According to this fitted line, how much did the farm population decline each year on the average during this period?

(c) Use the regression equation to predict the number of people living on farms in the year 2000. Does your prediction make sense? What went wrong?

WRITING PROJECTS

5.1 **Is math the key to success in college?** News articles often report cause-and-effect conclusions from observational studies. Here is the opening of a newspaper account of a College Board study of 15,941 high school graduates:[26]

> *Minority students who take high school algebra and geometry succeed in college at almost the same rate as whites, a new study says.*

> *The link between high school math and college graduation is "almost magical," says College Board President Donald Stewart, suggesting "math is the gatekeeper for success in college."*

> *"These findings," he says, "justify serious consideration of a national policy to ensure that all students take algebra and geometry."*

Write a careful critique of Mr. Stewart's reasoning. Are there lurking variables that might explain the association between taking several math courses in high school and success in college? Explain why requiring algebra and geometry may have little effect on who succeeds in college. Be sure that your explanation is understandable by someone who knows no statistics.

5.2 **Association is not causation.** Write a snappy, attention-getting article on the theme that "association is not causation." Use pointed but not-too-serious examples like those in Exercises 5.40, 5.41 (page 332), and 5.48 (page 334), or this one: there is an association between long hair and being female, but cutting a woman's hair will not turn her into a man. Be clear, but don't be technical. Imagine that you are writing for high school students.

5.3   **Toxic torts?** How to react to claims that a substance is toxic
when the statistical evidence is weak and experimental evidence
does not exist is a major issue for public policy. Bendectin
(Example 15, page 327) is one such case.

One point of view notes that our current legal structure
encourages lawsuits in the hope that the tragic condition of a
child with major birth defects will win a large award (for the
child's lawyers as well as for the child) from a sympathetic
jury even when there is no good evidence that Bendectin is
responsible. (Bendectin remains in use in Canada, where the
legal climate is different.) On the other hand, perhaps the public
is better protected when even a hint of risk is punished; scientific
standards of "good evidence" may be too strong when public
health is at risk.

Write an essay addressing these questions: What are the
benefits and the costs of a policy that punishes makers of
alleged toxic substances on even very weak evidence? What are
the benefits and the costs of insisting on strong evidence before
taking action? It isn't clear what public policy should be, but
understanding the costs and benefits helps lay out the issues.

CHAPTER 6

THE CONSUMER PRICE INDEX AND ITS NEIGHBORS

We all notice the high salaries paid to professional athletes. In the National Football League, for example, the average salary rose from $79,000 in 1980 to $737,000 in 1993. That's a big jump. Not as big as it first appears, however. *A dollar in 1993 did not buy as much as a dollar in 1980, so 1980 salaries cannot be directly compared with 1993 salaries.* The hard fact that the dollar has steadily lost buying power over time means that we must make an adjustment whenever we compare dollar values for different years. The adjustment is easy. What is not easy is measuring the changing buying power of the dollar. The government's Consumer Price Index (CPI) is the tool we need.

The CPI is a new kind of numerical description, an *index number*. In this chapter, we will look at index numbers and learn about the CPI, its use, and its weaknesses. The monthly CPI and the monthly unemployment rate are just two of the many "official statistics" produced by government statistical agencies. We will therefore look briefly at the current state of government statistics and at some ideas that help us interpret these regular series of measurements of our nation's economic and social condition.

▶ 1 INDEX NUMBERS

We can attach an index number to any numerical variable that we measure repeatedly over time. The essential idea of the index number

359

"Now this here's a genuine 1960 dollar. They don't make 'em like that anymore."

is to give a picture of changes in a variable much like that drawn by saying "The average cost of a hospital stay rose 52% between 1985 and 1990." That is, an index number describes the percent change from a base period.

> **Index number**
>
> An **index number** measures the value of a variable relative to its value at a base period. To find the index number for any value of the variable,
>
> $$\text{index number} = \frac{\text{value}}{\text{base value}} \times 100$$

EXAMPLE 1. Calculating an index number. A pound of oranges that cost 85 cents in 1980 cost $1.25 in 1995. The orange price index number in 1995, with 1980 as the base period, is

$$\text{index number} = \frac{125}{85} \times 100 = 147$$

The orange price index number for the base period, 1980, is

$$\text{index number} = \frac{85}{85} \times 100 = 100$$

The base period is essential to making sense of an index number, so be sure that it is stated. Because the index number for the base period is always 100, it is usual to identify the base period as 1980 by writing "1980 = 100." In news reports concerning the CPI, you will notice the mysterious equation "1982–84 = 100." That's shorthand for the fact that the years 1982 to 1984 are the base period for the CPI.

The recipe shows that an index number just gives the current value as a percent of the base value. An index of 140 means 140% of the base value, or a 40% increase from the base value. An index of 80 means that the current value is 80% of the base, a 20% decrease.

EXAMPLE 2. Your Mercedes SL600 sold for $85,000 in 1980 and for $125,000 in 1995. The Mercedes index number (1980 = 100) for 1995 is

$$\text{index number} = \frac{125,000}{85,000} \times 100 = 147$$

Comparing Examples 1 and 2 makes it clear that an index number measures only change relative to the base value. The orange price index and the Mercedes price index for 1995 are both 147. That says in plain language that the 1995 price was 147% of the 1980 price. That oranges rose from 85 cents to $1.25 and the Mercedes from $85,000 to $125,000 is information that is lost when we have only the index number.

FIXED MARKET BASKET PRICE INDEXES

It may seem that index numbers are little more than a plot to disguise simple statements in complex language. Why say "The Consumer Price

Index (1982–84 = 100) stood at 156.7 in June 1996" instead of "consumer prices rose 56.7% between the 1982–84 average and mid-1996?" In fact, the term "index number" usually means more than a measure of change relative to a base. It also tells us the kind of variable whose change we measure. That variable is a weighted average of several quantities, with fixed weights. Let's illustrate the idea by a simple price index.

EXAMPLE 3. The Mountain Man Price Index. Bill Smith lives in a cabin in the mountains and strives for self-sufficiency. He buys only salt, kerosene, and the services of a professional welder. Here are Bill's purchases in 1990, the base period. His cost, in the last column, is the price per unit multiplied by the number of units he purchased.

Good or service	1990 quantity	1990 price	1990 cost
Salt	100 pounds	$0.50/pound	$50.00
Kerosene	50 gallons	1.00/gallon	50.00
Welding	10 hours	14.00/hour	140.00
		Total cost =	$240.00

The total cost of Bill's collection of goods and services in 1990 was $240. To find the "Mountain Man Price Index" for 1995, we use 1995 prices to calculate the 1995 cost of this same collection of goods and services. Here is the calculation:

Good or service	**1990** quantity	**1995** price	1995 cost
Salt	100 pounds	$0.80/pound	$80.00
Kerosene	50 gallons	0.90/gallon	45.00
Welding	10 hours	17.50/hour	175.00
		Total cost =	$300.00

The same goods and services that cost $240 in 1990 cost $300 in 1995. So the Mountain Man Price Index (1990 = 100) for 1995 is

$$\text{index number} = \frac{300}{240} \times 100 = 125$$

The point of Example 3 is that we follow the cost of the *same* collection of goods and services over time. It may be that Bill refused to hire the welder in 1995 because his hourly rate rose sharply. No matter— the index number uses the 1990 quantities, ignoring any changes in Bill's purchases between 1990 and 1995. We call the collection of goods and services whose total cost we follow the *market basket*. The index number is then a *fixed market basket price index*.

Fixed market basket price index

A **fixed market basket price index** is an index number for the total cost of a fixed collection of goods and services.

The basic idea of a fixed market basket price index is that the weight given to each component (salt, kerosene, welding) remains fixed over time. The CPI is a fixed market basket price index, with several hundred items that represent all consumer purchases. Holding the market basket fixed allows a legitimate comparison of prices because we compare the prices of exactly the same items at each time. As we will see, it also poses severe problems for the CPI.

Government statistical offices publish many index numbers. Not all are price indexes, but all refer to weighted averages of a number of components with weights that remain fixed over time. Thus the term *index number* carries two ideas:

▶ It is a measure of the change of a variable relative to a base value.

▶ The variable is an average of many quantities, with the weight given to each quantity remaining fixed over time.

SECTION 1 EXERCISES

6.1   **The price of gasoline.** The average price of unleaded regular gasoline has fluctuated as follows:

1983	$1.24 per gallon
1987	$0.99 per gallon
1995	$1.22 per gallon

Give the gasoline price index numbers (1983 = 100) for 1983, 1987, and 1995.

6.2 **Air pollution.** Sulfur dioxide in the air is a pollutant that contributes to acid rain. The Environmental Protection Agency reports the amounts of several air pollutants emitted in the United States each year. In 1970, 31.1 million tons of sulfur dioxide were emitted. The figures for 1980 and 1990 were 25.8 and 22.3 million tons. Calculate an index number for sulfur dioxide emissions in each of these years, using 1980 as the base year. By what percent did sulfur dioxide emissions increase or decrease between 1970 and 1990?

6.3 **The price of orange juice.** The Department of Agriculture reports that the average retail price of a 12-ounce can of frozen orange juice has changed as follows:

1980	$0.88	1990	$2.02
1985	$1.32	1994	$1.55

Give the orange juice price index number (1985 = 100) for all four years.

6.4 Use your results for Exercise 6.3 to answer these questions.

(a) How many points did the orange juice price index rise from 1985 to 1990? What percent increase was this?

(b) How many points did the orange juice price index rise from 1980 to 1990? What percent increase was this?

You see that point increases in an index number are the same as percent increases only when we start from the base period.

6.5 The CPI (1982–84 = 100) averaged 130.7 in 1990 and 152.4 in 1995.

(a) How many points did the index gain between the 1982 to 1984 base period and 1990? What percent increase was this?

(b) How many points did the index gain between 1990 and 1995? What percent increase was this?

6.6 **Houston and New York.** The Bureau of Labor Statistics publishes separate consumer price indexes for major metropolitan areas in addition to the national CPI. The values of two of these indexes (1982–84 = 100) in April 1996 were:

141.5 for Houston
165.7 for New York

Can you conclude that consumer prices were higher in New York than in Houston in April 1996? Why or why not? What does a comparison of these two index numbers tell us?

6.7 **The cost of cable TV.** The Bureau of Labor Statistics tells us that the annual average index number (1982–84 = 100) for cable TV subscription rates was 197.4 in 1994. The annual average CPI (1982–84 = 100) for 1994 was 148.2.

(a) Explain exactly what the index number 197.4 tells us about the increase in cable TV rates up to 1994.

(b) A consumer group says that "Cable television rates have increased much faster than consumer prices in general." Is this true? How do you know?

6.8 **The Food Faddist Price Index.** A food faddist eats only steak, rice, and ice cream. In 1990, he buys:

Item	1990 quantity	1990 price
Steak	200 pounds	$5.45/pound
Rice	300 pounds	0.49/pound
Ice cream	50 gallons	5.08/gallon

After a visit from his mother, he adds oranges to his diet in 1994. Oranges cost $0.56/pound in 1990. Here are the food faddist's food purchases in 1994:

Item	1994 quantity	1994 price
Steak	175 pounds	$5.86/pound
Rice	325 pounds	0.53/pound
Ice cream	50 gallons	5.24/gallon
Oranges	100 pounds	0.66/pound

Find the fixed market basket Food Faddist Price Index (1990 = 100) for 1994.

6.9   **The Guru Price Index.** A guru purchases only olive oil, loincloths, and copies of the *Atharva Veda* from which to select mantras for his disciples. Here are the quantities and prices of his purchases in 1985 and 1995:

Item	1985 quantity	1985 price	1995 quantity	1995 price
Olive oil	20 pints	$2.50/pint	18 pints	$3.80/pint
Loincloth	2	$2.75 each	3	$2.80 each
Atharva Veda	1	$10.95	1	$12.95

From these data, find the fixed market basket Guru Price Index (1985 = 100) for 1995.

▶ 2 THE CONSUMER PRICE INDEX

The Consumer Price Index (CPI), published monthly by the Bureau of Labor Statistics (BLS), is a fixed market basket index that measures changes in the prices of consumer goods and services. The CPI is the most important of all U.S. index numbers.

What the CPI says

The CPI measures the cost of a market basket of goods and services that represent the typical purchases of urban households. When the CPI is 160, that means that it now takes $160 to buy what $100 would buy in the 1982 to 1984 base period.

Not only does the CPI make headlines as the most popular measure of inflation (the declining buying power of the dollar), but it directly affects the incomes of over 70 million Americans. This is so because many sources of income are "indexed." That is, they are tied to the CPI and automatically increase when the CPI increases. (The last year in which the CPI decreased was 1955, so we can ignore that unlikely possibility.) Social Security payments to 43 million people, food stamps received by 23 million people, and the benefits paid to 4 million military and government retirees are all adjusted as the CPI changes. Many union pay scales, the cost of school lunches, eligibility for welfare, and even the rate brackets for the federal income tax are all tied to the CPI.[1]

UNDERSTANDING THE CPI

What does the CPI measure? It is a fixed market basket price index covering the cost of a collection of about 400 goods and services. The CPI is computed just as in Example 3, though the arithmetic is a bit longer with 400 items. The real complications are not in the arithmetic, but in choosing the market basket from the thousands of goods and services available and in arriving at prices from the thousands of places we can buy these goods and services.

The official name for the common version of the CPI (there are others, but we will ignore them) is the Consumer Price Index for All Urban Consumers. The CPI market basket represents the purchases of people living in urban areas. The official definition of "urban" is broad, so that about 80% of the U.S. population is covered. But if you live on a farm, the CPI doesn't apply to you.

How is the market basket chosen? By an exercise in probability sampling. A sample of 29,000 households provided detailed information on their spending in the years 1982 to 1984. The BLS then divided the spending into categories and selected a sample of items to represent each category. For example, 73 specific items represent the "food and beverage" category. Both the sample of households and the sample of items are multistage stratified samples. The prices of the sample items get weights in the index that reflect how much households spend on the entire category that the items represent. An ongoing sample survey of household spending helps keep the weights up to date.

How are the prices determined? By more probability sampling. The BLS records the price of about 90,000 specific items each month. The retail outlets at which prices are measured are a sample of all outlets, chosen to represent not only geographic spread but the variety of stores people shop at. An ongoing sample survey learns if, for example, more people are buying computers at discount stores (such as Wal-Mart) or large specialty stores (such as CompUSA). The BLS records prices at different places as shopping habits change. A sample of about 8300 housing units produces data on the costs of homeowners and renters.

Does the CPI measure changes in my cost of living? No. First of all, the CPI can't measure "cost of living" because it uses a fixed market basket. You can change your cost of living by changing what you buy— if beef gets too expensive, you can eat less beef and more chicken (or more beans). A fixed market basket price index measures the cost of *living the same* from year to year—remember Bill Smith in Example 3. The CPI also leaves out taxes, which certainly affect your cost of living. Remember also that the CPI aims to measure changes in the cost of a market basket that is an "on the average" description of the spending of urban households. If you are a student, you spend more than the average on college tuition. If you are retired, you probably spend nothing on tuition and more than average on medical care. The CPI does not reflect your personal spending patterns.

Why is the base period 1982 to 1984? The BLS changes the base period every ten years or so. The 1982–84 base replaced the previous 1967 base in 1987. The base period may have changed again by the time you read this. Changing the base period changes the index numbers themselves, but not the results of comparisons based on them. Changes in the base usually coincide with major changes in the market basket and its weights. They may also save the government some embarrassment. For example, the CPI (1982–84 = 100) for June 1996 was 156.7. It would have been 469.0 on the previous 1967 = 100 basis. Did you want to be reminded that in 1996 it took \$469 to buy goods and services that cost \$100 in 1967?

THE WOES OF THE CPI

The CPI is important. In particular, it has a major impact on government spending. When the CPI goes up, government spending on Social Security, food stamps, and much else automatically goes up also. If the CPI goes up too fast, spending goes up too fast. In fact, most economists think that the CPI does go up too fast. That is, it overstates the true decline in the dollar's buying power. In 1995, Federal Reserve chairman Alan Greenspan estimated that the CPI overstates inflation by somewhere between 1/2% and 1 1/2% per year. How can all that probability sampling produce an upward bias?

The fault lies in the very idea of a fixed market basket price index. The world keeps changing, so that the market basket items and their weights rapidly cease to reflect actual consumer spending. As we have already hinted, the BLS in fact is constantly updating the index. Critics say it doesn't update fast enough. Let's look at some of the problems.

Changes in quality. A 1997 car is not the same as a 1984 car. So the "new car" item in the CPI market basket is different each year. We think that 1997 cars are better than 1984 cars—and if they are better, they are worth more. The BLS must decide each year how much of the rise in car prices is due to the better quality of the product, and how much is a genuine price hike. The 1993 car models, for example, cost (according to the BLS) $668.80 more than the 1992 models. But $310.96 of this increase paid for better quality, mostly the introduction of air bags. Only the remaining $357.84 counts as a price increase in the CPI. Cars are easy. What about computers? Or TV sets? Or even the "New! Improved!" shampoo? The BLS has to decide. Careful decisions take some time, so changes in the CPI lag behind actual quality improvements. That creates some upward bias because part of the higher price is buying better quality.

New products create similar problems. CDs replace LPs, stereo replaces mono and in turn gives way to home theater. Leaded gasoline is replaced by unleaded. New prescription drugs appear every month. *Changing buying habits* as consumers switch to cheaper products or to outlets that sell the same product at lower prices also outdate the market basket and the price sample. The BLS must update both constantly. Despite constant effort and much statistical competence, the BLS is always running to keep up with changes. The result is that the CPI goes up too fast. How much too fast is hard to judge—after all, the samples that undergird the CPI are our best source of information about consumer spending. The BLS says less than 1/2% per year. Critics say as much as 1 1/2%. Any adjustment in the CPI beyond what the BLS considers justified may, as often happens, be a political decision. Budget-balancers want the CPI to grow more slowly. Influential groups, especially Social Security recipients, profit when it goes up too fast. Social Security recipients usually win such fights.

USING THE CPI

A dollar in 1993 is not the same as a dollar in 1980. The decrepit 1993 dollar could buy much less than the more robust 1980 dollar. An income earned in 1980 looks misleadingly small now unless we restate it in present-day dollars.

To make legitimate comparisons of dollar amounts from different years, we must use dollars that have the same buying power. We do this by converting all dollar amounts into dollars of the same year. You will find tables in the *Statistical Abstract,* for example, with headings such as "Median Household Income, in Constant (1995) Dollars." That table has restated all incomes in dollars that will buy as much as the dollar would buy in 1995. Watch for the term *constant dollars* and for phrases like *real income.* They mean that all dollar amounts represent the same buying power.

We can use the CPI to adjust dollar amounts for the effects of inflation over time. Table 6-1 gives annual average CPIs to help our work. As Figure 6-1 shows, the CPI has generally increased throughout the twentieth century, and it rose steeply after 1973. Here is the recipe for converting dollars of one year into dollars of another year.

TABLE 6-1   Annual average Consumer Price
Index, 1982–84 = 100

Year	CPI	Year	CPI	Year	CPI
1915	10.1	1970	38.8	1985	107.6
1920	20.0	1975	53.8	1986	109.6
1925	17.5	1976	56.9	1987	113.6
1930	16.7	1977	60.6	1988	118.3
1935	13.7	1978	65.2	1989	124.0
1940	14.0	1979	72.6	1990	130.7
1945	18.0	1980	82.4	1991	136.2
1950	24.1	1981	90.9	1992	140.3
1955	26.8	1982	96.5	1993	144.5
1960	29.6	1983	99.6	1994	148.2
1965	31.5	1984	103.9	1995	152.4

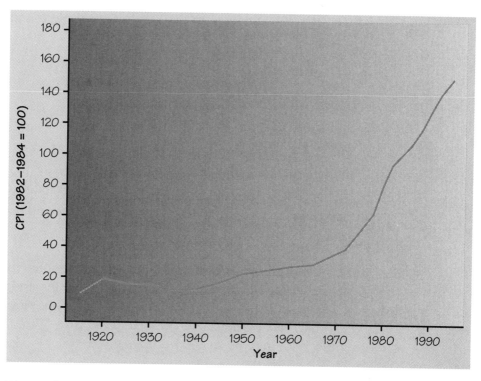

Figure 6-1 Annual average CPIs, 1915 to 1995. Except for the depression period in the 1930's, the CPI has steadily increased (and the dollar has lost buying power) over time.

Adjusting for changes in buying power

To convert an amount in dollars at time A to the amount with the same buying power at time B,

$$\text{Dollars at time B} = \text{Dollars at time A} \times \frac{\text{CPI at time B}}{\text{CPI at time A}}$$

Notice that the CPI for the year you are *going to* appears on the top in the ratio of CPIs in this recipe. Here are some examples.

EXAMPLE 4. **Salaries of professional athletes.** The average salary of National Football League players rose from $79,000 in 1980 to $737,000 in 1993. How big was the increase in real terms?

Let's convert the 1980 average into 1993 dollars. Table 6-1 gives the annual average CPIs that we need.

$$1993 \text{ dollars} = 1980 \text{ dollars} \times \frac{1993 \text{ CPI}}{1980 \text{ CPI}}$$

$$= \$79,000 \times \frac{144.5}{82.4}$$

$$= \$138,538$$

That is, it took $138,538 in 1993 to buy what $79,000 would buy in 1980. In fact, NFL players averaged $737,000 in 1993. So current players earn much more than 1980 players even after adjusting for the fact that the dollar buys less now.

EXAMPLE 5. Stagnant incomes. For a more serious example, let's leave the pampered world of professional athletes and look at the incomes of ordinary families. The median weekly earnings of all families that had earners was $400 in 1980 and $707 in 1993. Changing the 1980 income to 1993 dollars gives

$$1993 \text{ dollars} = \$400 \times \frac{144.5}{82.4} = \$701$$

Real family incomes were essentially the same in 1993 as in 1980, despite the fact that many more families had two earners in 1993.

Let's look at individuals rather than families. The average hourly earnings of workers in private industry were $6.66 in 1980 and $10.83 in 1993. That dollar increase is an illusion. Again restating the 1980 earnings in 1993 dollars,

$$1993 \text{ dollars} = \$6.66 \times \frac{144.5}{82.4} = \$11.68$$

Private industry workers earned more in real terms in 1980 than in 1993.

The calculations in Example 5 point to the most serious problem facing the American economy as we approach the end of the century. Comparing *real* incomes rather than just dollars earned is essential to seeing how serious the problem is. The real incomes of typical workers and families have stopped growing, while the gap between the incomes of better-educated workers and those with less education has increased dramatically. Why this has happened, and what we should do about it, are intensely controversial.

SECTION 2 EXERCISES

6.10   **Real earnings.** An article in the *New York Times* (January 17, 1989) reported that the earnings of workers without a college education were falling ever farther behind those of college graduates. The article stated that the real earnings of male college graduates between 25 and 34 years of age increased by 9% between 1979 and 1987. The real earnings of male high school graduates in the same age group fell 9% during this period, and the real earnings of high school dropouts fell 15%.

Explain carefully what is meant by "real earnings" in this news report.

6.11   **Real wages.** In one of the many reports on stagnant incomes in the United States, we read this: "Practically every income group faced a decline in real wages during the 1980s. However, workers at the 33rd percentile experienced a 14 percent drop in the real wage, workers at the 66th percentile experienced only a 6 percent drop, and workers in the upper tail of the distribution experienced a 1 percent wage increase."[2]

(a)   What is meant by "the 33rd percentile of the income distribution"?

(b)   What does "real wages" mean?

6.12   **Joe DiMaggio.** Yankee center fielder Joe DiMaggio was paid $32,000 in 1940 and $100,000 in 1950. Express his 1940 salary in 1950 dollars. By what percent did DiMaggio's real income change in the decade?

6.13   **Classic cars.** The "muscle cars" of the 1960s became popular with collectors in the late 1980s, perhaps because the collectors were teens in the 1960s. A 1965 Pontiac GTO hardtop sold new for $3100. The car was worth $17,000 in 1989. Restate the 1965 price in 1989 dollars to see how much value the GTO gained in real terms.

6.14   **False prosperity?** A news item in a local paper, comparing the county's economy in 1980 and 1990, says, "Our median family income increased from $20,554 to $36,073, a 75 percent increase." Explain why "a 75 percent increase" is misleading. What was the percent change in real median family income?

6.15   When Julie started college in 1990, she set a goal of earning $25,000 per year in her first job after graduation. Julie graduated in 1995. How much must she earn to have the same buying power that $25,000 had in 1990?

6.16   **The minimum wage.** The federal government sets the minimum hourly wage that employers can pay a worker. Labor wants a high minimum wage, but many economists argue that too high a minimum makes employers reluctant to hire workers with low skills and so increases unemployment. Here is information on changes in the federal minimum wage, in dollars per hour:

Year	1955	1960	1965	1970	1975
Minimum wage	$1.00	$1.00	$1.25	$1.60	$2.10

Year	1980	1985	1990	1995
Minimum wage	$3.10	$3.35	$3.80	$4.25

Use annual average CPIs from Table 6-1 to restate the minimum wage in constant 1955 dollars. Make two line graphs on the same axes, one showing the actual minimum wage during these years and the other showing the minimum wage in constant dollars. Explain carefully to someone who knows no statistics what your graphs show about the history of the minimum wage. (Perhaps because of the trend your calculations show, Congress agreed to increase the minimum wage to $5.15 per hour in 1997.)

6.17   **College tuition.** Tuition for Indiana residents at Purdue University has increased as follows (use tuition at your own college if you have those data):

Year	1977	1979	1981	1983	1985
Tuition	$820	$933	$1158	$1432	$1629

Year	1987	1989	1991	1993	1995
Tuition	$1816	$2032	$2324	$2696	$3056

Use annual average CPIs from Table 6-1 to restate the tuition in constant 1977 dollars. Make two line graphs on the same axes, one showing actual dollar tuition for these years and the other showing constant dollar tuition. Then explain to someone who knows no statistics what your graphs show.

6.18 The Ford Model T sold for $950 in 1910 and for $290 in 1925. Obtain the current value of the CPI, and use it to restate these prices in current dollars.

6.19 What would be the CPI for 1995 if the base period were 1950? [You can find this by changing $100 in 1950 dollars into 1995 dollars. The resulting number of 1995 dollars is the 1995 price index (1950 = 100).] What would the current CPI be if the base period were 1950?

6.20 **Item weights in the CPI.** Items representing "food and beverages" make up 17.84% of the total CPI. Items representing "entertainment" make up 4.38%. Why do food and beverages get greater weight in the CPI than entertainment?

6.21 **Woes of the CPI.** Here are some examples of how the CPI might overestimate price increases. Comment on each example. Which can be corrected by the ongoing sample survey of consumer spending because the overestimates are due to out-of-date information? Which require the BLS to adjust prices to take improved quality into account?

(a) Exterior house-paint prices in the CPI are up, but many buyers have switched from oil paints to latex water-base paints, a different product.

(b) Exterior house-paint prices are up, but new paints cover better so that less paint is needed to paint a house. New paints also last longer, so we need to buy them less often.

(c) Exterior house-paint prices are up, but buyers have switched from small hardware stores (higher prices) to discount stores (lower prices). So the price actually paid is not up as much as the hardware store price.

(d) Exterior house-paint prices are up, but new paints are easier to use and to clean up and so are more convenient.

6.22 **Local CPIs.** In addition to the national CPI, the BLS publishes separate CPIs for 29 large metropolitan areas. These local CPIs are considerably less precise (that is, they have considerably more sampling variation) than the national CPI. Explain why this is so.

6.23 "In the second recent dose of surprisingly high inflation figures, the Labor Department reported today that the Consumer Price Index rose two-tenths of 1 percent in February ..." (*New York Times,* March 20, 1991). Most news reports, like this one, tell us the percent change in the CPI rather than the actual value of the index number. Why do you think this is the usual practice?

▶ 3 ISSUES: THE PLACE OF GOVERNMENT STATISTICS

Modern nations run on statistics. Economic data in particular guide government policy and inform the decisions of private business and individuals. We have already seen how a single number, the monthly CPI, directly influences the incomes of 70 million Americans as well as the direction of financial markets and the shape of economic policy. Another number, the monthly unemployment rate, can decide the fortunes of a president seeking reelection.

Price indexes and unemployment rates, along with many other less publicized series of data, are produced by government statistical offices. Some countries have a single statistical office, such as Statistics Canada. Others attach smaller offices to various branches of government. The United States is an extreme case: there are 72 federal statistical offices, with relatively weak coordination among them. The Census Bureau (in the Department of Commerce) and the Bureau of Labor Statistics (in the Department of Labor) are most important, but you may at times sample the products of the Bureau of Economic Analysis, the National Center for Health Statistics, the Bureau of Justice Statistics, or others in the federal government's collection of statistical agencies.

Few topics are less stimulating than the question of how to organize government statistical agencies. Topics that are not stimulating are sometimes important, however. Citizens have a stake in accurate and timely government statistics that are not influenced by political pressure. A 1993 ranking of government statistical agencies by the heads of these agencies in several nations recently put Canada at the top, with the U.S. tied with Britain and Germany for sixth place.[3] The top spots generally went to countries with a single, independent statistical office. In 1996, Britain combined its main statistical agencies to form a new Office for National Statistics. American government statistics remains fragmented.

"Yes sir, I know that we have to know where the economy is going. But do we have to publish the statistics so that everyone else does too?"

THE PERILS OF GOVERNMENT STATISTICIANS: MONEY

What do citizens need from their government statistical agencies? First of all, data that are *accurate, timely,* and *keep up with changes in society and the economy.* Producing accurate data quickly demands considerable resources. We have seen the elaborate sampling programs that produce the unemployment rate and the CPI. The major U.S. statistical offices have a good reputation for accuracy and lead the world in getting data to the public quickly. Their record for keeping up with changes is much less good. The struggle to adjust the CPI for changing buying habits and changing quality is a minor problem next to the failure of U.S. economic statistics to keep up with trends such as the shift from manufacturing to services as the center of economic activity. Business organizations have expressed strong dissatisfaction with the overall state of our economic data.

Much of the difficulty stems from lack of money. The Current Population Survey sample of 60,000 households and the CPI sample of 90,000 retail prices are expensive. In the years after 1980, reducing federal spending was a political priority. Government statistical agencies lost staff and cut programs. Lower salaries made it hard to attract the best economists and statisticians to government, and constant sniping from politicians attacking "bureaucrats" made government service less attractive. As a result, efforts to modernize the measurement of such things as the gross national product (GNP) fell far behind schedule. (The GNP is the total value of goods and services produced by U.S. residents, a key measure of the overall level of economic activity.) Most economists and statisticians agree that U.S. government statistics are losing touch with the changing economy.

More dollars would help and are justified by the need of economic policy-makers for information about such things as the GNP and workers' productivity. There are, however, other issues. Just what data should the government produce? In particular, should the government produce data that are used mainly by private business rather than by the government's own policy makers? Perhaps such data should be either turned over to private concerns or produced only for those who are willing to pay. This is a question of political philosophy rather than statistics, but it helps determine what level of government statistics we want to pay for.

THE PERILS OF GOVERNMENT STATISTICIANS: FRAGMENTATION

Freedom from political influence is as important to government statistics as accuracy and timeliness. When a statistical office is part of a government ministry, it can be influenced by the needs and desires of that ministry. The Census Bureau is in the Department of Commerce, which serves business interests. The BLS is in the Department of Labor. Thus business and labor each have "their own" statistical office. The professionals in the statistical offices successfully resist direct political interference—a poor unemployment report is never delayed until after an election, for example. But indirect influence is clearly present. The BLS must compete with other Department of Labor activities for its budget, for example. Moreover, major agencies such as the Census and the BLS are overseen by different congressional committees, which have their own priorities and have sometimes interfered with cooperation between the agencies.

The 1996 reorganization of Britain's statistical offices was prompted in part by a widespread feeling that political influence was too strong. The details of how unemployment is measured in Britain were changed many times in the 1980s, for example, and almost all the changes had the effect of reducing the reported unemployment rate—just what the government wanted to see.

I favor a single "Statistics USA" office not attached to any other government ministry, as in Canada. Such unification might also help the money problem by eliminating duplication. It would at least allow a central decision about which programs deserve a larger share of limited resources. Unification is unlikely, but stronger coordination of the many federal statistical offices could achieve many of the same ends. In the words of Janet Norwood, the highly respected head of the BLS from 1979 to 1991:[4]

> *We have lost a lot of ground over the last few decades as we have steadily weakened coordination of the system and refused*

to provide the agencies with resources needed to incorporate new concepts and statistical knowledge. These are not new issues. More than a dozen studies of the statistical system, some over 100 years old, considered these problems. They all came to the same conclusion: we must either strengthen coordination of our decentralized system or we must build an entirely new centralized one. Most of the studies opted for continuation of decentralization but with much stronger coordination.

THE QUESTION OF SOCIAL STATISTICS

National economic statistics are well established with the government, the media, and the public. The government also produces much data on social issues such as education, health, housing, and crime. Social statistics are less complete than economic statistics. We have good data about how much money is spent on food, but less information about how many people are poorly nourished. Social data are also less carefully produced than economic data. Economic statistics are generally based on larger samples, are compiled more often, and are published with a shorter time lag. The reason is clear: economic data are used by the government to guide economic policy month by month. Social data help us understand our society and address its problems, but are not needed for short-term management.

Aside from their statistical quality, official social statistics have several other shortcomings in the eyes of social scientists. The government is often reluctant to collect data on sensitive issues. After first deciding to undertake a sample survey asking people about their sexual behavior, in part to guide AIDS policy, the government backed away. Instead, it funded a much smaller survey of 3452 adults by the University of Chicago's National Opinion Research Center (NORC). The study concluded, for example, that AIDS is likely to remain largely confined to the groups already heavily infected (homosexuals and intravenous drug users) because these groups have limited sexual contact with the rest of the population. The larger sample originally planned by the government would have provided more complete and more precise information.[5]

Government social data largely confine themselves to "objective" questions. "Have you been a victim of a crime during the past year?"

and "When did you last visit a doctor?" are questions the government is willing to ask. Social scientists would like to add "subjective" or "opinion" information to improved versions of the "objective" social statistics now available. Politicians as well as social scientists want to follow changes in the values Americans hold, their degree of satisfaction with their jobs, how much they are afraid of crime, and so on. A well-designed sample survey, regularly repeated, would provide fascinating data.

That data would be fascinating is not reason enough for the government to act, however. I have already pointed out that the government purpose of social data is not so obvious as the role of economic data in setting economic policy. Moreover, many people don't want the government to ask about their sexual behavior or religion. They feel safer answering questions from the University of Chicago than from Uncle Sam. Finally, it isn't clear just how we should measure such things as "satisfaction with the quality of life." Perhaps the government should avoid messy subjects that would annoy some citizens and do not serve the immediate needs of running the country.

Yet, in a generally prosperous nation, issues such as the quality of life are important to citizens. If we feel that our quality of life is slipping, we expect our government to do something about it. How can we get accurate information about social issues, collected consistently over time, and yet not entangle the government with sex, religion, and other touchy subjects?

THE NORC GENERAL SOCIAL SURVEY

The solution in the United States has been government funding of university social surveys. In particular, the National Science Foundation has long funded NORC's General Social Survey (GSS).[6] In operation since 1972, the GSS belongs with the Current Population Survey and the samples that undergird the CPI on any list of the most important sample surveys in the United States. The core questions are asked each year so that we can track changes over time.

Like other major sample surveys, the GSS uses a complex multi-stage stratified sampling design. The population is all adults who live in households (not institutions) and can be interviewed in English. Roughly

1500 people are interviewed each year. The GSS includes both objective and subjective items. Respondents are asked about their job security, their job satisfaction, their satisfaction with their city, their friends, and their families. They answer questions about race, religion, and sex. Many Americans would object if the government asked whether they had seen an X-rated movie in the past year, but they reply when the GSS asks this question.

This indirect system of government funding of a university-based sample survey fits the American feeling that the government itself should not be unduly invasive. It also insulates the survey from most political pressure. Alas, the government's budget-cutting extends to the GSS, which now describes itself as an "almost annual" survey because lack of funds has prevented taking samples in some years. The GSS is, I think, a bargain.

SECTION 3 EXERCISES

6.24   **Saving money?** One way to cut the cost of government statistics is to reduce the sizes of the samples. We might, for example, cut the Current Population Survey from 60,000 households to 20,000. Explain clearly, to someone who knows no statistics, why such cuts reduce the accuracy of the resulting data.

6.25   **The General Social Survey.** The General Social Survey places much emphasis on asking many of the same questions year after year. Why do you think it does this?

6.26   **Measuring quality.** Changing economic conditions afflict economic statistics with some of the same measurement problems that trouble social statistics. For example, the quantity of a manufacturing firm's output may now be less important than the quality of that output. Quantity is easy to measure, while quality is much tougher. Consider automobiles: suggest how you might measure the quality of cars in order to compare different manufacturers and to track changes in quality over time.

6.27   **Measuring the impact of crime.** We wish to include, as part of a set of social statistics, measures of the amount of crime and of the impact of crime on people's attitudes and activities. Suggest some possible measures in each of the following categories:

(a) Statistics to be compiled from official sources.

(b) "Objective" information to be collected using a sample survey of citizens.

(c) "Subjective" information on opinions and attitudes to be collected using a sample survey.

6.28 **Crime data.** The two primary collections of official statistics on crime in America are the FBI's annual *Crime in the United States* and the Bureau of Justice Statistics' annual *Criminal Victimization in the United States*. The FBI bases its data on the Uniform Crime Reports filed by law enforcement agencies. The Bureau of Justice Statistics reports results of the National Crime Victimization Survey. You can find a short description of each source in the introduction to the "Law Enforcement" section in the *Statistical Abstract*. Briefly describe these two sources. What kinds of data do they contain? How are the data collected? How frequently are they collected?

6.29 **Statistical agencies.** Write a short description of the work of one of these government statistical agencies. You can find material in the library, or at each agency's site on the World Wide Web. (The addresses given were correct as of mid-1996, but may change. A search for the agency's name will produce the current address.)

(a) Bureau of Economic Analysis (Department of Commerce, http://www.bea.doc.gov).

(b) Census Bureau (Department of Commerce, http://www.census.gov).

(c) Bureau of Labor Statistics (Department of Labor, http://www.bls.gov).

▶ 4 UNDERSTANDING CHANGE OVER TIME

The monthly CPI, the monthly unemployment rate, and most other economic statistics are produced at regular time intervals. Indeed, the *change* in prices or unemployment from month to month is often the center of attention rather than the actual value of the CPI or the unemployment rate.

> ### Time series
>
> A sequence of measurements of the same variable made at different times is called a **time series.** Usually (but not always) the variable is measured at regular intervals of time, such as monthly.

The variable measured in a time series may be any of the many kinds we have studied: counts (such as the number of persons employed), rates (such as the unemployment rate), or index numbers (such as the CPI). The idea of a time series is that each observation records both the value of the variable and the time when the observation was made. Time series data are often used to predict future values of the variable in question. The goal of the economic time series produced by the government is in part to tell us where we are but also to suggest where we are going. Predicting the future is always risky business, and statistical time series do not remove the risk. The interpretation of time series is complicated by the fact that several types of movement are going on together. Let us look at each type separately.

SEASONAL VARIATION

The unemployment rate peaks every year in January or February. It reaches its lowest level each year in either June or October. There is no mystery about this pattern. Unemployment rises in January when outdoor work slows down in the north due to winter weather and Christmas sales help are laid off. It would cause confusion (and perhaps political trouble) if the government's official unemployment rate jumped every January. The BLS knows about how much it expects unemployment to rise in January, so it adjusts the published data for this expected change. The published unemployment rate goes up only if actual unemployment rises more than expected. We can then see the underlying changes in the employment situation without being confused by regular seasonal changes.

> **Seasonal variation, seasonal adjustment**
>
> A pattern in a time series that repeats itself at known regular intervals of time is called **seasonal variation.** Many time series are **seasonally adjusted.** That is, the expected seasonal variation is removed before the data are published.

EXAMPLE 6. Politicians and statisticians. Most government economic statistics are seasonally adjusted. President Reagan, eager to claim that the economy was turning up, said in an April 1982 speech that unemployment had fallen by 88,000 in March. But the BLS had just announced an *increase* of 98,000 in the number of people out of work. The president used unadjusted figures. The BLS unemployment data are seasonally adjusted to take account of the fact that employment usually increases from February to March. "The statisticians in Washington have funny ways of counting," said Mr. Reagan.

TRENDS AND CYCLES

Once seasonal variation is removed, we can follow our usual strategy for examining data by looking for the overall pattern in a time series. *Trends* and *cycles* are long-term patterns in a time series.

> **Trends and cycles**
>
> A **trend** in a time series is a persistent long-term rise or fall. **Cycles** in time series are up-and-down movements of irregular strength and duration.

EXAMPLE 7. The price of fresh oranges. Figure 6-2(a) is a line graph of the index number (1982–84 = 100) for the monthly price of fresh oranges from the beginning of 1986 to mid-1996. The data are from the BLS samples that produce the CPI; they are not seasonally adjusted. There is a clear upward trend. That is, orange prices were generally rising during this period. Orange prices also display seasonal variation. They rise in the fall each year and then drop in the winter when oranges are harvested in Florida. Let's estimate the trend and the seasonal variation and see if they explain orange prices well.

We can estimate the trend by drawing a line on the graph. The line in Figure 6-2(a) is the least-squares regression line of orange price on time as the explanatory variable. Estimating seasonal variation by averaging the regular variation seen in many years' data is more complicated than drawing a trend line, but most statistical computer packages will do it for us. The estimated seasonal variation is the same each year because seasonal variation is a pattern that repeats each year. Figure 6-2(b) shows the estimated seasonal variation added to the trend line. The

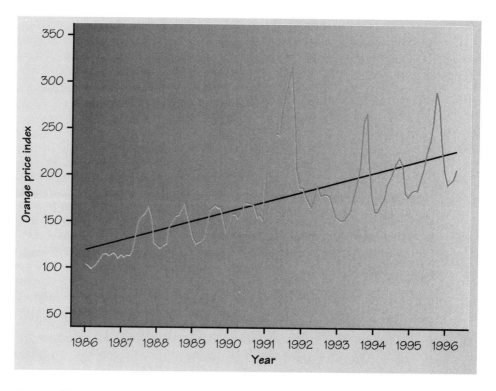

Figure 6-2(a) A time series and its component parts. The index number for the price of fresh oranges, with a line that shows the increasing trend.

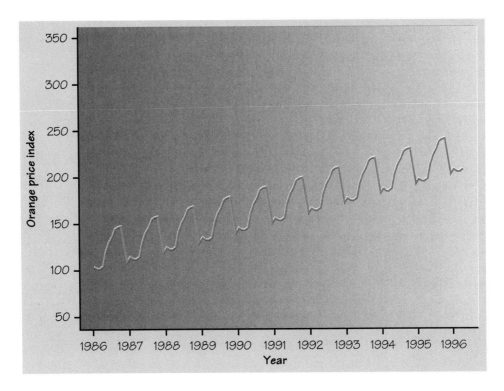

Figure 6-2(b) The estimated trend and seasonal variation in orange prices.

time series would look like this if only trend and seasonal variation were present.

PREDICTING THE FUTURE?

Do trend and seasonal variation together explain orange prices well enough to enable us to predict future prices? If we subtract both the trend and the seasonal variation from the original orange prices, we get Figure 6-2(c). This is a graph of the residual variation in orange prices that is not explained by either the long-term upward trend or the annual seasonal variation. We call these unexplained variations *erratic fluctuations*. Figure 6-2(c) shows quite a bit of erratic fluctuation, and several major deviations from the pattern set by the trend and

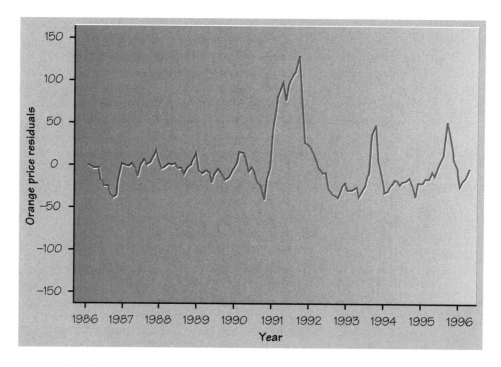

Figure 6-2(c) The erratic fluctuations in orange prices that remain after trend and seasonal variation are removed.

seasonal variation. The large jump in orange prices in 1991, for example, resulted from a freeze in Florida and could not be predicted. It is clear that trend and seasonal variation do not allow us to predict orange prices well.

Orange prices are simple compared with the national economy. *Business cycles,* the irregular up-and-down movements of the economy, are of great interest to investors and politicians, and to ordinary citizens whose lives are affected by good and bad times.

EXAMPLE 8. **Good times and bad times.** Business cycles are reflected in most economic statistics. Let's look at a familiar time series that is of particular concern to individual workers, the unemployment rate. Figure 6-3 is a line graph of the monthly unemployment rate from the beginning of 1970 to mid-1996.

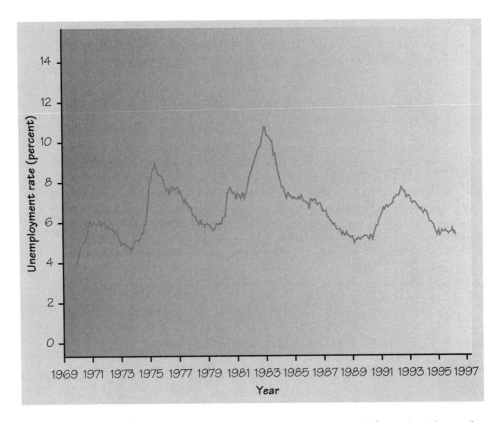

Figure 6-3 Monthly unemployment rate, 1970 to April 1996. Each tick on the "Year" axis marks the beginning of a year. The data are not seasonally adjusted. The cycles show periods of prosperity and economic recessions.

There is very little long-term trend in the unemployment rate. There are, however, strong cycles that show the effect of business cycles on unemployment. When times are good, unemployment is low. Economic recessions are marked by high unemployment. You can see that unemployment reached peaks in about 1971, 1975, 1983, and 1992. The recession of 1982–1983 was especially serious. The milder 1991–1992 recession helped Bill Clinton defeat the incumbent George Bush in the 1992 presidential election.

If President Bush's advisors had been able to predict the 1991–1992 recession and take corrective action, Mr. Bush might well have been reelected. It is very difficult to predict the future course of the economy. Not only are business cycles irregular, but erratic fluctuations are large. Storms, strikes, oil embargoes, and other unpredictable occurrences large and small affect the economy. Does a three-month rise in unemployment mean that the business cycle has peaked and is about to turn down? Perhaps it is only a "pause" (to quote the president's chief economic advisor on one such occasion) caused by lack of rain in Nebraska or a strike at Ford Motor Company. We cannot see the future clearly, even with the aid of a statistical record of the past.

A poor record of success does not prevent the foolhardy from continued attempts to predict the future. Economists often fail despite sophisticated statistics, foiled by the complexity of the economy and by

"Isn't it fascinating that you were ruined by a business cycle, while I was ruined by an erratic fluctuation?"

erratic outside shocks. More simple-minded predictions often discover coincidences and call them explanations.

> EXAMPLE 9. **Predicting presidential elections.** A writer in the early 1960s remarked on the great success of a simple method for predicting presidential elections: just choose the candidate with the longer name.[7] In the 22 elections from 1876 to 1960, this method failed only once. (In two cases the longer name correctly predicted the winner in the popular vote, but the winner then lost in the Electoral College. In 1916, both names were the same length so the method did not apply. Nonetheless, the record is impressive.)
>
> I hope that the writer didn't bet the family silver in later elections after coming up with this clever idea. The eight elections from 1964 to 1992 presented seven tests of the "long name wins" method (the 1980 candidates, Reagan and Carter, had names of the same length). The longer name lost five of the seven.

SECTION 4 EXERCISES

6.30   Which of these time series do you expect to show a clear trend? Will the trend be upwards or downwards? (All data are recorded annually.)

(a)   The percent of students entering a university who bring a typewriter with them.

(b)   The percent of students entering a university who bring a personal computer with them.

(c)   The percent of adult women who do not work outside the home.

6.31   You examine the average temperature in Chicago each month for many years. Do you expect this time series to show seasonal variation? Describe the kind of seasonal variation you expect to see.

6.32   **Gasoline prices.** Figure 6-4 is a line graph of the average monthly price of regular unleaded gasoline from 1986 to mid-1996. The prices are collected as part of the sampling that

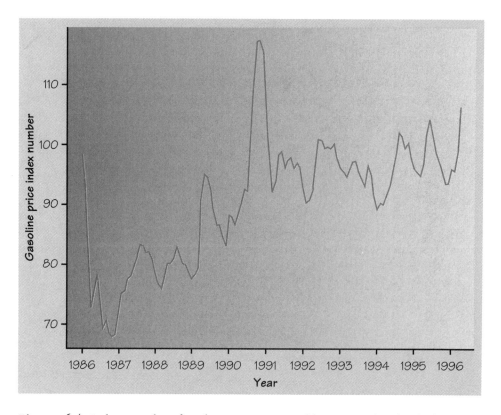

Figure 6-4 Index number for the average monthly price of unleaded regular gasoline from 1986 to mid-1996, from CPI data.

produces the CPI. The graph shows the index number (1982–84 = 100).

(a)   There is clear seasonal variation in gasoline prices. At what time of the year do gasoline prices regularly increase? Why do you think this happens?

(b)   Describe the trend in gasoline prices over the decade shown in the graph.

(c)   Describe the effect of the 1990 Gulf War on gasoline prices.

6.33   **The sunspot cycle.** Figure 6-5 is a line graph of the mean annual number of sunspots on the sun's visible face for years between 1912 and 1965. As the graph shows, the number of sunspots follows a quite regular cycle.

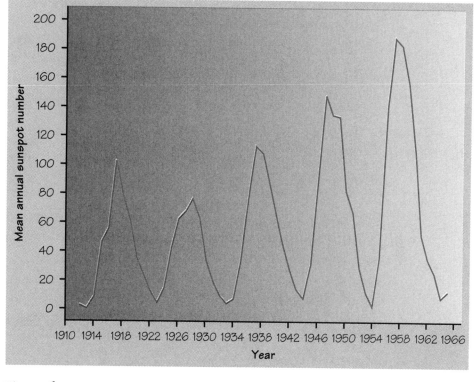

Figure 6-5 Mean annual sunspot numbers for the years 1912 to 1965. These data show the sunspot cycle.

(a)   Note from the graph the years in which the sunspot cycle reached its minimum. Then find the lengths of the five complete cycles that appear in the graph by finding the number of years between these minima.

(b)   Estimate the average length of the sunspot cycle by the mean of the five lengths.

6.34   The sales at your new gift shop in December are double the November value. Should you conclude that your shop is growing more popular and will soon make you rich? Explain your answer.

6.35   **Counting employed people.** A news article says, "More people were working in America in June than in any month since the end of 1990. If that's not what you thought you read in the papers last week, don't worry. It isn't. The report that employment plunged in June, with nonfarm payrolls declining by 117,000, helped to persuade the Federal Reserve to cut interest rates yet again. . . . In reality, however, there were 457,000 more people employed in June than in May."[8] What explains the difference between the fact that employment went up by 457,000 and the official report that employment went down by 117,000?

6.36   The BLS publishes the CPI both unadjusted and seasonally adjusted. There is some seasonal variation owing to weather, holidays, and so forth, so these two versions of the CPI often differ slightly.

(a)   If you want to follow general price trends in the economy, would you use the seasonally adjusted or unadjusted version of the CPI? Why?

(b)   If you have a labor contract with an escalation clause tied to the CPI, you want your wages to keep pace with the actual prices you must pay. Which version of the CPI would you use for this purpose, and why?

6.37   **More sunspots.** Figure 6-6 is a plot of the annual average number of sunspots on the sun's visible face from 1610 to 1989. The regular sunspot cycle that you explored in Exercise 6.33 is the most obvious feature of this time series, but there are other features in the longer run.

(a)   By tracing the curve of the sunspot peaks over many cycles, I think I can see a longer cycle superimposed on the sunspot cycle. Comment on this suggestion.

(b)   Does this time series show any striking noncyclical phenomena? Describe any you notice.

6.38   A rich eccentric once founded an institute to seek out cycles in natural and human affairs and use the regularity of cycles for prediction. The institute noted after studying thousands of time series that there had been, for example, a quite regular 5.9-year cycle in both cotton prices and the abundance of grouse. Would you be willing to risk your money in speculation on cotton prices on the basis of this cycle? What do you think explains the similar behavior of cotton prices and abundance of grouse?

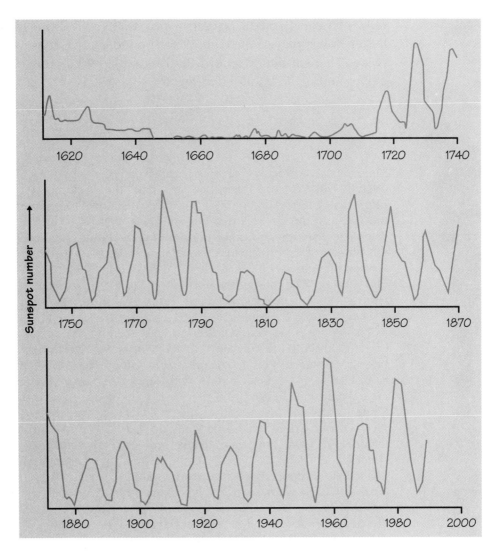

Figure 6-6 Mean annual sunspot numbers from 1610 to 1989. (Adapted from Peter V. Foukal, "The variable sun," *Scientific American,* February 1990, page 36.)

6.39   **The Super Bowl indicator.** One of the best known though least serious methods of predicting the stock market is the Super Bowl indicator. If the winner of the Super Bowl, played in January each year, is a team from the old National Football League, stocks will rise that year. If a team that came out of the old American Football League wins, stocks will drop. In the 23 Super Bowls played through 1989, the indicator was wrong only twice. That's a better record than most. How has this indicator performed in 1990 and later years?

NOTES

1. These data, correct as of 1996, appear in the BLS web site. The overall URL is http://stats.bls.gov, and the CPI home page is http://stats.bls.gov/cpihome.htm. More detailed information about the CPI and the surveys that sustain it, as well as about the Current Population Survey and other important data sources, can be found in the Bureau of Labor Statistics' *Handbook of Methods.* The handbook is revised regularly. The latest edition appeared in 1992.
2. G. J. Borjas, "The internationalization of the U.S. labor market and the wage structure," in *Federal Reserve Bank of New York Economic Policy Review,* Volume 1, Number 1 (1995), pp. 3–8. The quote appears on p. 3. This entire issue is devoted to articles seeking to explain stagnant earnings and the income gap. The consensus: we don't know.
3. "The good statistics guide," *Economist,* September 13, 1993, p. 65.
4. S. E. Fienberg, "A conversation with Janet Norwood," *Statistical Science,* Volume 9 (1994), pp. 574–590. The quote appears on p. 575.
5. Some results and discussion of the NORC survey appear in T. Lewin, "So, now we know what Americans do in bed. So?" *New York Times,* October 9, 1994. For more detail, see R. T. McNeil, et al., *Sex in America: A Definitive Survey,* Little, Brown and Company, 1994, or the scholarly report, E. O. Laumann, et al., *The Social Organization of Sexuality: Sexual Practices in the United States,* University of Chicago Press, 1994.
6. For general information, see J. A. Davis and T. W. Smith, *The NORC General Social Survey,* Sage Publications, 1992.
7. The writer is D. A. Borgman, *Language on Vacation,* Scribner's, 1965.
8. F. Norris, "Market Watch" column, *New York Times,* July 15, 1992.
9. From a news item in *Consumer Reports* magazine, June 1992, p. 392.

REVIEW EXERCISES

6.40   **Bowling and golfing.** As part of a report on trends in recreation, you decide to present index numbers to compare changes in different recreational activities. Here are data on participation in golf and bowling, from the *Statistical Abstract*. Give the index numbers (1980 = 100) for both sports for 1985, 1990, and 1993. Which sport has grown faster? Do the index numbers tell you which sport has more participants?

Sport	1980	1985	1990	1993
Bowlers (millions)	72.0	67.0	71.0	79.0
Golfers (thousands)	15,112	17,520	27,800	24,600

6.41 **New car prices.** According to the Motor Vehicle Manufacturers Association, the retail prices of new cars rose by 68% between 1982 and 1992. In contrast, the price of a new car used to calculate the CPI rose by only 24% in the same period.[9] Why doesn't the CPI include the full amount of the rise in the cost of new cars?

6.42 **Babe Ruth's salary.** In 1930, Babe Ruth earned $80,000, an astounding amount in those days. In fact, it was more than the president was paid. Asked about that, Ruth said, "I had a better year." What does the Babe's salary amount to in 1995 dollars?

6.43 **Phoning Japan.** A three-minute telephone call to Japan cost $30 in 1934, when AT&T opened the first radio channel across the Pacific. The same call cost $6.34 in 1964 on the first sea-bottom copper cable. By 1989, you could talk to Japan for three minutes for $3.78 via communications satellite. Restate the 1934 and 1964 prices in 1989 dollars to see the price decrease in real terms. (Use the 1935 and 1965 entries in Table 6-1.)

6.44 **The price of gold.** Some people recommend that investors buy gold "to protect against inflation." Here are the prices of an ounce of gold at the end of the year for years between 1983 and 1995. Make a graph that shows how the price of gold changed in real terms over this period. Would an investment in gold have protected against inflation by holding its value in real terms?

Year	1983	1985	1987	1989	1991	1993	1995
Gold price	$385	$329	$486	$403	$354	$391	$392

6.45   Sketch line graphs of time series with each of these characteristics. Mark your time axis in years. Be sure to include some erratic fluctuations.

(a)   A strong downward trend, but no seasonal variation.

(b)   Seasonal variation each year, but no clear trend.

(c)   A strong downward trend with yearly seasonal variation.

(d)   Yearly seasonal variation, no clear trend, and irregular cycles about 3 years long.

6.46   Explain why you need many years' data for an economic time series such as the unemployment rate in order to feel comfortable estimating the seasonal variation and any long-term trend.

WRITING PROJECTS

6.1   **Military spending.** Here are data on U.S. spending for national defense for fiscal years between 1940 and 1995, from the *Statistical Abstract*. The units are millions of dollars (this is real money).

Year	1940	1945	1950	1955	1960	1965
Spending	1,660	82,965	13,724	42,729	48,130	50,620

Year	1970	1975	1980	1985	1990	1995
Spending	81,692	86,509	133,995	252,748	299,331	271,600

Write an essay that describes the changes in military spending in real terms during this period from just before World War II until the end of the cold war. Do the necessary calculations and write a brief description that ties military spending to the major military events of this period: World War II (1941–1945), the Korean War (1950–1953), the Vietnam War (roughly 1964–1975), and the end of the cold war after the fall of the Berlin Wall in 1989.

6.2   **Many consumer price indexes?** In addition to the main CPI that applies to all urban consumers (CPI-U), the BLS publishes another CPI that uses different weights to represent the spending of only urban wage earners and clerical workers (CPI-W). The

CPI-W was the original version of the CPI. When the CPI-U was introduced as a better index for most purposes, political pressure from labor unions kept the CPI-W in place for use in indexing wages paid under union contracts.

If unions can have their own CPI, why not others? In particular, some members of Congress want to set up a price index for older consumers. This index would have a market basket that reflects the spending patterns of older citizens, such as their high spending on medical care. It would replace the general CPI to index Social Security payments. Briefly discuss the pros and cons of this proposal, and express your opinion.

PART III

DRAWING CONCLUSIONS FROM DATA

You have designed your experiment, assembled the subjects and assigned them at random, dealt with the myriad details, enforced double-blind, and made your measurements. You have graphed and tabled your data, averaged and correlated them. They seem to point to a clear conclusion. But how strong is the evidence that the data offer for that conclusion? Will your evidence stand the scrutiny of nitpicking journal editors or unhappy special interests?

Your data, and almost all other data, show variable outcomes. Patients receiving the new treatment live a little longer on the average, but not all live longer than all the patients who got the old treatment. Students taught by the new method learn a little more on the average; the new variety of corn gives a little higher yield on the average. Could that "little on the average" just be chance variation? You need more statistics to argue convincingly that your effects really are due to the treatment. What you need is *statistical inference:* formal methods for drawing conclusions from data, taking into account the effects of chance variation.

Formal statistical inference is based on *probability*, the language of chance. Probability is a subject of interest to those who wish to understand such worldly pursuits as roulette and state lotteries as well as to students of the lofty subject of statistics. It is the subject of Chapter 7. Chapter 8 presents some of the concepts behind statistical inference. Our emphasis is on the reasoning of inference. We will leave details of the methods of inference to more traditional introductions to statistics.

PROBABILITY: THE LANGUAGE OF CHANCE

Even the rules of football agree that tossing a coin avoids favoritism. Favoritism in choosing subjects for a sample survey or allotting patients to treatment and placebo groups in a medical experiment is as undesirable as it is in awarding first possession of the ball in football. Statisticians therefore recommend probability samples and randomized experiments, which are fancy versions of tossing a coin. The deliberate use of chance is the central idea of statistical designs for producing data. Both tossing a coin and choosing an SRS are *random*.

> **Randomness**
>
> We call a phenomenon **random** if:
>
> ▶ The exact outcome is not predictable in advance.
>
> ▶ Nonetheless, there is a predictable long-term pattern that can be described by the distribution of the outcomes of very many trials.

The inventors of probability samples and randomized experiments in this century were not the first to notice that some phenomena are random in this sense. They were drawing upon a long history of the study of randomness and applying the results of that study to statistics.

THE ANCIENT HISTORY OF CHANCE

Randomness is most easily noticed in many repetitions of games of chance—rolling dice, dealing shuffled cards, spinning a roulette wheel. Chance devices similar to these have been used from remote antiquity to discover the will of the gods. The most common method of randomization in ancient times was "rolling the bones," tossing several *astragali*. The astragalus is a solid, quite regular bone from the heel of animals that, when thrown, will come to rest on any of four sides. (The other two sides are rounded.) Cubical dice, made of pottery or bone, came later, but even dice existed before 2000 B.C. Gambling on the throw of astragali or dice is, compared with divination, almost a modern development. There is no clear record of this vice before about 300 B.C. Gambling reached flood tide in Roman times, then temporarily receded (along with divination) in the face of Christian displeasure.[1]

Chance devices such as astragali have been used from the beginning of recorded history. Yet none of the great mathematicians of antiquity studied the regular pattern of many throws of bones or dice. Perhaps this is because astragali and most ancient dice were so irregular that each had a different pattern of outcomes. Or perhaps the reasons lie deeper, in the classical reluctance to engage in systematic experimentation.

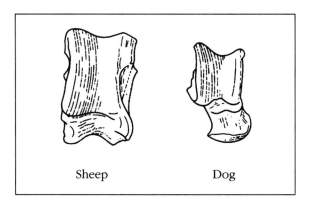

Sheep   Dog

Animal astragali, actual size. [Reproduced by permission of the publishers, Charles Griffin & Company, Ltd., of London and High Wycombe. From F. N. David, *Games, Gods, and Gambling,* 1962.]

Professional gamblers, who are not as inhibited as philosophers and mathematicians, did notice the regular pattern of outcomes of dice or cards and tried to adjust their bets to the odds of success. "How should I bet?" is the question that launched mathematical probability. The systematic study of randomness began (I oversimplify, but not too much) when seventeenth-century French gamblers asked French mathematicians for help in figuring out the "fair value" of bets on games of chance. *Probability theory,* the mathematical study of randomness, originated with Pierre de Fermat and Blaise Pascal in the seventeenth century and was well developed by the time statisticians took it over in the twentieth century. In this chapter we examine probability, but without mention of the actual mathematics that has grown from the first attempt of French mathematicians to aid their gambler friends.

Probability is now important for many reasons having little to do with either gambling or producing data. Many natural and artificial phenomena are random in the sense that they are not predictable in advance but have long-term patterns. The science of genetics grew from Gregor Mendel's observation that for given parents the characteristics of offspring are random, with long-run patterns that he began to uncover. The emission of particles from a radioactive source occurs randomly over time, with a pattern that helped to suggest the cause of radioactivity. Probability is used to describe phenomena in genetics, physics, and many other fields of study. Although we will not meet such applications here, the ideas of this chapter shed light on these fields as well.

▶ 1 WHAT IS PROBABILITY?

Probability begins with the observed fact that some phenomena are random. "Random" is not a synonym for "haphazard," but a description of a kind of order that emerges only in the long run. We often encounter the unpredictable side of randomness in our everyday experience, but we rarely see enough repetitions of the same random phenomenon to observe the long-term regularity that probability describes. The first five exercises for this section are intended to give you some experience with the somewhat mysterious phenomenon of randomness. Careful study of games of chance will also do the job, but at some risk to your financial health. Let's look first at a very simple random phenomenon.

EXAMPLE 1. Coin tossing. Toss a coin in the air. Will it land heads or tails? Sometimes it lands heads and sometimes tails. We cannot say what the next outcome will be. But if we toss a coin many times, a pattern emerges.

▶ The eighteenth-century French naturalist Count Buffon tossed a coin 4040 times. Result: 2048 heads, or proportion 2048/4040 = 0.5069 for heads.

▶ Around 1900, the English statistician Karl Pearson heroically tossed a coin 24,000 times. Result: 12,012 heads, a proportion of 0.5005.

▶ The English mathematician John Kerrich, while imprisoned by the Germans during World War II, tossed a coin 10,000 times. Result: 5067 heads, a proportion of 0.5067.

The example of coin tossing suggests how to use numbers to describe the long-run regularity of random phenomena. Figure 7-1 shows the results of tossing a coin 1000 times. For each number of tosses from 1 to 1000, I have plotted the proportion of those tosses that gave a head. The first toss was a head, so the proportion of heads starts at 1. The second toss was a tail, reducing the proportion of heads to 0.5 after two tosses. The next three tosses gave a tail followed by two heads, so the proportion of heads after five tosses is 3/5, or 0.6.

The proportion of heads is quite variable at first, but settles down as we make more and more tosses. Eventually this proportion gets close to 0.5 and stays there. We say that 0.5 is the *probability* of a head. The probability 0.5 appears as a horizontal line on the graph. This is the intuitive idea of probability. Probability 0.5 means "occurs half the time in a very large number of trials."

Probability

The **probability** of any outcome of a random phenomenon is the proportion of times the outcome would occur in a very long series of repetitions.

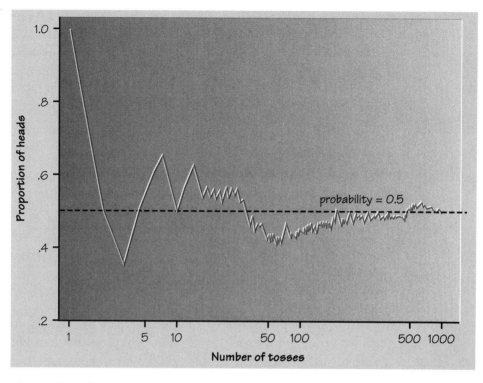

Figure 7-1 The proportion of tosses of a coin that give a head changes as we make more tosses. Eventually, however, the proportion approaches 0.5, the probability of a head.

When we say that a coin has probability 1/2 of coming up heads *on this toss,* we are applying to a single toss a measure of the chance of a head based on what would happen in a long series of tosses. This "definition" of probability as long-term proportion will not satisfy either a mathematician or a philosopher. It just gives us a language to describe the fact that long-term proportions often do settle down to fixed numbers.

THINKING ABOUT PROBABILITY

You may feel that it is obvious from the balance of a coin that the probability of a head is about 1/2. Such opinions are not always correct.

Take a penny and, instead of tossing it, hold it on end on a hard surface with the index finger of one hand and snap it with the other index finger. The coin will spin for several seconds and then fall with either heads or tails showing. A long series of trials reveals that the probability of a head in this random experiment is clearly different from 1/2. Moral: we defined probability *empirically,* that is, in terms of observations. Only by observation can we be reasonably sure of the approximate value of the probability of an outcome.

Of course, not all phenomena are random. Drop a coin from a fixed height and measure the time it takes to fall. You will get the same result (except for measurement errors) every time. The time is predictable, and you can take a physics course to learn how to predict it.

Other phenomena are unpredictable but display no long-term pattern. This is often because someone can intervene in the sequence of outcomes to disrupt the regularity on which randomness depends. If, for example, the operator of a roulette wheel has a brake that she can apply at will, she can prevent the proportion of "red" from settling down to a fixed probability in the long run. Thinking of the roulette operator watching the play and hitting the brake from time to time reminds us that probability as long-run proportion requires more than imagining a long series of repetitions. The repetitions must be *independent.* Independence means that the outcome of one trial gives no information about the outcome of any other trial. Spins of an honest roulette wheel are independent because the wheel doesn't remember where it stopped on previous spins. The operator of a dishonest wheel can remember previous outcomes and intervene to destroy independence.

Although many phenomena are not random, randomness is nonetheless widespread. The behavior of large groups of individuals is often as random as the behavior of many coin tosses or many random samples. Life insurance, for example, is based on the fact that deaths occur at random among many individuals.

> **EXAMPLE 2. The probability of dying.** We can't predict whether a particular person will die in the next year. But if we observe millions of people, deaths are random. The proportion of men aged 25 to 34 who will die next year is about 0.0021. This is the *probability* that a young man will die next year. For women that age, the probability of death is about 0.0007.

If an insurance company sells many policies to people aged 25 to 34, it knows that it will have to pay off on about 0.21% of the policies sold to men and on about 0.07% of the policies sold to women. It will charge more to insure a man because the probability of having to pay is higher.

PROBABILITY MODELS

We describe a random phenomenon by assigning probabilities to its outcomes. This is called giving a *probability model* for the phenomenon. The mathematics of probability begins by laying down the properties of all legitimate probability models. We will need only a few simple rules.

Probability rules

Any legitimate assignment of probabilities to the outcomes of a random phenomenon satisfies these rules:

A. Any probability is a number between 0 and 1.

B. All possible outcomes together must have probability 1.

An **event** is any set of outcomes. Probabilities of events obey these rules as well:

C. The probability that an event does not occur is 1 minus the probability that the event does occur.

D. If two events have no outcomes in common, the probability that one or the other occurs is the sum of their individual probabilities.

These rules are clearly true for proportions; for example, any proportion must be a number between 0 and 1. Although the rules are motivated by the idea of probability as long-run proportion, their purpose is to make it easier to work with probabilities. The simplest way to give a probability model is to assign a probability to each individual outcome. *Any assignment that satisfies Rules A and B is legitimate.* Find the probability of any event by adding the probabilities of its outcomes. Rules C and D are then automatically satisfied.

EXAMPLE 3. **Choosing at random.** Choose at random an American woman aged 25 to 29 years. That is, choose one woman in a way that gives all women an equal chance to be chosen. When we do this, the probability that the woman we choose is (say) married is just the proportion of all women aged 25 to 29 who are married. The Census Bureau tells us that the probabilities that we choose a woman who is single, married, widowed, or divorced are:

Outcome	Single	Married	Widowed	Divorced
Probability	0.352	0.577	0.003	0.068

Check that these probabilities satisfy Rules A and B. The probability that the woman we select is not married is the sum of the probabilities that she is single, widowed, or divorced:

Probability she is not married $= 0.352 + 0.003 + 0.068 = 0.423$

You can also use Rule C:

Probability she is not married $= 1 -$ probability she is married
$$= 1 - 0.577 = 0.423$$

EXAMPLE 4. **Rolling two dice.** Rolling two dice is a common way to lose money in casinos. There are 36 possible outcomes when we roll two dice and record the up faces in order (first die, second die). Figure 7-2 displays these outcomes. What probabilities should we assign to these outcomes?

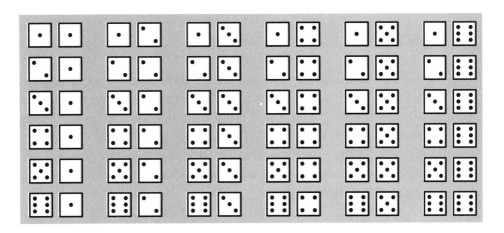

Figure 7-2 The possible outcomes of rolling two dice.

Casino dice are carefully made. Their spots are not hollowed out, which would give the faces different weights, but are filled with white plastic of the same density as the red plastic of the body. For casino dice it is reasonable to assign the same probability to each of the 36 outcomes in Figure 7-2. Because these 36 probabilities must have sum 1 (Rule B), each outcome must have probability 1/36.

We are interested in the sum of the spots on the up faces of the dice. What is the probability that this sum is 5? The event "roll a 5" contains these four outcomes:

Because each outcome has probability 1/36, the probability of rolling a 5 is

$$\frac{1}{36} + \frac{1}{36} + \frac{1}{36} + \frac{1}{36} = \frac{4}{36} = 0.111$$

The rules only tell us what probability models *make sense.* They don't tell us whether the assignment is *correct,* that is, whether it actually describes what happens in the long run. The probabilities in Example 4, for example, are correct for casino dice. Inexpensive dice with hollowed-out spots are not balanced and this assignment of probabilities does not describe their behavior.

SAMPLING DISTRIBUTIONS AGAIN

Choosing a random sample from a population and calculating a statistic such as the sample proportion or sample mean is certainly a random phenomenon. The sampling distribution of a statistic displays the regular pattern of values of the statistic in a large number of samples from the same population. That sounds like the idea of probability.

Sampling distributions

The sampling distribution of a statistic gives a probability model for the values of the statistic in repeated sampling.

EXAMPLE 5. An opinion poll. Ask an SRS of 1373 parents whether they would have children again. The proportion who say "Yes,"

$$\hat{p} = \frac{\text{number who say "Yes"}}{1373}$$

is a statistic. The values of $\hat{p}$ vary from sample to sample, not haphazardly but in a regular pattern described by the sampling distribution.

Figure 7-3 shows two versions of this sampling distribution in the case when the truth about the population is that 80% would say "Yes." The histogram displays the results of 1000 samples. The normal curve is a good approximation to the distribution that would emerge after many thousands of samples. It has mean 0.8 and standard deviation 0.011.

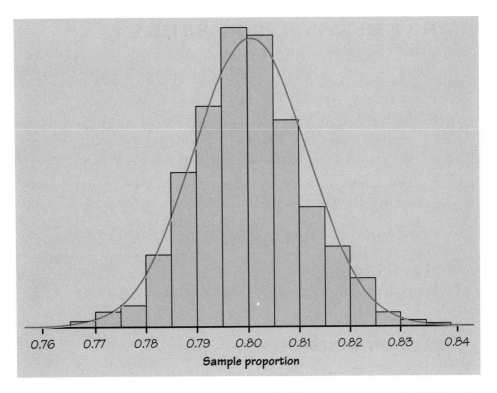

Figure 7-3 A sampling distribution. This distribution assigns probabilities that describe how often the statistic takes various values in very many samples from the same population.

> *The normal curve assigns probabilities to sets of outcomes.* For example, the "95" part of the 68–95–99.7 rule says that 95% of all samples will give a $\hat{p}$ falling within 0.022 of 0.8, that is, between 0.778 and 0.822. We now have more concise language for this fact: the probability is 0.95 that between 77.8% and 82.2% of the people in a sample will say "Yes."

Normal curves don't assign probabilities to individual outcomes like "Exactly 1126 of the 1373 people in the sample say 'yes'." They are only good approximations to the exact sampling distributions. Normal curves assign probabilities only to events such as "the sample proportion $\hat{p}$ takes a value between 0.778 and 0.822." We find the probabilities from the 68–95–99.7 rule or from Table B of percentiles of normal distributions. These probabilities satisfy Rules A to D.

HOW TO ASSIGN PROBABILITIES

Our brief tour of probability models has given us two ways to assign probabilities:

▶ Assign probabilities to every individual outcome in a way that satisfies Rules A and B. The probability of any event is the sum of the probabilities of the outcomes in that event.

▶ Use a normal curve to assign probabilities directly to sets of outcomes using the 68–95–99.7 rule or Table B.

Normal curves and sampling distributions will dominate the next chapter, where we turn to statistical inference. In this chapter we will seek to understand probability from examples of simpler models that assign a probability to each possible outcome.

SECTION 1 EXERCISES

7.1　　**Randomness: tossing a thumbtack.** Toss a thumbtack on a hard surface 100 times. How many times did it land with the point up? What is the approximate probability of landing point up?

7.2   **Randomness: random digits.** The table of random digits (Table A) was produced by a random mechanism that gives each digit probability 0.1 of being a 0. What proportion of the first 200 digits in the table are 0s? This proportion is an estimate of the true probability, which in this case is known to be 0.1.

7.3   **Randomness: spinning a coin.** Hold a penny upright on its edge under your forefinger on a hard surface, then snap it with your other forefinger so that it spins for some time before falling. Based on 50 spins, what is the probability of heads?

7.4   **Randomness: waiting for the first head.** When we toss a penny, experience shows that the probability (long-term proportion) of a head is close to 1/2. Suppose now that we toss the penny repeatedly until we get a head. What is the probability that the first head comes up in an odd number of tosses (1, 3, 5, and so on)? To find out, repeat this experiment 50 times, and keep a record of the number of tosses needed to get a head on each of your 50 trials.

(a)   From your experiment, estimate the probability of a head on the first toss. What value should we expect this probability to have?

(b)   Use your results to estimate the probability that the first head appears on an odd-numbered toss.

7.5   **Randomness: rolling dice.** Roll a pair of dice 100 times and record the sum of the spots on the upward faces in each trial. On what proportion of these 100 trials do the dice total 8? Compare your result with the probability calculated from the probability model for casino dice in Example 4.

7.6   You read in a book on poker that the probability of being dealt three of a kind in a five-card poker hand is 1/50. Explain in simple language what this means.

7.7   Probability is a measure of how likely an event is to occur. Match one of the probabilities that follow with each statement about an event. (The probability is usually a much more exact measure of likelihood than is the verbal statement.)

0, 0.01, 0.3, 0.6, 0.99, 1

(a)   This event is impossible. It can never occur.

(b)   This event is certain. It will occur on every trial of the random phenomenon.

(c)   This event is very unlikely, but it will occur once in a while in a long sequence of trials.

(d)   This event will occur more often than not.

7.8   **Colors of M&Ms.** Buy a bag of M&M candies and draw a candy at random from the bag. The candy can have any one of six colors. The probability of drawing each color depends on the proportion of each color among all candies made. The table below gives the probability that a randomly chosen M&M has each color. What must the probability for blue candies be to make this a legitimate assignment of probabilities? What is the probability that the candy drawn is red, yellow, or orange?

Color	Brown	Red	Yellow	Green	Orange	Blue
Probability	0.3	0.2	0.2	0.1	0.1	?

7.9   **Does he do his share?** The *New York Times* (August 21, 1989) reported a poll that interviewed a probability sample of 1025 women. The married women in the sample were asked whether their husband did his fair share of household chores. Here are the results:

Outcome	Probability
Does more than his fair share	0.12
Does his fair share	0.61
Does less than his fair share	?

These proportions give us a probability model for the random phenomenon of choosing a married woman at random and asking her opinion.

(a)   What must be the probability that the woman chosen says that her husband does less than his fair share? Why?

(b)   The event "I think my husband does at least his fair share" contains the first two outcomes. What is its probability?

7.10 Exactly one of Brown, Chavez, and Williams will be promoted to partner in the law firm that employs them all. Brown thinks that she has probability 0.25 of winning the promotion and that Williams has probability 0.2. What probability does Brown assign to the outcome that Chavez is the one promoted?

7.11 **Rolling dice.** Example 4 gives a probability model for rolling two dice. According to this model, what is the probability that the sum of the spots on the up faces is either 7 or 11?

7.12 **Roulette.** An American roulette wheel contains compartments numbered 1 through 36 plus 0 and 00. Of the 38 compartments, 0 and 00 are colored green, 18 of the others are red, and 18 are black. A ball is spun in the direction opposite to the wheel's motion, and bets are made on the number where the ball comes to rest. A simple wager is *red-or-black,* in which you bet that the ball will stop in, say, a red compartment. If the wheel is fair, all 38 compartments are equally likely.

(a) What is the probability of a red?

(b) In the red-or-black bet, green belongs to the house. What is the probability of green?

7.13 Figure 7-4 displays four assignments of probability to the outcomes of rolling a single die. Which, if any, is *correct* for this die can be discovered only by rolling the die many times. But some of the models are not *legitimate* assignments of probability. Which are legitimate and which are not, and why?

7.14 In each of the following cases, state whether or not the given assignment of probabilities to individual outcomes is legitimate in the sense of satisfying Rules A and B. If not, give specific reasons why not.

(a) When a coin is spun, $P(H) = 0.55$ and $P(T) = 0.45$.

(b) When two coins are tossed, $P(HH) = 0.4$, $P(HT) = 0.4$, $P(TH) = 0.4$, and $P(TT) = 0.4$.

(c) The mixture of colors for M&M's given in Exercise 7.8 is about to be replaced by a new mixture. Now there will be no blue candies, and the other five colors will have the same probabilities given in Exercise 7.8.

Outcome	Probability			
	Model 1	Model 2	Model 3	Model 4
⚀	1/7	1/3	1/3	1
⚁	1/7	1/6	1/6	1
⚂	1/7	1/6	1/6	2
⚃	1/7	0	1/6	1
⚄	1/7	1/6	1/6	1
⚅	1/7	1/6	1/6	2

Figure 7-4 Four probability models for rolling a die, for Exercise 7.13.

7.15 A six-sided die is rolled and comes to rest with 1, 2, 3, 4, 5, or 6 spots on the up face. Make an assignment of probabilities to the six outcomes that satisfies Rules A and B and that you think is close to correct for casino dice. (Assume that the die is balanced so that all faces are equally likely to come up. Probability models are often based on assuming balance or symmetry. In this case, observation supports the assumption.) According to your model, what is the probability that an odd face will come up?

7.16 When you toss two coins, the four possible outcomes are (head, head), (head, tail), (tail, head), and (tail, tail). Make an assignment of probabilities to these outcomes that you think is correct for balanced coins. What probability does your model give for the event that exactly one of the coins shows a head?

7.17 **Sampling distributions: Do you jog?** An opinion poll asks an SRS of 1500 adults, "Do you happen to jog?" Suppose (as is approximately correct) that the population proportion who jog is $p = .15$. In a large number of samples, the proportion $\hat{p}$ who answer "Yes" will be approximately normally distributed with mean

0.15 and standard deviation 0.009. Sketch this normal curve and use it to answer these questions:

(a) What percent of many samples will have a sample proportion who jog that is 0.15 or less? Explain clearly why this percent is the probability that $\hat{p}$ is 0.15 or less.

(b) What is the probability that $\hat{p}$ will take a value between 0.141 and 0.159? (Use the 68–95–99.7 rule.)

(c) Now use Rule C for probability: What is the probability that $\hat{p}$ does not lie between 0.141 and 0.159?

7.18 **Sampling distributions: Applying to college.** You ask an SRS of 1500 college students whether they applied for admission to any other college. Suppose that in fact 35% of all college students applied to colleges besides the one they are attending. (That's close to the truth.) The sampling distribution of the proportion $\hat{p}$ of your sample who say "Yes" is approximately normal with mean 0.35 and standard deviation 0.01. Sketch this normal curve and use it to answer these questions:

(a) Explain in simple language what the sampling distribution tells us about the results of our sample.

(b) What percent of many samples would have a $\hat{p}$ larger than 0.37? (Use the 68–95–99.7 rule.) Explain in simple language why this percent is the probability of an outcome larger than 0.37.

(c) What is the probability that your sample will have a $\hat{p}$ less than 0.33?

(d) Use Rule D: What is the probability that your sample result will be either less than 0.33 or greater than 0.35?

▶ 2 ISSUES: DEALING WITH CHANCE

The idea of probability seems straightforward. It answers the question "What would happen if we did this many times?" In fact, both the behavior of random phenomena and the idea of probability are a bit subtle. We meet chance behavior constantly, and psychologists tell us that we deal with it poorly.

THE MYTH OF SHORT-RUN REGULARITY

The idea of probability is that randomness is regular *in the long run.* Unfortunately, our intuition about randomness tries to tell us that random phenomena should also be regular in the short run. When they aren't, we look for some explanation other than chance variation. Our intuition is a very poor guide to understanding chance.[2]

> EXAMPLE 6. **What looks random?** Toss a coin six times and record heads (H) or tails (T) on each toss. Which of these outcomes is more probable?
>
> HTHTTH   TTTHHH
>
> Almost everyone says that HTHTTH is more probable, because TTTHHH does not "look random." In fact, both are equally probable. That heads and tails are equally probable says only that about half of a very long sequence of tosses will be heads. It doesn't say that heads and tails must come close to alternating in the short run. The coin has no memory. It doesn't know what past outcomes were, and it can't try to create a balanced sequence.

Example 6 noted that a *run* of three heads followed by a run of three tails does not look random. Runs of identical outcomes are much more likely to occur by chance than our intuition suggests. Suppose, for example, that you toss a coin 10 times. Runs of three straight heads or tails do not look random to most people. Yet the probability of a run of 3 or more consecutive heads or tails in 10 tosses is greater than 0.8.

> EXAMPLE 7. **The hot hand in basketball.** Belief that runs must result from something other than "just chance" influences behavior. If a basketball player makes several consecutive shots, both the fans and his teammates believe that he has a "hot hand" and is more likely to make the next shot. This is wrong. Careful study has shown that runs of baskets made or missed are no more frequent in basketball than would be expected if each shot is independent of the player's previous shots. Players perform consistently, not in streaks. If a player makes half her shots in the long run, her hits and misses behave just like tosses of a coin—and that means that runs of hits and misses are more common than our intuition expects.[3]

THE "LAW OF AVERAGES"

Six tosses of a coin are exactly as likely to give TTTTTT as HTHTTH. But if we get TTTTTT, isn't the next toss more likely to give a head than a tail because in the long run heads must appear half the time? Many people think that the "law of averages" says that future outcomes must compensate for an imbalance such as six straight tails.[4]

That's wrong. The seventh toss is still equally likely to produce a head or a tail. The coin has no memory. It doesn't know that the first

"So the law of averages doesn't guarantee me a girl after seven straight boys, but can't I at least get a group discount on the delivery fee?"

six outcomes were tails, and it can't try to get a head on the next toss to even things out. Of course, things do even out *in the long run*. After 10,000 tosses, the results of the first six tosses don't matter. They are overwhelmed by the results of the next 9994 tosses, not compensated for.

> **EXAMPLE 8.** **We want a boy.** Belief in this phony "law of averages" can lead to consequences close to disastrous. A few years ago, "Dear Abby" published in her advice column a letter from a distraught mother of eight girls. It seems that she and her husband had planned to limit their family to four children. When all four were girls, they tried again—and again, and again. After seven straight girls, even her doctor had assured her that "the law of averages was in our favor 100 to one." Unfortunately for this couple, having children is like tossing coins. Eight girls in a row is highly unlikely, but once seven girls have been born, it is not at all unlikely that the next child will be a girl—and it was.

The key to understanding why short random sequences need not look regular and why no law of averages demands that future outcomes compensate for early lack of balance lies in the phrase "the coin has no memory." That is, repeated tosses of a coin are *independent*.

Independence

Two random phenomena are **independent** if knowing the outcome of one does not change the probabilities for outcomes of the other.

Independence, like all aspects of probability, can be verified only by observing many repetitions. It is plausible that repeated tosses of a coin are independent (the coin has no memory), and observation shows that they are. It seems less plausible that successive shots by a basketball player are independent, but observation shows that they are at least very close to independent.

THE WOES OF GAMBLERS

Jimmy likes to gamble. He stands by the craps table in a casino, watching Jane roll the dice. On each throw, Jimmy can bet that Jane will win or that she will lose. He watches her win five straight times. What should Jimmy do?

Many gamblers follow the "hot hand" theory. If Jane has thrown five straight winners, she must have a hot hand. So Jimmy should bet that Jane will win again on her next throw. Other gamblers believe in the "law of averages." If Jane has won five times, she is due to lose. So Jimmy should bet against Jane.

Neither strategy makes sense, of course. The dice have no memory. Jane's throws are independent. She has the same probability of winning on each throw. What is more, the casino knows the probabilities and has set the payoffs so that in the long run it will take in more money than it pays out no matter how Jimmy and the other gamblers bet.

WHAT PROBABILITY DOESN'T SAY

There is, you may think, little difference between these statements:

(a) "In many tosses of a fair coin, the proportion of heads will be close to one-half."

(b) "In many tosses of a fair coin, the number of heads will be close to one-half the number of tosses."

Alas, (a) is true, but (b) is false. Statement (a) is what we mean by saying that the probability of a head is 1/2. In many tosses of a fair coin, however, the number of heads is certain to deviate more and more from one-half the number of tosses. To see why this is true, consider the following example.

> EXAMPLE 9. **100,000 tosses of a coin.** Suppose that we toss a coin 100,000 times and record the count of heads after 100, 1000,

10,000 and 100,000 tosses. Here are our results:

Number of tosses	Number of heads	Proportion of heads	Difference between number of heads and 1/2 number of tosses
100	51	0.51	1
1,000	510	0.51	10
10,000	5,100	0.51	100
100,000	51,000	0.51	1,000

There it is: the *proportion* of heads stays close to 1/2, while the *number* of heads departs more and more from one-half the number of tosses. (This exact outcome is unlikely, but it is typical of what happens in many repetitions.)

DOES PROBABILITY MEASURE RISK?

Scientists often use the probability of an unpleasant event to describe its risk. Individuals and society, however, seem to ignore probabilities. We overreact to some risks with very low probabilities while ignoring others that are much more probable.

EXAMPLE 10. **Asbestos in the schools.** High exposures to asbestos are dangerous. Low exposures, such as that experienced by teachers and students in schools where asbestos is present in the insulation around pipes, are not very risky. The probability that a teacher who works for 30 years in a school with typical asbestos levels will get cancer from the asbestos is around 15/1,000,000. The risk of dying in a car accident during a lifetime of driving is about 15,000/1,000,000. That is, driving regularly is 1000 times more risky than teaching in a school where asbestos is present.[5]

Risk does not stop us from driving. Yet the much smaller risk from asbestos launched massive cleanup campaigns and a federal requirement that every school inspect for asbestos and make the findings public.

Why do we take asbestos so much more seriously than driving? Why do we worry about very unlikely threats such as tornados and terrorists more than we worry about heart attacks?

▶ We feel safer when a risk seems under our control than when we cannot control it. We are in control (or so we imagine) when we are driving, but we can't control the risk from asbestos or tornados or terrorists.

▶ It is hard to comprehend very small probabilities. Probabilities of 15 per million and 15,000 per million are both so small that our intuition cannot distinguish between them. Psychologists have shown that we generally overestimate very small risks and underestimate higher risks. Perhaps this is part of the general weakness of our intuition about how probability operates.

▶ The probabilities for risks like asbestos in the schools are not as certain as probabilities for tossing coins. They must be estimated by experts from complicated statistical studies. Perhaps it is safest to suspect that the experts may have underestimated the level of risk.

We don't carry out probability calculations in reacting to risks. Our reactions do depend somewhat on the probability that something bad will happen, but they also depend on our psychological makeup and on social influences. As one writer noted, "Few of us would leave a baby sleeping alone in a house while we drove off on a 10-minute errand, even though car-crash risks are much greater than home risks."[6]

SECTION 2 EXERCISES

7.19 Ask several of your friends (at least 10 people) to choose a four-digit number "at random." How many of the numbers chosen start with 1 or 2? How many start with 8 or 9? (There is strong evidence that people in general tend to choose numbers starting with low digits.[7])

7.20 **Playing "pick four."** The "pick four" games in many state lotteries announce a four-digit winning number each day. The winning number is essentially a four-digit group from a table of random digits. You win if your choice matches the winning digits. The winnings are divided among all players who matched the winning digits. That suggests a way to get an edge.

(a)   The winning number might be, for example, either 2873 or 9999. Explain why these two outcomes have exactly the same probability. (It is 1 in 10,000.)

(b)   If you asked many people which outcome is more likely, they would strongly favor one of them. Use the information in this section to say which one and to explain why. You should therefore choose a number that people think is unlikely. You have the same chance to win, but you will win a larger amount with such numbers because fewer other people will choose them.

7.21   **Shaq's free throws.** The basketball player Shaquille O'Neal makes about half of his free throws over an entire season. In today's game, Shaq makes his first three free throws. The TV commentator says "Shaq's technique really looks good today." Explain why the claim that Shaq has improved his shooting technique is not justified.

7.22   Probability works not by compensating for imbalances but by swamping them. Suppose that the first six tosses of a coin give six tails and that tosses after that are exactly half heads and half tails. (Exact balance is unlikely, but it illustrates how the first six outcomes are swamped by later outcomes.) What is the proportion of heads after the first six tosses? What is the proportion of heads after 100 tosses if the last 94 produce 47 heads? What is the proportion of heads after 1000 tosses if the last 994 produce half heads? What is the proportion of heads after 10,000 tosses if the last 9994 produce half heads?

7.23   The baseball player Tony Gwynn gets a hit about 35% of the time over an entire season. After he has failed to hit safely in six straight at-bats, the TV commentator says "Tony is due for a hit by the law of averages." Is that right? Why?

7.24   A meteorologist, predicting a colder than normal winter, said, "First, in looking at the past few winters, there has been a lack of really cold weather. Even though we are not supposed to use the law of averages, we are due."[8] Do you think that the weather in successive winters is independent? Do you think that "due by the law of averages" makes sense in talking about the weather?

7.25 **An unenlightened gambler.**

(a)   A gambler knows that red and black are equally likely to occur on each spin of a roulette wheel. He observes five consecutive reds occur and bets heavily on black'at the next spin. Asked why, he explains that black is "due by the law of averages." Explain to the gambler what is wrong with this reasoning.

(b)   After hearing you explain why red and black are still equally likely after five reds on the roulette wheel, the gambler moves to a poker game. He is dealt five straight red cards. He remembers what you said, and assumes that the next card dealt in the same hand is equally likely to be red or black. Is the gambler right or wrong, and why?

7.26 **Choose your game.** You are gambling with a coin that has probability 1/2 of coming up heads on each toss. You are allowed to choose either 10 or 100 tosses.

(a)   On the first bet, you win if the proportion of heads is between 0.4 and 0.6. Should you choose 10 tosses or 100 tosses? Why?

(b)   On the second bet, you win if exactly half of the tosses are heads. Should you choose 10 tosses or 100 tosses? Why?

7.27 **Reacting to risks.** The probability of dying if you play high school football is about 10 per million each year you play. The risk of getting cancer from asbestos if you attend a school in which asbestos is present for 10 years is about 5 per million. If we ban asbestos from schools, should we also ban high school football? Briefly explain your position.

7.28 **Reacting to risks.** National newspapers such as *USA Today* and the *New York Times* carry many more stories about deaths from airplane crashes than about deaths from automobile crashes. Auto accidents kill about 40,000 people in the United States each year. Crashes of all scheduled air carriers have killed between 25 and 550 people per year in recent years.

(a)   Why do the news media give more attention to airplane crashes?

(b)   How does news coverage help explain why most people consider flying more dangerous than driving?

▶ 3 FINDING PROBABILITIES BY SIMULATION

Toss a coin 10 times. What is the probability of a run of 3 or more consecutive heads or tails? A couple plans to have children until they have a girl or until they have four children, whichever comes first. What is the probability that they will have a girl among their children? There are three methods we can use to answer probability questions like these:

"I've had it! Simulated wood, simulated leather, simulated coffee, and now simulated probabilities!"

1. Actually carry out the random phenomenon many times. That's slow and often awkward.

2. Start with a probability model and use mathematics. That's often hard. The probability of a run of 3 heads or 3 tails, for example, is too hard for most math majors.

3. Start with a probability model and use random digits to *simulate* many repetitions. This is quicker than repeating the real phenomenon (especially if we program a computer to do the simulation) and we can do problems too hard for our math to reach.

Simulation

Using random digits from a table or from computer software to imitate chance behavior is called **simulation**.

SIMULATION BASICS

Simulation is an effective tool for finding probabilities of complex events once we have a trustworthy probability model. We can use random digits to simulate many repetitions quickly. The proportion of repetitions on which an event occurs will eventually be close to its probability, so simulation can give good estimates of probabilities. The art of simulation is best learned from a series of examples.

> **EXAMPLE 11. Doing a simulation.** Toss a coin 10 times. What is the probability of a run of at least 3 consecutive heads or 3 consecutive tails?
>
> **Step 1. Give a probability model.** Our model for coin-tossing has two parts:
>
> ▶ Each toss has probabilities 0.5 for a head and 0.5 for a tail.
>
> ▶ Tosses are independent of each other.
>
> **Step 2. Assign digits to represent outcomes.** Digits in Table A of random digits will stand for the outcomes, in a way that matches the probabilities from Step 1. We know that each digit in Table A

has probability 0.1 of being any of 0, 1, 2, 3, 4, 5, 6, 7, 8, or 9, and that successive digits in the table are independent. Here is one assignment of digits for coin-tossing:

▶ One digit simulates one toss of the coin.

▶ Odd digits represent heads, even digits represent tails.

This works because the 5 odd digits give probability 5/10 to heads. Successive digits in the table simulate independent tosses.

Step 3. Simulate many repetitions. Ten digits simulate 10 tosses, so looking at 10 consecutive digits in Table A simulates one repetition. Read many groups of 10 digits from the table to simulate many repetitions. Be sure to keep track of whether or not the event we want (a run of 3 heads or 3 tails) occurs on each repetition.

Here are the first three repetitions, starting at line 101 in Table A. I underlined all runs of 3 or more heads or tails.

Digits	19223 95034	05756 28713	96409 12531
Heads/tails	HHTT<u>H</u> HHTHT	T<u>HHH</u>T TT<u>HHH</u>	HT<u>TTT</u>H HT<u>HHH</u>
Run of 3	YES	YES	YES

Continuing in Table A, I did 25 repetitions; 23 of them did have a run of 3 or more heads or tails. So we estimate the probability of a run by the proportion

$$\text{estimated probability} = \frac{23}{25} = 0.92$$

Of course, 25 repetitions are not enough to be confident that our estimate is accurate. Now that we understand how to do the simulation, we can tell a computer to do many thousands of repetitions. A long simulation (or hard mathematics) finds that the true probability is about 0.826.

Once you have gained some experience in simulation, setting up the probability model (Step 1) is usually the hardest part of the process. Although coin-tossing may not fascinate you, the model in Example 11 is typical of many probability problems because it consists of independent trials (the tosses) all having the same possible outcomes and probabilities. Shooting 10 free throws and observing the sexes of 10 children have similar models and are simulated in much the same way.

It may seem a bit shady to begin the process of finding a probability by assuming that we already know some other probabilities, but not even mathematics can give us something for nothing. The idea is to state the basic structure of the random phenomenon and then use simulation to move from this model to the probabilities of more complicated events. The model is based on opinion and past experience. If it does not correctly describe the random phenomenon, the probabilities derived from it by simulation will also be incorrect.

Step 2 (assigning digits) rests on the properties of the random digit table. Here are some examples of this step.

> **EXAMPLE 12. Assigning digits for simulation.**
>
> **(a)** Choose a person at random from a group of which 70% are employed. One digit simulates one person:
>
> $$0, 1, 2, 3, 4, 5, 6 = \text{employed}$$
>
> $$7, 8, 9 = \text{not employed}$$
>
> **(b)** Choose one person at random from a group of which 73% are employed. Now *two* digits simulate one person:
>
> $$00, 01, 02, \ldots, 72 = \text{employed}$$
>
> $$73, 74, 75, \ldots, 99 = \text{not employed}$$
>
> We assigned 73 of the 100 two-digit pairs to "employed" to get probability 0.73. Representing "employed" by 01, 02, ..., 73 would also be correct.
>
> **(c)** Choose one person at random from a group of which 50% are employed, 20% are unemployed, and 30% are not in the labor force. There are now three possible outcomes, but the principle is the same. One digit simulates one person:
>
> $$0, 1, 2, 3, 4 = \text{employed}$$
>
> $$5, 6 = \text{unemployed}$$
>
> $$7, 8, 9 = \text{not in the labor force}$$

MORE ELABORATE SIMULATIONS

The building and simulation of random models is a powerful tool of contemporary science, yet a tool that can be understood without advanced mathematics. What is more, several attempts to simulate random phenomena will increase your understanding of probability more than many pages of my prose. Having in mind these two goals of understanding simulation for itself and understanding simulation to understand probability, let us look at two more elaborate examples.

> EXAMPLE 13. **We want a girl.** A couple plans to have children until they have a girl or until they have four children, whichever comes first. What is the probability that they will have a girl among their children?
>
> **Step 1.** The probability model is the same as for coin-tossing:
>
> ▶ Each child has probability 0.5 of being a girl and 0.5 of being a boy.
>
> ▶ The sexes of successive children are independent.
>
> **Step 2.** Assigning digits is also easy. One digit simulates the sex of one child:
>
> $$0, \ 1, \ 2, \ 3, \ 4 = \text{girl}$$
> $$5, \ 6, \ 7, \ 8, \ 9 = \text{boy}$$
>
> **Step 3.** To simulate one repetition of this child-bearing strategy, read digits from Table A until the couple has either a girl or four children. Notice that the number of digits needed to simulate one repetition depends on how quickly the couple gets a girl. Here is the simulation, using line 130 of Table A. To interpret the digits, I have written G for girl and B for boy under them, have added space to separate repetitions, and under each repetition have written "+" if a girl was born and "−" if not.

690	51	64	81	7871	74	0
BBG	BG	BG	BG	BBBG	BG	G
+	+	+	+	+	+	+

951	784	53	4	0	64	8987
BBG	BBG	BG	G	G	BG	BBBB
+	+	+	+	+	+	−

In these 14 repetitions, a girl was born 13 times. Our estimate of the probability that this strategy will produce a girl is therefore

$$\text{estimated probability} = \frac{13}{14} = 0.93$$

Some mathematics shows that if our probability model is correct, the true probability of having a girl is 0.938. Our simulated answer came quite close. Unless the couple is unlucky, they will succeed in having a girl.

EXAMPLE 14. The Asian stochastic beetle. We are studying the Asian stochastic beetle.[9]

Step 1. Females of this insect have the following pattern of reproduction:

▶ 20% of females die without female offspring, 30% have one female offspring, and 50% have two female offspring.

▶ Different females reproduce independently.

What will happen to the population of Asian stochastic beetles: Will they increase rapidly, barely hold their own, or die out? Notice that we can ignore the male beetles in studying reproduction, as long as there are some around for certain essential purposes. Although we are studying only a single population, ecologists use probability models and simulation to study the interaction of several populations, such as predators and prey.

Step 2. To simulate the model we have given, use one digit to simulate one female:

$$0, 1 = \text{dies without female offspring}$$
$$2, 3, 4 = \text{has one female offspring}$$
$$5, 6, 7, 8, 9 = \text{has two female offspring}$$

Step 3. To answer the question "What is the future of the Asian stochastic beetle?" we will simulate the female descendents of several female beetles until they either die out or reach the fifth generation. Beginning at line 122 of the table of random digits,

13873   81598   95052   90908   73592

the first beetle dies without offspring (1). The second has one female offspring (3); she in turn has two female offspring (8); the first of these has two (7) and the second has one (3) female offspring. So the fourth generation of this family contains three female beetles.

We need a clearer way to record this simulation. Figure 7-5 displays the female descendents of seven female Asian stochastic beetles. Each family is followed to the fifth generation. The two families on the left are those we just met. The random digits beside these beetles in Figure 7-5 remind you how we used line 122 "from left to right in each generation of offspring." The fifth generation has 29 female beetles from the original 7. It is clear that the population of Asian stochastic beetles will increase rapidly until crowding, shortage of food, or increased predator populations change the reproductive pattern we have simulated.

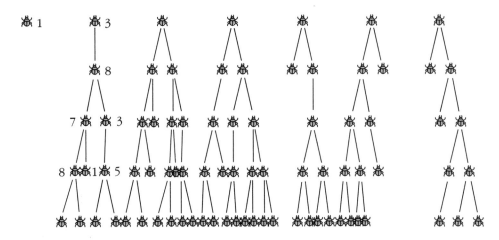

Figure 7-5 Simulation of the female offspring of seven female Asian stochastic beetles. The stimulation shows that the beetle population will increase rapidly.

SECTION 3 EXERCISES

7.29 An opinion poll selects adult Americans at random and asks them, "Which political party, Democratic or Republican, do you think is better able to manage the economy?" Explain carefully how you would assign digits from Table A to simulate the response of one person in each of the following situations.

(a) Of all adult Americans, 50% would choose the Democrats and 50% the Republicans.

(b) Of all adult Americans, 60% would choose the Democrats and 40% the Republicans.

(c) Of all adult Americans, 40% would choose the Democrats, 40% would choose the Republicans, and 20% are undecided.

(d) Of all adult Americans, 53% would choose the Democrats and 47% the Republicans.

7.30 Use Table A to simulate the responses of 10 independently chosen adults in each of the four situations of Exercise 7.29.

(a) For situation (a), use line 110.

(b) For situation (b), use line 111.

(c) For situation (c), use line 112.

(d) For situation (d), use line 113.

7.31 **A small opinion poll.** Suppose that 80% of a university's students favor abolishing evening exams. You ask 10 students chosen at random. What is the probability that all 10 favor abolishing evening exams?

(a) Give a probability model for asking 10 students independently of each other.

(b) Assign digits to represent the answers "Yes" and "No."

(c) Simulate 25 repetitions, starting at line 129 of Table A. What is your estimate of the probability?

7.32 **Missing free throws.** A basketball player makes 70% of her free throws in a long season. In a tournament game she shoots 5 free throws late in the game and misses 3 of them. The fans think she was nervous, but the misses may simply be chance. You will shed some light by estimating a probability.

(a)   Describe how to simulate a single shot if the probability of making each shot is 0.7. Then describe how to simulate 5 independent shots.

(b)   Simulate 50 repetitions of the 5 shots and record the number missed on each repetition. Use Table A starting at line 125. What is the approximate probability that the player will miss 3 or more of the 5 shots?

7.33   **Drilling for oil.** A wildcat oil driller estimates that the probability of finding a producing well when he drills a hole is 0.1. If he drills 10 holes without finding oil, he will be broke. What is the probability that he will go broke?

(a)   Give a probability model for drilling 10 holes independently of each other.

(b)   Explain how you will assign random digits to simulate drilling one hole. Then explain how to simulate drilling 10 holes.

(c)   Use Table A beginning at line 140 to simulate 20 repetitions of drilling 10 holes. Estimate from this simulation the probability that the wildcatter will go broke.

7.34   **Repeating an exam.** Elaine is enrolled in a self-paced course that allows three attempts to pass an examination on the material. She does not study and has probability 2/10 of passing on any one attempt by luck. What is Elaine's probability of passing on at least one of the three attempts? (Assume the attempts are independent because she takes a different examination on each attempt.)

(a)   Explain how you would use random digits to simulate one attempt at the exam. Elaine will of course stop taking the exam as soon as she passes.

(b)   Simulate 50 repetitions, starting at line 120 of Table A. What is your estimate of Elaine's probability of passing the course?

(c)   Do you think the assumption that Elaine's probability of passing the exam is the same on each trial is realistic? Why?

7.35 **A better model for repeating an exam.** A more realistic probability model for Elaine's attempts to pass an exam in the previous exercise is as follows: on the first try she has probability 0.2 of passing. If she fails on the first try, her probability on the second try increases to 0.3 because she learned something from her first attempt. If she fails on two attempts, the probability of passing on a third attempt is 0.4. She will stop as soon as she passes. The course rules force her to stop after three attempts in any case.

(a) Explain how to simulate one repetition of Elaine's tries at the exam. Notice that she has different probabilities of passing on each successive try.

(b) Simulate 50 repetitions and estimate the probability that Elaine eventually passes the exam. Use Table A starting at line 130.

7.36 **Gambling in ancient Rome.** Tossing four astragali was the most popular game of chance in Roman times. Many throws of a present-day sheep's astragalus show that the approximate probability distribution for the four sides of the bone that can land uppermost are:

Outcome	Probability
Narrow flat side of bone	1/10
Broad concave side of bone	4/10
Broad convex side of bone	4/10
Narrow hollow side of bone	1/10

The best throw of four astragali was the "Venus," when all the four uppermost sides were different.[10]

(a) Explain how to simulate the throw of a single astragalus. Then explain how to simulate throwing four astragali independently of each other.

(b) Simulate 25 throws of four astragali. Estimate the probability of throwing a "Venus."

7.37 **Winning a lottery.** The first state lotteries were much simpler than the present games. The original Pennsylvania State Lottery worked as follows: you choose a six-digit number. A weekly winning number is drawn at random, so every six-digit number has the same chance to be drawn. Suppose that you choose the number 123456. You win

$50,000	if the winning number is	123456
$2,000	if the winning number is	X23456 or 12345X
$200	if the winning number is	XX3456 or 1234XX
$40	if the winning number is	XXX456 or 123XXX

where X stands for any nonmatching number.

(a) Explain how to simulate one play of this lottery by using Table A to draw the weekly winning number.

(b) Simulate 20 plays of the lottery and record the weekly winning numbers drawn.

(c) On which plays did your number 123456 win a prize? What were the values of the prizes won?

(d) Based on your 20 trials, what is your estimate of the probability of winning a prize if you hold one ticket in this lottery?

7.38 **Lottery probabilities.** From your experience with random digits, you can find the exact value of the probability of winning the Pennsylvania lottery as it is given in Exercise 7.37. Again suppose that you chose number 123456.

(a) How many different six-digit numbers are there?

(b) How many of these will pay you each of

$$\$50,000 \qquad \$2000 \qquad \$200 \qquad \$40$$

if drawn as the weekly winning number? (Be careful!)

(c) Now find the total number of winning numbers that will pay you a prize. From this and part (a), find the probability of winning a prize.

7.39 **Beetle populations again.** Females of the benign boiler beetle have the following reproductive pattern:

▶ 40% die without female offspring, 40% have one female offspring, and 20% have two female offspring.

▶ Different females reproduce independently.

(a)   Explain how you would use random digits to simulate the number of offspring of a single female benign boiler beetle.

(b)   Simulate the family trees to the fifth generation of enough of these beetles to decide whether the population will definitely die out, will definitely grow rapidly, or appears to barely hold its own. (Simulate the offspring of at least five beetles.)

7.40   **Two safety systems.** A nuclear reactor is equipped with two independent automatic shutdown systems to shut down the reactor when the core temperature reaches the danger level. Neither system is perfect. System A shuts down the reactor 90% of the time when the danger level is reached. System B does so 80% of the time. The reactor is shut down if *either* system works.

(a)   Explain how to simulate the response of system A to a dangerous temperature level.

(b)   Explain how to simulate the response of system B to a dangerous temperature level.

(c)   Both systems are in operation simultaneously. Combine your answers to (a) and (b) to simulate the response of both systems to a dangerous temperature level. Explain why you cannot use the same entry in Table A to simulate both responses.

(d)   Now simulate 100 trials of the reactor's response to an emergency of this kind. Estimate the probability that it will shut down. This probability is higher than the probability that either system working alone will shut down the reactor.

7.41   **Playing craps.** The game of craps is played with two dice. The player rolls both dice and wins immediately if the outcome (the sum of the faces) is 7 or 11. If the outcome is 2, 3, or 12, the player loses immediately. If he rolls any other outcome, he continues to throw the dice until he either wins by repeating the first outcome or loses by rolling a 7.

(a)   Explain how to simulate the roll of a single fair die. (Hint: just use digits 1 to 6 and ignore the others.) Then explain how to simulate a roll of two fair dice.

(b)   Use Table A beginning at line 114 to simulate three plays of craps. Explain at each throw of the dice what the result was.

(c)   Now that you understand craps, simulate 25 plays and estimate the probability that the player wins.

▶ 4 THE HOUSE EDGE: EXPECTED VALUES

Gambling on chance outcomes goes back to ancient times. Both public and private lotteries were common in the early years of the United States. After disappearing for a century or so, government-run gambling reappeared in 1964, when New Hampshire caused a furor by introducing a lottery to raise public revenue without raising taxes. The furor subsided quickly as larger states adopted the idea, until almost all states outside the South now sponsor lotteries. State lotteries made gambling acceptable as entertainment. Casinos and other legal gambling establishments, once found only in Nevada and Atlantic City, had spread to 24 states by 1996. Over half of all adult Americans have gambled legally. What makes a bet good or bad?

> **EXAMPLE 15. Comparing bets.** Suppose you are offered this choice of bets, each costing the same: bet A pays $10 if you win and you have probability 1/2 of winning. Bet B pay $10,000 and offers probability 1/10 of winning.

"What kind of childish nonsense are you working on now?"

You would very likely choose B even though A offers a better chance to win, because B pays much more if you win. It would be foolish to decide which bet to make just on the basis of the probability of winning. How much you can win is also important. *When a random phenomenon has numerical outcomes, we are concerned with their amounts as well as with their probabilities.*

What will be the average payoff of these two bets in many plays? The probabilities tell us the long-run proportions of plays on which each outcome occurs. Bet A produces $10 half the time in the long run and nothing half the time. So the average payoff will be

$$\left(\$10 \times \frac{1}{2}\right) + \left(\$0 \times \frac{1}{2}\right) = \$5$$

Bet B, on the other hand, pays out $10,000 on 1/10 of all bets in the long run. Bet B's average payoff is

$$\left(\$10{,}000 \times \frac{1}{10}\right) + \left(\$0 \times \frac{9}{10}\right) = \$1000$$

If you can place many bets, you should certainly choose B.

Here is a general definition of the kind of "average outcome" we used to compare the two bets in Example 15.

Expected value

The **expected value** of a random phenomenon that has numerical outcomes is found by multiplying each outcome by its probability and then summing over all possible outcomes.

In symbols, if the possible outcomes are $a_1, a_2, \ldots, a_k$ and their probabilities are $p_1, p_2, \ldots, p_k$, the expected value is

$$\text{expected value} = a_1 p_1 + a_2 p_2 + \ldots + a_k p_k$$

State lotteries have introduced ever more gimmicks and special prizes in an attempt to keep public interest high, and these make computation of the expected value of a ticket difficult. Let us therefore go back a few years and study the original, uncluttered New York State Lottery.

EXAMPLE 16. The original New York lottery. The New York State Lottery awards, for each one million tickets sold,

$$
\begin{array}{rl}
1 & \$50{,}000 \text{ prize} \\
9 & \$5000 \text{ prizes} \\
90 & \$500 \text{ prizes} \\
900 & \$50 \text{ prizes}
\end{array}
$$

If you buy one ticket, your probability of winning $50,000 is 1/1,000,000, and so on. There were 1000 winning tickets in all, so your probability of winning nothing is 999,000/1,000,000. The expected value of your winnings is

$$
(\$50{,}000)\left(\frac{1}{1{,}000{,}000}\right) + (\$5000)\left(\frac{9}{1{,}000{,}000}\right) + (\$500)\left(\frac{90}{1{,}000{,}000}\right)
$$

$$
+ (\$50)\left(\frac{900}{1{,}000{,}000}\right) + (\$0)\left(\frac{999{,}000}{1{,}000{,}000}\right) = \$0.185
$$

Lottery tickets cost 50 cents, which seems a high price to pay for an expected return of 18.5 cents.

THE LAW OF LARGE NUMBERS

The definition of expected value says that it is an average of the possible outcomes, but an average in which outcomes with higher probability count more. The expected value is the average outcome in another sense as well.

The law of large numbers

If a random phenomenon with numerical outcomes is repeated many times independently, the mean value of the actually observed outcomes approaches the expected value.

The law of large numbers is closely related to the idea of probability. In many independent repetitions, the proportion of each possible outcome will be close to its probability, and the average outcome obtained will be close to the expected value. These facts express the long-run regularity of chance events. They are the true version of the "law of averages."

The law of large numbers explains why gambling, which is a recreation or an addiction for individuals, is a business for a casino. The "house" in a gambling operation is not gambling at all. The average winnings of a large number of customers will be quite close to the expected value. The house has calculated the expected value ahead of time and knows what its take will be in the long run. There is no need to load the dice or stack the cards to guarantee a profit. Casinos concentrate on inexpensive entertainment and cheap bus trips to keep the customers flowing in. If enough bets are placed, the law of large numbers guarantees the house a profit. Life insurance companies operate much like casinos—they bet that the people who buy insurance will not die. Some do die, of course, but the insurance company knows the probabilities and relies on the law of large numbers to predict the average amount it will have to pay out. Then the company sets its premiums high enough to guarantee a profit.

THINKING ABOUT EXPECTED VALUES

As with probability, it is worth exploring a few fine points about expected values and the law of large numbers.

How large is a large number? The law of large numbers says that the actual average outcome of many trials gets closer to the expected value as more trials are made. It doesn't say how many trials are needed to guarantee an average outcome close to the expected value. That depends on the *variability* of the random outcomes.

The more variable the outcomes, the more trials are needed to ensure that the mean outcome is close to the expected value. Games of chance must be quite variable if they are to hold the interest of gamblers. Even a long evening in a casino has an unpredictable outcome. Gambles with extremely variable outcomes, like state lottos with their very large but very improbable jackpots, require impossibly large numbers of trials to ensure that the average outcome is close to the expected value. Though most forms of gambling are less variable than lotto, the layman's answer to the applicability of the law of large numbers is usually that the house plays often enough to rely on it, but you don't. Much of the psychological allure of gambling is its unpredictability for the player. The business of gambling rests on the fact that the result is not unpredictable for the house.

Is there a winning system? Serious gamblers often follow a system of betting in which the amount bet on each play depends on the outcome of previous plays. You might, for example, double your bet on each spin of the roulette wheel until you win—or, of course, until your fortune is exhausted. Such a system tries to take advantage of the fact that you have a memory even though the roulette wheel does not. Can you beat the odds with a system? No. Mathematicians have established a stronger version of the law of large numbers that says that if you do not have an infinite fortune to gamble with, your average winnings (the expected value) remains the same as long as successive trials of the game (such as spins of the roulette wheel) are independent. Sorry.

THE STATE OF LEGALIZED GAMBLING

Is a lottery ticket a good bet? Tickets in the old New York Lottery cost 50 cents. As Example 16 shows, the expected value of a ticket was 18.5 cents. The law of large numbers says that New York paid out very close to 18.5 cents per ticket sold. That is, the state paid out a bit less than 40% of the amount wagered by ticket buyers and kept over 60%. Competition has since sweetened the pot a bit. State lotteries typically pay out about half the dollars bet. Another 15% goes for advertising and expenses, and the remaining 35% flows into the state's treasury. Most states also pay grand prize winners over time, usually 20 years. Because money earns interest over time, a jackpot advertised as $10 million actually costs the state only about $4.8 million. By way of comparison, casinos in Las Vegas or Atlantic City pay out 85% to 95% of the amount wagered, depending on which game you choose.

State lotteries are a poor bet. Professional gamblers and statisticians avoid them, not wanting to waste money on so bad a bargain. Politicians avoid them for reasons nicely put by Nelson Rockefeller when he was governor of New York: "I'm afraid I might win." Almost everyone else in lottery states plays. Surveys usually show that about 80% of adult residents have purchased at least one lottery ticket. Why is such a poor bet so popular? Lack of knowledge of how poor the bet is plays

"I think the lottery is a great idea. If they raised taxes instead, we'd have to pay them."

a role. So does heavy advertising by the state. But the major attraction is probably the lure of possible wealth, no matter how unlikely the jackpot is. Many people find a dollar a week a fair price for the entertainment value of imagining themselves rich. As Carmen Brutto of Harrisburg, Pennsylvania, said in a newspaper interview, "My chances of winning a million are better than my chances of earning a million."[11]

Are casinos worth the trouble? States license casino gambling for much the same reasons that they run lotteries: gambling raises state revenues without raising taxes. The state keeps its 35% of lottery wagers, and it taxes casinos to get its share of their take. Casinos paid more than $1.4 billion in state and local taxes in 1994. States also hope that casinos will attract tourists from outside the state.

Gambling has its disadvantages as a revenue source. Dollars from gambling are small relative to state budgets. They vary from year to year, making planning difficult. As gambling has become more widespread, competition has grown, so that attracting customers requires heavy advertising and a constant flow of new attractions. In the case of lotteries, the states are in the business of trying to convince their citizens to gamble. Low-income people are more likely than the rich to buy lottery tickets. Should states really try to persuade their poorer citizens to make bets, especially bets as bad as lottery tickets?

Casinos, more than lotteries, bring social problems with them. The number of addicted gamblers climbs, bringing bankruptcies, broken marriages, and neglected children. Crime may also increase. One University of Illinois study estimated that costs to social and police agencies raise state expenses by about three times the amount that gambling brings into the state treasury.

By the mid-1990s, a backlash against gambling was setting in. Between 1994 and 1996, only two more states approved legalized gambling, while 22 others considered gambling but turned it down. Some major new casinos lost money when the crowds did not arrive. Congress considered, against heavy lobbying by gambling interests, holding hearings on the social costs of gambling. For the first time since New Hampshire started it all in 1964, the future growth of legalized gambling is uncertain.

FINDING EXPECTED VALUES BY SIMULATION

How can we calculate expected values in practice? You know the mathematical recipe, but that requires that you start with the probability of each outcome. Expected values too difficult to compute in this way can be found by simulation. The procedure is as before: give a probability model, use random digits to imitate it, and simulate many repetitions. By the law of large numbers, the average outcome of these repetitions will be close to the expected value.

> **EXAMPLE 17. We want a girl, again.** A couple plans to have children until they have a girl or until they have four children, whichever comes first. Example 13 estimated the probability that they will have a girl among their children. Now we ask a different question: How many children, on the average, will couples who follow this plan have?
>
> The simulation is exactly as in Example 13. The only difference is that we keep track of the number of children in each repetition

and average these to estimate the expected value. Here are the same 14 repetitions we simulated in Example 13, with the number of children noted:

690 51 64 81 7871 74 0 951 784 53 4 0 64 8987
BBG BG BG BG BBBG BG G BBG BBG BG G G BG BBBB
 3   2   2   2   4   2   1   3   3   2   1   1   2   4

The mean outcome is

$$\overline{x} = \frac{3 + 2 + \cdots + 4}{14} = \frac{32}{14} = 2.3$$

We estimate that if many couples follow this plan, they will average 2.3 children each. This simulation is too short to be trustworthy. Math or a long simulation shows that the actual expected value is 1.8 children.

A SUMMING UP

Probability and expected values give us a language to describe randomness. Random phenomena are not haphazard or chaotic any more than random sampling is haphazard. Randomness is instead a kind of order in the world, a long-run regularity as opposed to either chaos or a determinism that fixes events in advance. When randomness is present, probability answers the question "How often in the long run?" and expected value answers the question "How much in the long run?" The two answers are tied together by the definition of expected value in terms of probabilities.

It appears that randomness is embedded in the way the world is made. Albert Einstein reacted to the growing emphasis on randomness in physics by saying "I cannot believe that God plays dice with the universe." Lest you have similar qualms, I remind you again that randomness is not chaos but a kind of order. Our immediate concern, however, is randomness that we arrange—not God's dice, but Reno's. In particular, statistical designs for producing data are founded on deliberate randomizing. The order thus introduced into the data is the basis for statistical inference, as we have noticed repeatedly and will study more thoroughly in the next chapter. If you understand probability, statistical inference is stripped of mystery. That may console you as you contemplate the remark of the great economist John Maynard Keynes on long-term orderliness: "In the long run, we are all dead."

SECTION 4 EXERCISES

7.42   **The numbers racket.** The numbers racket is a well-entrenched illegal gambling operation in the poorer areas of most large cities. One version works as follows: you choose any one of the 1000 three-digit numbers 000 to 999 and pay your local numbers runner a dollar to enter your bet. Each day, one three-digit number is chosen at random and pays off $600. What is the expected value of a dollar bet on the numbers?

7.43   **Making decisions.** The psychologist Amos Tversky did many studies of our perception of chance behavior. I cited some of his work in Section 2. In its obituary of Tversky (June 6, 1996), the *New York Times* cited the following example.

(a)   Tversky asked subjects to choose between two public health programs that affect 600 people. One has probability 1/2 of saving all 600 and probability 1/2 that all 600 will die. The other is guaranteed to save exactly 400 of the 600 people. Find the expected number of people saved by the first program.

(b)   Tversky then offered a different choice. One program has probability 1/2 of saving all 600 and probability 1/2 of losing all 600, while the other will definitely lose exactly 200 lives. What is the difference between this choice and that in (a)?

(c)   Given option (a), most subjects choose the second program. Given option (b), most subjects choose the first program. Do the subjects appear to use expected values in their choice? Why do you think the choices differ in the two cases?

7.44   **The Connecticut lottery.** The Connecticut State Lottery (in its original simple form) awarded at random, for each 100,000 50-cent tickets sold,

18	$200 prizes
120	$25 prizes
270	$20 prizes

What is the expected value of the winnings on one ticket in this lottery? What percent of the money bet did the state pay out as winnings?

7.45 **Sizes of households.** A household is a group of people living together, regardless of their relationship to each other. Sample surveys such as the Current Population Survey select a random sample of households. Here is the probability model for the number of people living in a randomly chosen American household:

Household size	1	2	3	4	5	6	7
Probability	0.251	0.321	0.171	0.154	0.067	0.022	0.014

These probabilities are the proportions for all households in the country. They give the probabilities that a single household chosen at random will have each size. (The very few households with more than 7 members are placed in the 7 group.) Check that this is a legitimate assignment of probabilities by verifying that it satisfies Rules A and B of Section 1. Then find the expected size of a randomly chosen household.

7.46 **Beetle reproduction.**

(a) What is the expected number of female offspring produced by a female Asian stochastic beetle? (See Example 14 on page 431 for this insect's reproductive pattern.)

(b) What is the expected number of female offspring produced by a female benign boiler beetle? (See Exercise 7.39 on page 436 for the reproductive pattern.)

(c) Use the law of large numbers to explain why the population should grow if the expected number of female offspring is greater than 1 and die out if this expected value is less than 1. Do your expected values in parts (a) and (b) confirm the results of the simulations of these populations done in Section 3?

7.47 **Life insurance.** A life insurance company sells a term insurance policy to a 21-year-old male that pays $20,000 if the insured dies within the next five years. The probability that a randomly chosen male will die each year can be found in mortality tables. The company collects a premium of $100 each year as payment for the insurance. The amount that the company earns on this policy is $100 per year, less the $20,000 that it must pay if the insured dies. Here are the probabilities for the possible amounts the company can earn. Calculate the expected value of its earnings.

Age at death	21	22	23	24	25	≥ 26
Earnings	−$19,900	−$19,800	−$19,700	−$19,600	−$19,500	$500
Probability	0.0018	0.0019	0.0019	0.0019	0.0019	0.9906

7.48 **Understanding insurance.** It would be quite risky for you to insure the life of a 21-year-old friend under the terms of the previous exercise. With high probability your friend would live, and you would gain $500 in premiums; but if he were to die, you would lose almost $20,000. Explain carefully why selling insurance is not risky for an insurance company that insures many thousands of 21-year-old men.

7.49 We play a game by reading a pair of random digits from Table A. If the two digits are the same (for example, 2 and 2), you win $10. If they differ by one (for example, 1 and 0, 7 and 8, or 9 and 0—note that we say 0 and 9 differ by one), you win $5. Otherwise, you lose $3. What is your expected outcome in this game?

7.50 **Green Mountain Numbers.** Green Mountain Numbers, a part of the Vermont state lottery, offers a choice of several bets. In each case, the player chooses a three-digit number. The lottery commission announces the winning number, chosen at random, at the end of the day. Find the expected winnings for a $1 bet for each of the following options in this lottery:

(a) The "triple" pays $500 if your chosen number exactly matches the winning number.

(b) The "box" pays $83.33 if your chosen number has the same digits as the winning number, in any order. (Assume that you chose a number having three distinct digits.)

7.51 **An expected rip-off?** A "psychic" runs the following ad in a magazine:

Expecting a baby? Renowned psychic will tell you the sex of the unborn child from any photograph of the mother. Cost $10. Moneyback guarantee.

This may be a profitable con game. Suppose that the psychic simply replies "girl" to all inquiries. In the worst case, everyone who has a boy will ask for her money back. Find the expected value of the psychic's profit by filling in the table below.

Sex of child	Probability	The psychic's profit
Male		
Female		

7.52 **The Pennsylvania lottery.** Use the probability distribution of prizes for the Pennsylvania State Lottery (Exercise 7.37 on page 435) to compute the expected value of the winnings from one ticket in that lottery.

7.53 **We really want a girl.** Let's modify Example 17 a bit. If a couple keeps having children until they get a girl, what is the expected number of children they will have? Here is the same question in a different guise: If a fair coin is tossed repeatedly, what is the expected number of trials required to obtain the first head? Design a simulation to answer this question. Then simulate 25 repetitions. What is your estimate of the expected value?

7.54 **Stochastic beetles simulation.** Simulate the offspring (one generation only) of 100 female Asian stochastic beetles. What is your estimate of the expected number of offspring of one such beetle, based on this simulation? Compare the simulated value with the exact expected value you found in Exercise 7.46(a). Explain how your results illustrate the law of large numbers.

7.55 OK, friends, I've got a little deal for you. We have a fair coin (heads and tails each have probability 1/2). Toss it twice. If two heads come up, you win right there. If you get any result other than two heads, I'll give you another chance: toss the coin twice more, and if you get two heads you win. (Of course, if you fail to get two heads on the second try, I win.) Pay me a dollar to play. If you win, I'll give you your dollar back plus another dollar.

(a)   Explain how to simulate one play of this game. Start by simulating two tosses of a fair coin.

(b)   You have only two possible monetary outcomes, $-\$1$ if you lose and $\$1$ if you win. Simulate 50 plays, using Table A starting at line 125. Use your simulation to estimate the expected value of the game.

7.56   **Your state's lottery.** If your state has a lottery, find out what percent of the money bet is returned to the bettors in the form of prizes. What percent of the money bet is used by the state to pay lottery expenses? What percent is net revenue to the state? For what purposes does the state use lottery revenue?

Notes

1. More detail can be found in the opening chapters of F. N. David, *Games, Gods and Gambling,* Charles Griffin and Co., 1962. The historical information given here comes from this excellent and entertaining book.

2. See A. Tversky and D. Kahneman, "Belief in the law of small numbers," *Psychological Bulletin,* Volume 76 (1971), pp. 105–110, and other writings of these authors for a full account of our misperception of randomness.

3. R. Vallone and A. Tversky, "The hot hand in basketball: on the misperception of random sequences," *Cognitive Psychology,* Volume 17 (1985), pp. 295–314.

4. For a discussion and amusing examples, see A. E. Watkins, "The law of averages," *Chance,* Volume 8 (Spring 1995), pp. 28–32.

5. Estimated probabilities from R. D'Agostino, Jr. and R. Wilson, "Asbestos: the hazard, the risk, and public policy," in K. R. Foster, D. E. Bernstein, and P. W. Huber (eds.), *Phantom Risk: Scientific Inference and the Law,* The MIT Press, 1994, pp. 183–210. See also the similar conclusions in B. T. Mossman, et al., "Asbestos: scientific developments and implications for public policy," *Science,* Volume 247 (1990), pp. 294–301.

6. The quotation is from R. J. Zeckhauser and W. K. Viscusi, "Risk within reason," *Science,* Volume 248 (1990), pp. 559–564.

7. T. Hill, "Random-number guessing and the first digit phenomenon," *Psychological Reports,* Volume 62 (1988), pp. 967–971.
8. Quoted in the article by Watkins cited in Note 4.
9. Stochastic beetles are well known in the folklore of simulation, if not in entomology. They are said to be the invention of Arthur Engle of the School Mathematics Study Group.
10. Information from the book by F. N. David cited in Note 1.
11. Quoted in an article by F. J. Prial in the *New York Times,* February 17, 1976.

REVIEW EXERCISES

7.57 **Playing "heads or tails."** In the game of *heads or tails,* Betty and Bob toss a coin four times. Betty wins a dollar from Bob for each head, and pays Bob a dollar for each tail. That is, Betty wins or loses the difference between the number of heads and the number of tails. For example, if there are 1 head and 3 tails, Betty loses $2. You can check that Betty's possible outcomes are

$$-4, \ -2, \ 0, \ 2, \ 4$$

Assign probabilities to these outcomes by playing the game 20 times and using the proportions of the outcomes as estimates of the probabilities. If possible, combine your trials with those of other students to obtain long-run proportions that are closer to the probabilities. What do you estimate the expected value of Betty's winnings to be?

7.58 **The distribution of blood types.** All human blood can be typed as one of O, A, B, or AB, but the distribution of the types varies a bit with race. Here is the distribution of the blood type of a randomly chosen black American.

Blood type	O	A	B	AB
Probability	0.49	0.27	0.20	?

(a) If this is to be a legitimate assignment of probabilities, what must be the probability of type AB blood?

(b) What is the probability that the person chosen does not have type O blood?

(c)   Does it make sense to speak of the expected value of blood type? Why?

7.59   A bridge deck contains 52 cards, four of each of the 13 face values ace, king, queen, jack, ten, nine, ..., two. You deal a single card from such a deck and record the face value of the card dealt. Give an assignment of probabilities to these outcomes that should be correct if the deck is thoroughly shuffled. Give a second assignment of probabilities that is legitimate (that is, satisfies Rules A and B) but differs from your first choice. Then give a third assignment of probabilities that is *not* legitimate, and explain what is wrong with this choice.

7.60   **Personal probabilities.** Las Vegas Zeke, when asked to predict the Atlantic Coast Conference basketball champion, follows the modern practice of giving probabilistic predictions. He says, "North Carolina's probability of winning is twice Duke's. North Carolina State and Virginia each have probability 0.1 of winning, but Duke's probability is three times that. Nobody else has a chance." Has Zeke given a legitimate assignment of probabilities to the 8 teams in the conference? (These are personal probabilities: they express Zeke's informed opinion about how the race would come out if the season could be repeated very many times.)

7.61   **How far do fifth graders go?** A study selected a sample of fifth-grade pupils and recorded how many years of school they eventually completed. Based on this study we can give the following assignment of probabilities for the years of school that will be completed by a randomly chosen fifth grader.

Years	4	5	6	7	8	9	10	11	12
Probability	0.010	0.007	0.007	0.013	0.032	0.068	0.070	0.041	0.752

(a)   Verify that this is a legitimate assignment of probabilities.

(b)   What outcomes make up the event "The student completed at least one year of high school?" (High school begins with the 9th grade.) What is the probability of this event?

(c)   What is the expected number of years of school completed?

7.62 **What are the odds?** Gamblers often express chance in terms of *odds* rather than probability. Odds of *A* to *B* against an outcome means that the probability of that outcome is $B/(A+B)$. So "odds of 5 to 1" is another way of saying "probability 1/6."

(a) The odds against being dealt three of a kind in a five-card poker hand are about 49-to-1. What is the probability of being dealt three of a kind?

(b) The probability of being killed in an auto accident in a lifetime of driving is about 0.015. What are the odds against being killed?

7.63 **Social mobility.** Sociologists have studied whether children move out of their parents' occupational class. The overall result can be expressed in terms of probabilities. If a father has a white-collar occupation, here are the probabilities for the occupational class of his adult son:

Professional	0.2
White collar	0.5
Blue collar	0.2
No steady occupation	0.1

(a) Explain how to simulate the occupational class of a randomly selected son of a white-collar father.

(b) If we follow five sons of white-collar fathers, and these sons progress independently of each other, what is the probability that at least two of the five will have professional occupations? Use Table A, beginning at line 101, to simulate 50 repetitions of the five sons' occupations and estimate this probability.

7.64 **A grocery store prize game.** A grocery chain runs a prize game based on giving each customer a ticket that may award a prize when a box on the ticket is scratched. The chain is required to reveal the probabilities of winning. For customers who visit the store once each week during the game, these probabilities are:

Amount won	Probability
$250	0.01
$100	0.05
$10	0.25

(a)   What is the probability of winning nothing?

(b)   What is the expected value of the amount won?

(c)   Describe how to simulate the winnings of a single customer who visits the store once each week.

(d)   Three friends who each visit the store once a week agree to pool their winnings. Explain how you would simulate their three outcomes.

(e)   Simulate 50 repetitions of the three friends' winnings, using Table A starting at line 140. From your results, estimate their expected total winnings and also the probability that they will win at least $100.

7.65   **The birthday problem.** A famous example in probability theory shows that the probability that at least two people in a room have the same birthday is already greater than 1/2 when 23 people are in the room. The probability model is:

▶   The birth date of a randomly chosen person is equally likely to be any of the 365 dates of the year.

▶   The birth dates of different people in the room are independent.

Explain carefully how you would simulate the birth dates of 23 people to see if any two have the same birthday. Do the simulation once, using line 139 of Table A. (*Comment*: It is easiest to let three-digit groups stand for the birth dates of successive people, so that 001 is January 1 and 365 is December 31. Ignore leap years and skip groups that don't label birth dates. The simulation is too lengthy to repeat many times without a computer, but in principle you can find the probability of matching birthdays by routine repetition. The birthday problem is too hard for most of your math-major friends to solve, so it shows the power of simulation.)

7.66   **Rolling two dice.** Example 4 (page 409) gives a probability model for rolling two casino dice and recording the number of spots on each of the two up faces. Suppose that instead we just count the spots on the two up faces and record the sum. The possible outcomes are now 2, 3, 4, ..., 12. Follow the method of Example 4 to find the probability model for this random phenomenon. Then use the probabilities to find the expected value.

7.67 **Keno.** Keno is a favorite game in casinos. Balls numbered 1 to 80 are tumbled in a machine as the bets are placed, then 20 of the balls are chosen at random. Players select numbers by marking a card. A bewildering variety of Keno bets are available. Here are two of the simpler bets. Give the expected winnings for each.

(a) A $1 bet on "Mark 1 number" pays $3 if the single number you mark is one of the 20 chosen; otherwise you lose your dollar.

(b) A $1 bet on "Mark 2 numbers" pays $12 if both your numbers are among the 20 chosen. The probability of this is about 0.06. Is Mark 2 a more or a less favorable bet than Mark 1?

7.68 **Stock prices.** The behavior of common stock prices is often described by probability models. The stock of Random Enterprises behaves as follows: each day it either gains $1 or loses $1, independent of earlier price movements. Mr. Smart buys a share for $10. As soon as it goes up to $11, he sells. After 5 days, he must sell anyway because he needs to use the money. Mr. Smart can gain $1 or lose as much as $5, depending on the stock's price changes over the five-day period. Describe how to simulate his gain or loss, then simulate 50 repetitions. Based on your simulation, estimate the probability that Mr. Smart will finish with a gain. Also estimate the expected value of his gain. (Take losses to be negative gains; for example, a loss of $2 is a gain of −$2.)

WRITING PROJECTS

7.1 **Reacting to risks.** "Few of us would leave a baby sleeping alone in a house while we drove off on a 10-minute errand, even though car-crash risks are much greater than home risks." Take it as a fact that the probability that the baby will be injured in the car is very much higher than the probability of any harm at home in the same time period. Would you leave the baby alone? Explain your reasons in a short essay. If you would not leave the baby alone, be sure to explain why you choose to ignore the probabilities.

7.2   **Lotteries, pro and con.** What are the arguments in favor of state-run lotteries as a means of financing state government? What are the arguments against lotteries? Explain why you support or oppose such lotteries.

CHAPTER 8

INFERENCE: CONCLUSIONS WITH CONFIDENCE

To *infer* means "to draw a conclusion." *Statistical inference* offers us methods for drawing conclusions from data. We have, of course, been drawing conclusions from data all along. What is new in formal inference is that we use probability to say how confident we are that our conclusions are correct. Probability allows us to take chance variation into account and so to correct our judgment by calculation. Here are two examples of how probability can correct our judgment.

> **EXAMPLE 1. Was the draft lottery fair?** In the Vietnam War years, a lottery determined the order in which men were drafted for army service. The lottery assigned draft numbers by choosing birth dates in random order. A random lottery should produce draft numbers that have no systematic relationship with the order of birth. A scatterplot of the results of the first draft lottery (Figure 8-1) shows no clear relationship. When we calculate the correlation between birth date and draft number, however, we find that $r = -0.226$. That is, men born later in the year tended to get lower draft numbers. Is this small correlation evidence that the lottery was not truly random?
>
> Our unaided judgment can't tell, because any two variables will have some correlation, just by chance. We therefore do a probability calculation, and find that a correlation this far from zero has probability less than 0.001 in a truly random lottery. Because a correlation as strong as that observed would almost never occur in a random lottery, there is strong evidence that the draft lottery was unfair.

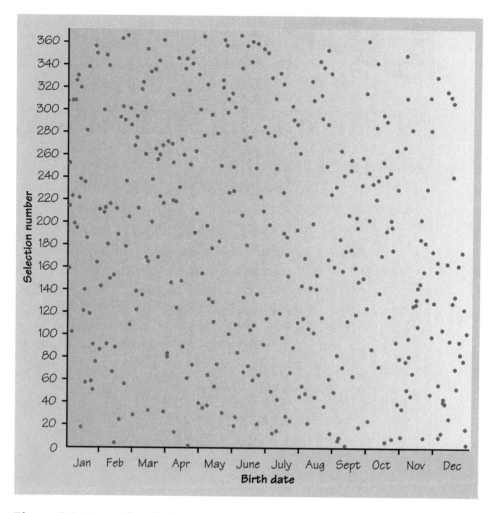

Figure 8-1 Scatterplot of the association between birth dates and draft numbers in the first Vietnam-era draft lottery. When we label birth dates from 1 (January 1) to 366 (December 31) and draft numbers from 1 (drafted first) to 366 (drafted last), the correlation is $r = -0.226$.

EXAMPLE 2. Does vitamin C prevent flu? Probability calculations can also protect us from jumping to a conclusion when only chance variation may be at work. Consider a small randomized comparative double-blind experiment to study whether taking vitamin C every day will reduce flu. We assign 20 subjects at random to vitamin C and another 20 to a placebo. At the end of the winter, 12 of the placebo group and 8 of the vitamin C group have had flu. Can we conclude that flu will be less frequent among all people if they take vitamin C?

Again, do a probability calculation. A difference this large or larger between the results for two groups of 20 subjects would occur about one time in five simply because of chance variation. An effect that could so easily be just chance is not convincing.

Drawing conclusions in mathematics is a matter of starting from a hypothesis and using logical argument to prove without doubt that the conclusion follows. This is a *deductive argument* from hypothesis to consequences. Empirical science argues in almost the reverse order. If vitamin C prevents flu, we would expect the vitamin C group to have less flu than the placebo group. The vitamin C group did have less flu, so this is evidence in favor of vitamin C's effect. This is an *inductive argument* from consequences back to a hypothesis. Inductive arguments do not produce proof. The good health of the vitamin C group *might* be due to something other than the vitamin C they took.

Statistical inference uses probability to say how strong an inductive argument is. In Example 1, the outcome of the draft lottery gives strong evidence for the hypothesis that the lottery was not truly random. In Example 2, the outcome of the experiment gives only weak evidence for the hypothesis that vitamin C prevents flu. You have no doubt heard that "statistics cannot prove anything." True enough. Neither can any other kind of inductive argument. Outside mathematics there is no proof. But inductive arguments can be quite convincing, and statistical arguments are often among the most convincing.

Because statistical inference is based on probability, it is most trustworthy when the data are produced by a probability sample or randomized comparative experiment. The deliberate use of chance in producing the data guarantees that the rules of probability apply to the results. Probability tells us what would happen in many repetitions of the experiment or survey. That is, statistical inference asks "What would happen if we did this many times?" This entire chapter is a working out of that idea.[1]

Here is a distinction we first met in Chapter 1 that you absolutely must keep in mind when thinking about statistical inference.

> **Parameters and statistics**
>
> A **parameter** is a number that describes the *population.* For example, the proportion of the population having some characteristic of interest is a parameter that we call *p.* In a statistical inference problem, population parameters are fixed numbers, but we do not know their values.
>
> A **statistic** is a number that describes the *sample data.* For example, the proportion of the sample having some characteristic of interest is a statistic that we call $\hat{p}$. Statistics change from sample to sample. We use the observed statistics to get information about the unknown parameters.

▶ 1 ESTIMATING WITH CONFIDENCE

The first type of inference we will think about has a simple goal: use a sample statistic to estimate the numerical value of an unknown population parameter.

EXAMPLE 3. A worried senator. Senator Bean wants very much to know what percent of the voters in his state plan to vote for him in the election, now only a month away. This unknown proportion of the population of voters is a *parameter p.* He therefore commissions a poll. Being rich, he can afford a genuine SRS of 1000 registered voters. Of these voters, 570 say that they plan to vote for Bean. That's a reassuring 57% of those polled. This observed proportion of the sample is a *statistic,* $\hat{p} = 0.57$.

But wait. Bean and his pollster know very well that another SRS of 1000 voters would no doubt produce a different response—perhaps 59% or 55%. Or perhaps even 51% or (horrors) 49%. How should the senator interpret the 57% sample result?

If you advised Senator Bean to demand a confidence statement, you may join the chorus of the wise. A confidence statement attaches a margin of error to that 57% estimate and also states how confident we can be that the true proportion of voters who favor Bean falls within the margin of error. Confidence statements are based on the distribution of values of the sample proportion $\hat{p}$ that would occur if many independent SRSs were taken from the same population. This is the *sampling distribution* of the statistic $\hat{p}$.

SAMPLING DISTRIBUTIONS

The sampling distribution of a statistic answers the question, "What would happen if we did this many times?" Figure 8-2 shows the idea. The population contains the proportion p of Bean supporters. Draw an SRS of 1000 voters and find the proportion $\hat{p}$ who support Bean. Draw another SRS and get another $\hat{p}$. Keep on drawing samples and recording the sample proportions $\hat{p}$ that result. If you draw samples forever, the distribution of values of $\hat{p}$ will be close to a normal curve centered at p. This is the sampling distribution of the statistic $\hat{p}$. In the language of probability, the sampling distribution gives the *probability model* for $\hat{p}$. That is, it gives the probabilities of the possible values of $\hat{p}$.

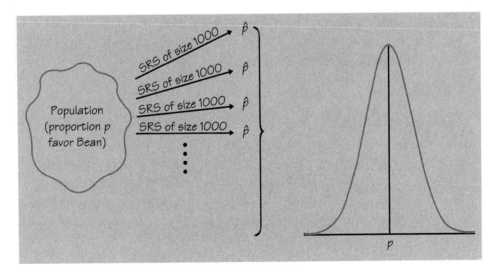

Figure 8-2 The idea of a sampling distribution. Take many samples from the same population. Record the value of a statistic (the sample proportion $\hat{p}$) for each sample. The distribution of these values is the sampling distribution of the statistic.

EXAMPLE 4. **A sampling distribution.** Suppose that in fact 55% of the several million voters in Bean's state plan to vote for him. What would be the pattern (sampling distribution) of the sample proportion $\hat{p}$ favoring Bean in many independent SRSs of size 1000?

We could discover this pattern by a long simulation. Fortunately, it is also possible to discover the pattern by mathematics, and I will tell you the answer. The probability distribution of the sample proportion $\hat{p}$ favoring Bean in an SRS of size 1000 when 55% of the population favor Bean is very close to the normal distribution with mean 0.55 and standard deviation 0.015. If you don't believe me, go and simulate several thousand SRSs of size 1000 from such a population and make a histogram of the several thousand values of $\hat{p}$ you get. That histogram will look very much like the normal curve with mean 0.55 and standard deviation 0.015.

Figure 8-3 shows the sampling distribution of the sample proportion $\hat{p}$ in Senator Bean's case. This distribution shares the properties of all normal curves. For example, the "95" part of the 68–95–99.7 rule says:

> *The probability is 0.95 that an SRS of size 1000 from this population will have a $\hat{p}$ between 0.52 and 0.58 (that is, within two standard deviations of the mean).*

THE REASONING OF CONFIDENCE INTERVALS

Senator Bean is becoming impatient. We are telling him how $\hat{p}$ behaves for a known population proportion p. He wants to know what he can say about p, the *unknown* fraction of voters who are for him, once he has taken a sample and observed $\hat{p} = 0.57$. Patience, senator. We're coming to that. Consider this line of argument.

A. When we choose an SRS of size 1000 from a large population of which a proportion p favor Bean, the proportion $\hat{p}$ of the sample who favor Bean has approximately the normal distribution with mean equal to p and standard deviation 0.015. The mean is p

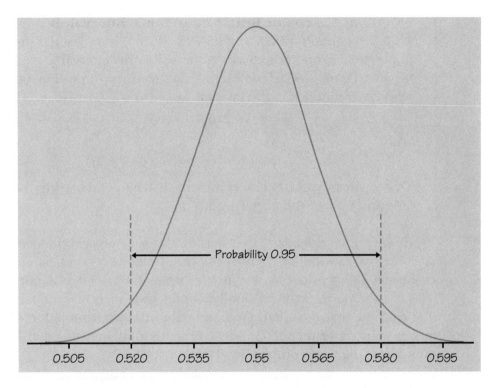

Figure 8-3 The sampling distribution of the proportion $\hat{p}$ of an SRS of size 1000 who favor Senator Bean. The truth about the population is that proportion $p = 0.55$ favor Bean. The sampling distribution is therefore centered at 0.55. The probability is 0.95 that $\hat{p}$ falls within two standard deviations of p.

because the SRS has no bias. The standard deviation is fixed by the size of the sample.*

B. The "95" part of the 68–95–99.7 rule says that the probability is 0.95 that $\hat{p}$ will be within 0.03 (two standard deviations) of its mean p.

C. That is exactly the same as saying that the probability is 0.95 that the interval from $\hat{p} - 0.03$ to $\hat{p} + 0.03$ captures the unknown p. We say that the interval $\hat{p} \pm 0.03$ is a *95% confidence interval* for the parameter p.

---

*Actually, the standard deviation changes when p changes. But for p anywhere between $p = 0.3$ and $p = 0.7$, the standard deviation of $\hat{p}$ is within 0.001 of 0.015. You can find more detailed information in Section 2.

EXAMPLE 5. Senator Bean's confidence interval. Bean's SRS of 1000 voters had $\hat{p} = 0.57$. In 95% of all such samples, the unknown population proportion p is captured by the interval $\hat{p} \pm 0.03$. So Senator Bean is 95% confident that the true proportion of voters who favor him lies in the interval between

$$\hat{p} - 0.03 = 0.57 - 0.03 = 0.54$$

and

$$\hat{p} + 0.03 = 0.57 + 0.03 = 0.60$$

That is, Bean can be 95% confident that he is favored by between 54% and 60% of the population.

Be sure you understand the basis for the senator's confidence. There are only two possibilities. *Either* the true p lies between 0.54 and 0.60, *or* Bean's SRS was one of the few samples for which $\hat{p}$ is not within 0.03 of the true p. Only 5% of all samples give such inaccurate results. It is possible that Bean had the bad luck to draw a sample for which $\hat{p}$ misses p by more than 0.03, but over many drawings this will happen only 5% of the time (probability 0.05).

Confidence intervals

A **95% confidence interval** is an interval obtained from the sample data by a method that in 95% of all samples will produce an interval that captures the true population parameter.

We call 95% the **confidence level**. It is the probability that the method gives an interval that captures the true parameter. **95% confidence** is short for "I got this result by a method that gives correct results 95% of the time."

Figure 8-4 illustrates how confidence intervals behave. The normal curve at the top of the figure is the sampling distribution of $\hat{p}$. As we take many samples, the actual values of $\hat{p}$ vary according to this distribution. The values of $\hat{p}$ observed in 25 samples appear as dots below the curve, together with the confidence intervals $\hat{p} \pm 0.03$ that extend out 0.03 on either side of the observed $\hat{p}$. The true population proportion p is marked by the vertical line. Although the intervals vary from sample to sample, all but one of these samples gave a confidence interval that captures the true p. To say that these are 95% confidence intervals is

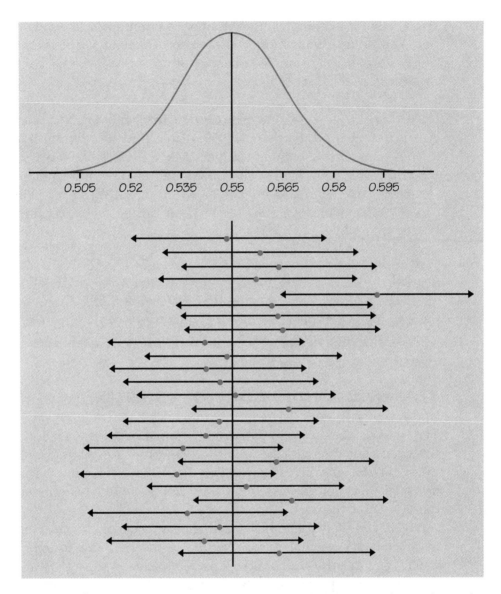

Figure 8-4 The behavior of confidence intervals in repeated sampling. The intervals are 95% confidence intervals for a population proportion $\hat{p}$ from 25 SRSs. In the long run, 95% of such intervals will contain the true value of p.

just to say that the interval covers the true p in 95% of all samples, and misses in only 5%. Be sure you understand that this 95% and 5% refer to what would happen if we continued to take samples forever. In a

small number of samples, the number of confidence intervals that fail
to cover the true p may be a bit more or less than 5% of the samples.
In Figure 8-4, for example, one out of 25, or 4%, of the confidence
intervals fails to contain p.

> **EXAMPLE 6. A 99.7% confidence interval.** Bean is a cautious
> man. A method that is right 95% of the time and wrong 5% of the
> time is not good enough for him. Very well, let's use the "99.7" part
> of the 68–95–99.7 rule. The probability is 0.997 that $\hat{p}$ falls within
> 0.045 (three standard deviations) of its mean p. So $\hat{p} \pm 0.045$ is a
> 99.7% confidence interval for p. Bean can be 99.7% confident that
> the true p falls between
>
> $$\hat{p} - 0.045 = 0.57 - 0.045 = 0.525$$
>
> and
>
> $$\hat{p} + 0.045 = 0.57 + 0.045 = 0.615$$

Now he's smiling. A method that is correct 997 times in 1000 in the
long run (probability 0.997) estimates that he is safely ahead.

THINKING ABOUT CONFIDENCE INTERVALS

The statement that we are 95% confident that between 54% and 60% of
the voters favor Senator Bean is shorthand for "We got these numbers
by a method that gives correct results 95% of the time." That's the
essential idea. As always, some fine points should be pondered. To
these we now turn.

What confidence does not mean. The language of statistical inference
uses a fact about what would happen in the long run to express our
confidence in the results of a single sample. For Senator Bean's poll,
a method with probability 95% of being right estimated that he was
favored by between 0.54 and 0.60 of the voters in this state. Be careful:
we *cannot* say that the probability is 95% that the true p falls between
0.54 and 0.60. All we can say is that we got the interval 0.54 to 0.60
by a method that covers the true p in 95% of all possible samples.
This particular interval from this particular sample either does or does
not cover p—there is no randomness left once we have chosen one
particular sample, so probability does not make sense.

High confidence is not free. Why would anyone use a 95% confidence interval when 99.7% confidence is available? Look again at Senator Bean. His 95% confidence interval was $\hat{p} \pm 0.03$, while 99.7% confidence required a wider interval, $\hat{p} \pm 0.045$. It is generally true that there is a tradeoff between the confidence level and the width of the interval. To obtain higher confidence from the same sample, we must be willing to accept a larger margin of error (a wider interval).

Larger samples give shorter intervals. If we want high confidence *and* a short interval, we must take a larger sample. The precision of a sample statistic increases as the sample size increases. That is, the standard deviation of the sampling distribution gets smaller. That means that for a fixed level of confidence, the confidence interval grows ever shorter as the sample size increases.

It's a poor cook who uses the same recipe in every meal. The Gallup poll's probability sampling method is such that when a sample of size 1000 is taken, we can be 95% confident that the population proportion is within 4 points of Gallup's sample proportion. (Table 1-1 on page 40 gives this information in more detail.) In more formal language, $\hat{p} \pm 0.04$ is a 95% confidence interval for the population proportion p based on a sample size 1000 drawn by Gallup's probability sampling method.

The 95% confidence interval for an SRS of size 1000 is $\hat{p} \pm 0.03$. That recipe is wrong for the Gallup poll because Gallup does not use an SRS. Gallup's sampling method is less precise (has a wider 95% confidence interval) than an SRS of the same size. That is the price paid for the convenience of Gallup's multistage sampling design. The recipe for a confidence interval depends on how the data were collected. Section 2 gives more detailed recipes for use when you have an SRS. You can't use these recipes when the data are not an SRS.

Confidence intervals are used whenever statistical methods are applied, and some of the recipes are complicated indeed. But the idea of 95% confidence is always the same: the recipe employed catches the true parameter value 95% of the time when used repeatedly.

SECTION 1 EXERCISES

8.1 **The schools' most serious problem.** The report of a sample survey of 1500 adults says, "With 95% confidence, between 27%

and 33% of all American adults believe that drugs are the most serious problem facing our nation's public schools." Explain to someone who knows no statistics what the phrase "95% confidence" means in this report.

8.2   **A pre-election poll.**   In 1976, a closely contested presidential election pitted Jimmy Carter against Gerald Ford. A poll taken immediately before the election showed that 51% of the sample intended to vote for Carter. The polling organization announced that they were 95% confident that the sample result was within ± 2 points of the true percent of all voters who favored Carter.

(a)   Explain in plain language to someone who knows no statistics what "95% confident" means in this announcement.

(b)   The poll showed Carter leading. Yet the polling organization said the election was too close to call. Explain why.

8.3   Suppose that Senator Bean's SRS of size 1000 produced 520 voters who plan to vote for Bean.

(a)   What is the 95% confidence interval for the population proportion p who plan to vote for Bean?

(b)   What is the 99.7% confidence interval for p?

(c)   Based on the sample results, can Bean be confident that he is leading the race?

8.4   An SRS of 1000 graduates of a university showed that 54% earned at least $50,000 a year. Give a 95% confidence interval for the proportion of all graduates of the university who earn at least $50,000 a year.

8.5   If Senator Bean took an SRS of only 25 voters, the probability distribution of the sample proportion $\hat{p}$ who favor him would be (very roughly) normal with mean equal to the population proportion p and standard deviation 0.1.

(a)   What property of this distribution shows that $\hat{p}$ is an unbiased estimate of p?

(b)   A sample of size 25 gives less precision than a sample of size 1000. How is this reflected in the distributions of $\hat{p}$ for the two sample sizes?

p ± .2 is a 95% CI for the proportion of Sea Bean voters in the pop.

2 × .1

(c)   For the SRS of size 25, what is the number m such that the interval from $\hat{p} - m$ to $\hat{p} + m$ is a 95% confidence interval for p? Explain your answer.   *95% of all samples*

(d)   Suppose Senator Bean found 17 of 25 voters in an SRS favoring him. Give a 95% confidence interval for the proportion of all voters who favor Bean. $\hat{p} = \frac{17}{25} = .68$ *(.48, .88) is a 95%...*

8.6   Spinning a coin.   Hold a penny upright on its edge under your forefinger on a hard surface, then snap it with your other forefinger so that it spins for some time before falling. Do this 25 times. You have taken an SRS of size 25 from the population of all spins of this coin. The probability p that the spinning coin will fall heads is the population proportion of heads.

(a)   What is the sample proportion $\hat{p}$ of heads for your 25 trials?

(b)   Part (c) of the previous exercise derives a 95% confidence interval for p based on a sample of size 25. What is your 95% confidence interval for the probability p that the spinning coin will fall heads?

(c)   Your interval is quite wide. How could you get a shorter 95% confidence interval?

8.7   Tossing a thumbtack.   If you toss a thumbtack on a hard surface, what is the probability that it will it land point up? Estimate this probability p by tossing a thumbtack 100 times. The 100 tosses are an SRS of size 100 from the population of all tosses. The proportion of these 100 tosses that land point up is the sample proportion $\hat{p}$.

(a)   The sampling distribution of $\hat{p}$ for an SRS of size 100 is approximately normal with mean p and standard deviation 0.05. Sketch a normal curve and mark on the axis the region where the central 95% of all values of $\hat{p}$ would fall if you repeated your 100 tosses many times.

(b)   Give the recipe for a 95% confidence interval for p, the probability that the tack lands point up, based on 100 tosses.

(c)   What is the 95% confidence interval from your 100 tosses? Explain in simple language your conclusion about the probability that a thumbtack lands point up.

8.8   In Senator's Bean's SRS of 1000 voters, what is the number m such that the interval from $\hat{p} - m$ to $\hat{p} + m$ is a 68% confidence interval for p? Explain your answer.

8.9   **Simulating confidence intervals.** We are going to simulate the performance of a 68% confidence interval for p based on an SRS of size 25.

(a)   Using the information given in Exercise 8.5, explain why the interval from $\hat{p} - 0.1$ to $\hat{p} + 0.1$ is a 68% confidence interval for p.

(b)   Suppose that in fact $p = 0.6$ (that is, 60% of all the voters favor Bean). Explain how to use Table A to simulate drawing an SRS of 25 voters.

(c)   Starting at line 101 of Table A, simulate drawing 10 SRSs of size 25 from this population. For each sample, compute the sample proportion $\hat{p}$ who favor Bean. Then compute the 68% confidence interval $\hat{p} - 0.1$ to $\hat{p} + 0.1$ for each sample.

(d)   Make a drawing like Figure 8-4 that shows your 10 confidence intervals. How many of the intervals covered the true $p = 0.6$? How many failed to cover p?

8.10   On hearing of the poll mentioned in Exercise 8.2, a nervous politician asked, "What is the probability that over half the voters prefer Carter?" A statistician replied that this question not only can't be answered from the poll results, it doesn't even make sense to talk about such a probability. Explain why.

8.11   **A Gallup poll.** Table 1-1 on page 40 is a table of margins of error for samples of several sizes drawn by the Gallup poll's probability sampling procedure. We can use that table to give 95% confidence intervals.

A Gallup poll survey of 1028 adults finds that 678 favor a constitutional amendment that would permit organized prayer in public schools. Give a 95% confidence interval for the proportion of all American adults who support such an amendment.

8.12   **A Gallup poll.** A Gallup poll survey of 1540 adults finds that 15% jog regularly. Use Table 1-1 on page 40 to give a 95%

confidence interval for the proportion p of all adults who jog regularly.

8.13   Comparing confidence intervals.

(a)   If you have an SRS of size 1000 from a population of which the proportion p have some characteristic:

   $\hat{p} \pm 0.015$ is a 68% confidence interval for p.

   $\hat{p} \pm 0.03$ is a 95% confidence interval for p.

   $\hat{p} \pm 0.045$ is a 99.7% confidence interval for p.

What general fact about confidence intervals does a comparison of these three intervals illustrate?

(b)   For the same population:

$\hat{p} \pm 0.2$ is a 95% confidence interval for samples of size 25.

$\hat{p} \pm 0.1$ is a 95% confidence interval for samples of size 100.

$\hat{p} \pm 0.03$ is a 95% confidence interval for samples of size 1000.

What general fact about confidence intervals does a comparison of these three intervals illustrate?

▶ 2 CONFIDENCE INTERVALS FOR PROPORTIONS AND MEANS*

Although the idea of a confidence interval remains ever the same, specific recipes vary greatly. The form of a confidence interval depends first on the parameter you wish to estimate—a population proportion, or mean, or median, or whatever. The second influence is the design of the sample or experiment. Estimating a population proportion from a stratified sample requires a different recipe than if the data come from an SRS. The sampling design and the parameter to be estimated determine the form of the confidence interval. The final details depend on the sample size and the confidence level you choose. The two recipes in this section are quite useful, but in comparison with the statistician's full array of confidence intervals for all occasions, these

---

*This section is more technical than the rest of the book.

two resemble a tool kit containing only a chisel and a roofing square. These are useful tools, but only sometimes.

CONFIDENCE INTERVALS FOR A POPULATION PROPORTION

When we select a large SRS from a population, the sampling distribution of the sample proportion $\hat{p}$ is close to a normal distribution. This normal sampling distribution has mean equal to the population proportion p because $\hat{p}$ is unbiased as an estimator of p. When the sample size n is 1000 and p is between 0.3 and 0.7, the standard deviation of $\hat{p}$ is close to 0.015. By the 68–95–99.7 rule, $\hat{p} \pm (2)(0.015)$ is a 95% confidence interval for p. The same reasoning leads to the conclusion that whenever we know the standard deviation of the sampling distribution of $\hat{p}$, a 95% confidence interval for p is

$$\hat{p} \pm (2)(\text{standard deviation of } \hat{p})$$

because 95% of the probability in the normal distribution of $\hat{p}$ falls within two standard deviations of the mean p.

THE SAMPLING DISTRIBUTION OF $\hat{p}$

By mathematics we can discover the standard deviation of the normal sampling distribution of $\hat{p}$. Here is the full story.

Sampling distribution of a sample proportion

Choose an SRS of size n from a large population with population proportion p having some characteristic of interest. Let $\hat{p}$ be the proportion of the sample having that characteristic. Then:

▶ The sampling distribution of $\hat{p}$ is **approximately normal** when the sample size n is large.

▶ The **mean** of the sampling distribution is equal to p.

▶ The **standard deviation** of the sampling distribution is

$$\sqrt{\frac{p(1-p)}{n}}$$

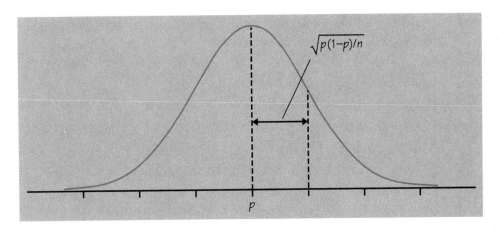

Figure 8-5 The sampling distribution of a sample proportion $\hat{p}$. The distribution of $\hat{p}$ is approximately normal for large samples.

Figure 8-5 illustrates this sampling distribution. The standard deviation of $\hat{p}$ depends on the true p and on the sample size n. For example, when $n = 1000$ and $p = 0.5$, the standard deviation of $\hat{p}$ is

$$\sqrt{\frac{(0.5)(0.5)}{1000}} = \sqrt{0.00025} = 0.0158$$

When $p = 0.3$ (or 0.7), the standard deviation is

$$\sqrt{\frac{(0.3)(0.7)}{1000}} = \sqrt{0.00021} = 0.0145$$

These more exact results lie behind the statement in Section 1 that for p between 0.3 and 0.7, the standard error of $\hat{p}$ in an SRS of size 1000 is close to 0.015.

We now know that a more accurate recipe for a 95% confidence interval for p, taking p and n into account, is

$$\hat{p} \pm 2\sqrt{\frac{p(1-p)}{n}}$$

We can't use this formula because we do not know p. (If we did know p, we would not need to settle for 95% confidence!) When n is large, $\hat{p}$ is quite close to p. So, at the cost of further approximation, we can use the estimated standard deviation formed by replacing p by $\hat{p}$ in the recipe. The final version of the 95% confidence interval for p is

$$\hat{p} \pm 2\sqrt{\frac{\hat{p}(1 - \hat{p})}{n}}$$

CONFIDENCE INTERVALS WITH OTHER CONFIDENCE LEVELS

All of this started with the fact that in any normal distribution there is probability 0.95 within two standard deviations of the mean. What if we want a 90% confidence interval, or a 75% confidence interval? For any probability C between 0 and 1, there is a number z^* such that any

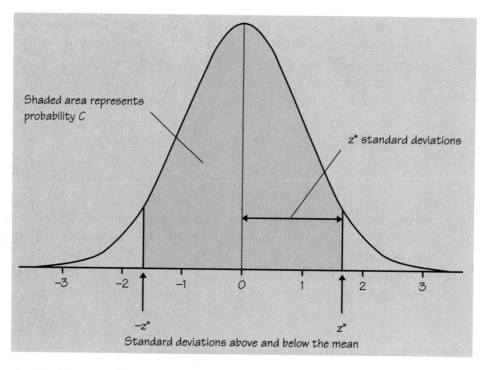

Figure 8-6 Critical values of the normal distributions. In any normal distribution, there is probability C within z^* standard deviations of the mean.

normal distribution has probability C within z^* standard deviations of the mean. Figure 8-6 shows how the probability C and the number z^* are related.

Table C in the back of the book gives the numbers z^* for various choices of C. These numbers are called **critical values** of the normal distributions. Table C shows that any normal distribution has probability 0.90 within ± 1.64 standard deviations of its mean. The table also shows that any normal distribution has probability 0.95 within ± 1.96 standard deviations of its mean. The 68–95–99.7 rule uses 2 in place of the critical value $z^* = 1.96$. That is good enough for practical purposes, but the table gives the more exact value.

From Figure 8-6 we see that with probability C, the sample proportion $\hat{p}$ takes a value within z^* standard deviations of p. That is just to say that with probability C, the interval extending z^* standard deviations on either side of the observed $\hat{p}$ captures the unknown p. Using the estimated standard deviation of $\hat{p}$ produces the following recipe.

Confidence interval for a population proportion

Choose an SRS of size n from a population of individuals of which proportion p have some characteristic of interest. When n is large, an approximate level C confidence interval for p is

$$\hat{p} \pm z^* \sqrt{\frac{\hat{p}(1 - \hat{p})}{n}}$$

where z^* is the critical value for probability C from Table C.

This recipe is valid only when our data are an SRS from the population of interest. Even then it is only approximately correct, for two reasons. First, the sampling distribution of $\hat{p}$ is only approximately normal. Second, the estimated standard deviation of $\hat{p}$ is only approximately equal to the exact standard deviation $\sqrt{p(1 - p)/n}$. Both of these approximations improve as the sample size n increases. For samples of size 100 and larger, the recipe given is quite accurate. It is often used for sample sizes as small as 25 or 30. If you have a small sample, or if your sample is not an SRS, please visit your friendly local statistician for advice.

EXAMPLE 7. Senator Bean's confidence intervals.   Senator Bean, our acquaintance from Section 1, took an SRS of 1000 registered voters. Of these, 570 supported the Senator in his bid for re-election. So

$$\hat{p} = \frac{570}{1000} = 0.57$$

and the estimated standard deviation of $\hat{p}$ is

$$\sqrt{\frac{\hat{p}(1 - \hat{p})}{n}} = \sqrt{\frac{(0.57)(0.43)}{1000}} = 0.0157$$

A 95% confidence interval for the proportion p of all registered voters who support Bean is therefore

$$\hat{p} \pm z^* \sqrt{\frac{\hat{p}(1 - \hat{p})}{n}} = 0.57 \pm (1.96)(0.0157)$$

$$= 0.57 \pm 0.031$$

or 0.54 to 0.60. This is the same result we found in Section 1. If Bean insists on 99% confidence, we must use the critical value $z^* = 2.58$ for confidence level $C = 0.99$. The interval is

$$\hat{p} \pm z^* \sqrt{\frac{\hat{p}(1 - \hat{p})}{n}} = 0.57 \pm (2.58)(0.0157)$$

$$= 0.57 \pm 0.041$$

or 0.53 to 0.61. As usual, higher confidence exacts its price in the form of a larger margin of error.

The $\sqrt{n}$ in the recipe for our confidence interval tells us exactly how the sample size n affects the margin of error. Because n appears in the denominator, the margin of error gets smaller as n gets larger. Because it is the square root of n that appears, a sample four times larger is needed to give a margin of error half as large. Senator Bean's SRS of 1000 voters gave a margin of error of about ± 0.04 with 99% confidence; to reduce the margin of error to ± 0.02, the senator would need an SRS of 4000 voters. The square root of n increases much more slowly than n itself. This mathematical fact means that small margins of error require large samples and therefore lots of money. Sorry about that.

CONFIDENCE INTERVALS FOR A POPULATION MEAN

What is the mean number of hours your college's first-year students study each week? What is the mean number of beers they drink each week? We often want to estimate the mean of a population. To distinguish the population mean (a parameter) from the sample mean $\bar{x}$, we write the population mean as μ, the Greek letter mu. We use the mean $\bar{x}$ of an SRS to estimate the unknown mean μ of the population.

Like the sample proportion $\hat{p}$, the sample mean $\bar{x}$ from a large SRS has a sampling distribution that is close to normal. Because the sample mean of an SRS is an unbiased estimator of μ, the sampling distribution of $\bar{x}$ has μ as its mean. The standard deviation of $\bar{x}$ depends on the standard deviation of the population, which is usually written as σ (the Greek letter sigma). By mathematics we can discover the following fact.

Sampling distribution of the sample mean

Choose an SRS of size n from a large population having mean μ and standard deviation σ. Let $\bar{x}$ be the mean of the sample. Then:

▶ The sampling distribution of $\bar{x}$ is **approximately normal** when the sample size n is large.

▶ The **mean** of the sampling distribution is equal to μ.

▶ The **standard deviation** of the sampling distribution is $\sigma/\sqrt{n}$.

Figure 8-7 compares the distribution of a single observation with the distribution of the mean $\bar{x}$ of 10 observations. Both have the same center, but the distribution of $\bar{x}$ is less spread out. That averages are less variable than individual observations is an important statistical fact. (In Figure 8-7, the distribution of individual observations is normal. If that is true, then the sampling distribution of $\bar{x}$ is exactly normal for any size sample, not just approximately normal for large samples. The comparison of spreads, however, does not depend on normality. Averages of several observations are always less variable than individual observations.)

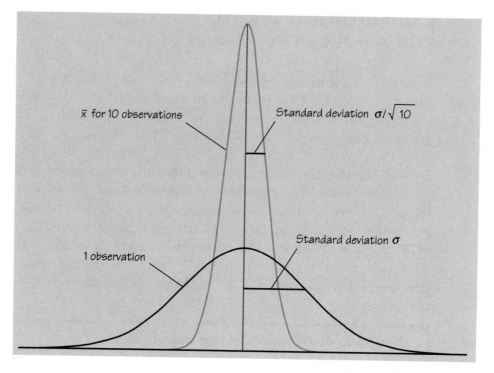

Figure 8-7 The sampling distribution of the mean $\overline{x}$ of 10 observations compared with the distribution of a single observation. If individuals have standard deviation σ, the mean of 10 individuals has standard deviation $\sigma/\sqrt{10}$.

The standard deviation of $\overline{x}$ depends on both σ and the sample size n. We know n, but not σ. When n is large, the sample standard deviation s is close to σ and can be used to estimate it. The estimated standard deviation of $\overline{x}$ is therefore $s/\sqrt{n}$. Now we can find confidence intervals for μ just as for p.

Confidence interval for a population mean

Choose an SRS of size n from a population of individuals having mean μ. When n is large, an approximate level C confidence interval for μ is

$$\overline{x} \pm z^* \frac{s}{\sqrt{n}}$$

where z^* is the critical value for probability C from Table C.

The cautions cited in estimating p apply here as well. The recipe is valid only when an SRS is drawn and the sample size n is reasonably large. And the margin of error again decreases only at a rate proportional to $\sqrt{n}$ as the sample size n increases.

EXAMPLE 8. Bacteria counts in milk. Exercise 4.17 on page 230 reports the count of coliform bacteria per milliliter in each of 100 specimens of milk. Suppose that these specimens are an SRS of the milk sold in east coast groceries. We want a 90% confidence interval for the mean coliform count μ of all milk sold in the region.

First compute the sample mean and standard deviation for this set of data. The results are

$$\overline{x} = 5.88 \quad s = 2.02$$

The 90% confidence interval is therefore

$$\overline{x} \pm z^* \frac{s}{\sqrt{n}} = 5.88 \pm (1.64)\frac{2.02}{\sqrt{100}}$$

$$= 5.88 \pm 0.33$$

We are 90% confident that the mean coliform count in the entire population of milk falls between 5.55 and 6.21 per milliliter.

SECTION 2 EXERCISES

8.14   **A poll of voters.** You are the polling consultant to a member of Congress. An SRS of 500 registered voters finds that 28% name "environmental problems" as the most important issue facing the nation. Give a 90% confidence interval for the proportion of all voters who hold this opinion. Then explain carefully to the member of Congress what your conclusion tells us about voters' opinions.

8.15   In the setting of the previous exercise:

(a)   Give a 75% confidence interval and a 99% confidence interval. How does the confidence level affect the width of the interval?

(b)   Suppose that the sample result $\hat{p} = 0.28$ had come from an SRS of 100 persons or an SRS of 4000 persons. Give a 90%

confidence interval in both cases. How does the sample size affect the width of the interval?

8.16 Risky sexual behavior. How common is behavior that puts people at risk of AIDS? The National AIDS Behavioral Surveys interviewed a random sample of 2673 adult heterosexuals. Of these, 170 said they had multiple sexual partners.[2]

(a) Give a 99% confidence interval for the proportion of all adult heterosexuals who have multiple partners. Then explain carefully, to someone who knows no statistics, what the confidence interval tells us about sexual behavior.

(b) What type of nonsampling error is likely to affect the results of this survey? Does the confidence interval in (a) take nonsampling errors into account?

8.17 The congressman you advise receives 1310 pieces of mail on pending gun control legislation. Of these, 86% oppose the legislation. Having learned from you about estimating with confidence, he asks you for an analysis of these opinions. What will you tell him?

8.18 Crowds at Yellowstone Park. The Forest Service is considering additional restrictions on the number of vehicles allowed to enter Yellowstone National Park. To assess public reaction, the Service asks a simple random sample of 150 visitors if they favor the proposal. Of these, 89 say "Yes." Give a 95% confidence interval for the proportion of all visitors to Yellowstone who favor the restrictions. Are you 95% confident that more than half are in favor? Explain your answer.

8.19 Count Buffon's coin. The eighteenth-century French naturalist Count Buffon tossed a coin 4040 times. He got 2048 heads. Give a 95% confidence interval for the probability that Buffon's coin lands heads. Are you confident that this probability is not 1/2? Why?

8.20 We want to be rich. In a recent year, 73% of first-year college students responding to a national survey identified "being very well-off financially" as an important personal goal. A state university finds that 132 of an SRS of 200 of its first-year students say that this goal is important. Give a 95% confidence interval for the proportion of all first-year students at the university who would identify being well-off as an important personal goal. Are you confident that the proportion at this university is different than the national proportion, 73%? Why?

8.21 How large a sample? The congressman you advise knows from preliminary polls that about half the registered voters in his district favor his re-election. He wants to commission a poll that will estimate this proportion accurately. You decide to take an SRS large enough to get a 95% confidence interval with margin of error ± 0.02. How large a sample must you take? (Hint: the margin of error is $\pm 1.96 \sqrt{p(1-p)/n}$, and p is close to 0.5. You must find n to make this margin of error 0.02.)

8.22 Scores on the American College Testing (ACT) college admissions examination vary normally with mean $\mu = 18$ and standard deviation $\sigma = 6$. The range of reported scores is 1 to 36.

(a) What range contains the middle 95% of all individual scores?

[handwritten: $18 \pm 2 \times 6$ gives middle 95% of pop]

(b) If the ACT scores of 25 randomly selected students are averaged, what range contains the middle 95% of the averages $\bar{x}$?

[handwritten: $6/\sqrt{25} = 6/5 = 1.2$ $18 \pm 2 \times 1.2$ $(15.6, 20.4)$]

8.23 More credit-card spending? A bank wonders whether dropping the annual credit card fee for customers who charge at least $2400 in a year will increase the amount charged on its credit cards. The bank makes this offer to an SRS of 200 of its credit card customers. It then compares the amount these customers charge this year with the amount that they charged last year. The mean increase is $332, and the standard deviation is $108.

(a) Give a 99% confidence interval for the mean amount of the increase.

(b) Unfortunately, the study is poorly designed—the result doesn't tell us whether the new policy caused the increase. Explain why. Then outline a design that will tell the bank what it wants to know.

8.24 Blood pressure. A randomized comparative experiment studied the effect of diet on blood pressure. Researchers divided 54 healthy white males at random into two groups. One group received a calcium supplement, the other a placebo. At the beginning of the study, the researchers measured many variables on the subjects. The paper reporting the study gives $\overline{x} = 114.9$ and $s = 9.3$ for the seated systolic blood pressure of the 27 members of the placebo group.

(a)   Give a 95% confidence interval for the mean blood pressure of the population from which the subjects were recruited.

(b)   The recipe you used in part (a) requires an important assumption about the 27 men who provided the data. What is this assumption?

8.25   **How long do presidents live?**   Table 4-3 on page 228 gives the ages at death of the American presidents. The mean is $\overline{x} = 69.14$ years and the standard deviation is $s = 11.28$ years. It *makes no sense* to use these statistics to find a 95% confidence interval for the mean age at death μ of all men who have been American presidents. Why is this so?

8.26   **Incomes.**   The income of the householder was one of the items included on only 17% of the forms in the 1990 census. Suppose (alas, it is too simple to be true) that the households who answer this question are an SRS of the households in each district. In Middletown, a city of 40,000 persons, the 1990 census asked 2621 householders their 1989 income. The mean of the responses was $\overline{x} = \$23,453$ and the standard deviation was $\$8721$. Give a 99% confidence interval for the 1989 mean income of Middletown householders.

8.27   **Standard errors.**   The estimated standard deviation of a statistic is often called a *standard error*. For example, the standard error of $\overline{x}$ from an SRS of size n is $s/\sqrt{n}$. When the standard error is given, you can compute confidence intervals for means and proportions from complex sample designs without knowing the formula that led to the standard error.

The Bureau of Labor Statistics estimates that there are 6,000,000 unemployed adult men in the labor force this month. This estimate is based on the Current Population Survey, a complex multistage sample. The BLS tells us that the standard error of the estimate is 122,000. Give a 90% confidence interval for the number of unemployed adult men.

▶ 3 STATISTICAL SIGNIFICANCE

Statistical inference uses sample data to draw conclusions about the population that the data represent. The reasoning of inference works by asking "What would happen if we did this many times?" The answers are in terms of probabilities, based on sampling distributions. A confidence interval estimates the value of an unknown parameter. It says, "If we took many samples, most of our intervals would capture the true parameter value." Replace the vague "most" by a probability and you have a numerical measure of our confidence that we have caught the true value.

The other major type of formal inference is the *test of significance.* The purpose of a statistical test is to assess the evidence provided by the data against some claim about a parameter. A test says, "If we took many samples and the claim were true, we would rarely get a result like this." Observing a result that would rarely occur if a claim were true is evidence that the claim is not true. Replace the vague "rarely" by a probability, and you have a numerical measure of our confidence in the evidence that the data give us. Here is an example of this reasoning at work.

> **EXAMPLE 9. Is the coffee fresh?** People of taste are supposed to prefer fresh-brewed coffee to the instant variety. On the other hand, perhaps many coffee-drinkers just want their caffeine fix. A skeptic claims that only half of all coffee-drinkers prefer fresh coffee. Let's do an experiment to test this claim.
>
> Each of 50 subjects tastes two unmarked cups of coffee and says which they prefer. One cup in each pair contains instant coffee, the other, fresh-brewed coffee. The statistic that records the result of our experiment is the proportion $\hat{p}$ of the sample who say they like the fresh-brewed coffee better. We find that 36 of our 50 subjects choose the fresh coffee. That is,
>
> $$\hat{p} = \frac{36}{50} = 0.72$$
>
> To make a point, let's compare our outcome $\hat{p} = 0.72$ with another possible result. If only 28 of the 50 subjects like the fresh coffee

better than instant coffee, the sample proportion is

$$\hat{p} = \frac{28}{50} = 0.56 = 56\%$$

Surely 72% is stronger evidence against the skeptic's claim than 56%. But how much stronger? Is even 72% in favor in a *sample* convincing evidence that a majority of the *population* prefer fresh coffee? Statistical tests answer these questions.

THE REASONING OF TESTS OF SIGNIFICANCE

A. The skeptic claims that only half of all coffee-drinkers prefer fresh-brewed coffee. That is, he claims that the population proportion p is only 0.5. *Suppose for the sake of argument that this claim is true.* From mathematics or a long simulation, find the sampling distribution of the sample proportion $\hat{p}$ for our 50 subjects if in fact $p = 0.5$. Figure 8-8 displays this distribution. It is approximately normal with mean 0.5 and standard deviation 0.0707.

B. Locate the observed $\hat{p}$ on this sampling distribution and ask, "If we repeated the study many times, how often would the outcome favor fresh coffee this strongly?" You can see from Figure 8-8 that a sample proportion as large as $\hat{p} = 0.56$ would occur quite often just by chance, but that a $\hat{p}$ as large as 0.72 would rarely occur.

C. In fact, the probability that 56% or more of the sample favor fresh coffee is 0.20. Because a result this strong would happen one time in five just by chance, it is not good evidence against the skeptic's claim. The probability that 72% or more of the sample prefer fresh coffee *if in fact only half the population do so* is only 0.001. That is, we would almost never find 72% or more of the sample favoring fresh coffee if the skeptic is right. This is convincing evidence that he is wrong, and that the true population p is greater than 0.5.

Be sure you understand why this evidence is convincing. There are two possible explanations of the fact that 72% of our subjects prefer fresh to instant coffee:

(a)   The skeptic is correct ($p = 0.5$), and by bad luck a very unlikely outcome occurred.

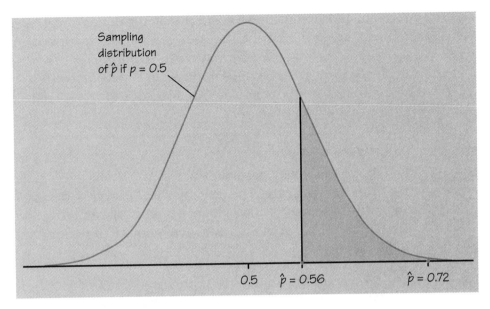

Figure 8-8 Suppose that just half of all coffee-drinkers prefer fresh coffee over instant coffee. If we ask 50 people which they prefer, the proportion favoring fresh coffee will have this sampling distribution. We would often observe $\hat{p} = 0.56$ just by chance, but would rarely see a sample proportion as large as 0.72. The shaded area represents the probability of an outcome 0.56 or greater.

> (b) In fact the population proportion favoring fresh coffee is greater than 0.5, so the sample outcome is about what would be expected.

We cannot be certain that explanation (a) is untrue. Our taste test results *could* be due to chance alone. But the probability that such a result would occur by chance is so small (0.001) that we are quite confident that explanation (b) is right. The basic reasoning of a statistical test is that results that would be surprising if a claim were true give us evidence that the claim is not true. Examples 1 and 2 at the beginning of this chapter are other examples of this reasoning.

HYPOTHESES AND *P*-VALUES

Tests of significance refine (and perhaps hide) this basic reasoning. In most studies, we hope to show that some definite effect is present in the population. In Example 9, we suspect that a majority of coffee-drinkers prefer fresh-brewed coffee. A statistical test begins by supposing for the sake of argument that the effect we seek is *not* present. We then look for evidence against this supposition and in favor of the effect we hope to find. The first step in a test of significance is to state a claim that we will try to find evidence *against*.

Null hypothesis H_0

The statement being tested in a test of significance is called the **null hypothesis**. The test of significance is designed to assess the strength of the evidence against the null hypothesis. Usually the null hypothesis is a statement of "no effect" or "no difference."

The term "null hypothesis" is abbreviated as H_0, which is read "H-nought." It is a statement about the population and so must be stated in terms of a population parameter. In Example 9, the parameter is the proportion p of all coffee-drinkers who prefer fresh to instant coffee. The null hypothesis is

$$H_0 : p = 0.5$$

The statement we hope or suspect is true instead of H_0 is called the **alternative hypothesis** and is abbreviated by H_a. In Example 9, the alternative hypothesis is that a majority of the population favor fresh coffee. In terms of the population parameter, this is

$$H_a : p > 0.5$$

A test of significance assesses the strength of the evidence against the null hypothesis in terms of probability. If the observed outcome is surprising under the supposition that the null hypothesis is true, but is more probable if the alternative hypothesis is true, that outcome is evidence against H_0 in favor of H_a. Outcomes are surprising if they are far from the null hypothesis in the direction given by the alternative hypothesis. It would be surprising to find 38 of 50 subjects favoring fresh coffee if in fact only half of the population feel this way.

The probability that describes our confidence that H_0 is false and H_a is true is the probability of getting an outcome at least as far as the actually observed outcome from what we would expect when H_0 is true. What counts as "far from what we would expect" depends on H_a

as well as H_0. In the taste test, the probability we want is the probability that $\hat{p}$ is 0.72 or larger. This probability is very small (0.001), so our result would be very surprising if $p = 0.5$ were true.

P-value

The probability, computed assuming that H_0 is true, that the test statistic would take a value as extreme or more extreme than that actually observed is called the ***P*-value** of the test. The smaller the *P*-value is, the stronger is the evidence against H_0 provided by the data.

In practice, most statistical tests are carried out by computer software that calculates the *P*-value for us. It is usual to report the *P*-value in describing the results of studies in many fields. You should therefore understand what *P*-values say even if you don't do statistical tests yourself, just as you should understand what "95% confidence" means even if you don't roll your own confidence intervals.

> **EXAMPLE 10. Count Buffon's coin.** The French naturalist Count Buffon (1707–1788) tossed a coin 4040 times. He got 2048 heads. The sample proportion of heads is
>
> $$\hat{p} = \frac{2048}{4040} = 0.507$$
>
> That's a bit more than one-half. Is this evidence that Buffon's coin was not balanced? This is a job for a significance test.
>
> The null hypothesis says that the coin is balanced ($p = 0.5$). We did not suspect a bias in a specific direction before we saw the data, so the alternative hypothesis is just "the coin is not balanced." That is,
>
> $$H_a : p \neq 0.5$$
>
> We need the sampling distribution of $\hat{p}$, *assuming for the sake of argument that the null hypothesis $p = 0.5$ is true*. Mathematics or a long simulation shows that this distribution is very close to normal, with mean 0.5 and standard deviation 0.008. Figure 8-9 shows the sampling distribution.
>
> Locate the observed $\hat{p} = 0.507$ on the sampling distribution. It is clear from Figure 8-9 that this $\hat{p}$ is not a surprising result for 4040 tosses of a balanced coin. What is the *P*-value?

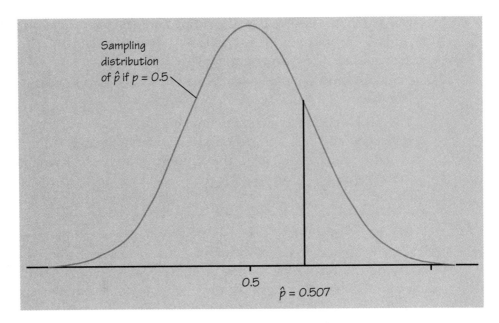

Figure 8-9 Suppose that the coin that Count Buffon tossed 4040 times was fair ($p = 0.5$). The sampling distribution of the proportion of heads in 4040 tosses is then close to normal with mean 0.5 and standard deviation 0.008. Buffon's result, $\hat{p} = 0.507$, could easily occur by chance in tossing a fair coin.

Because the alternative hypothesis allows p to lie on either side of 0.5, values of $\hat{p}$ far from 0.5 in either direction provide evidence against H_0 and in favor of H_a. The P-value is therefore the probability that the observed $\hat{p}$ lies as far from 0.5 *in either direction* as the observed $\hat{p} = 0.507$. Figure 8-10 shows the P-value. It is $P = 0.37$.

In plain language, a truly balanced coin would give a result this far or farther from 0.5 in 37% of all repetitions of Buffon's trial. His result gives no reason to think that his coin was not balanced.

The alternative $H_a : p > 0.5$ in Example 9 is a **one-sided alternative** because the effect we seek evidence for says that the population proportion is greater than one-half. The alternative $H_a : p \neq 0.5$ in Example 10 is a **two-sided alternative.** Whether the alternative is one-sided or two-sided determines whether sample results extreme in one or both directions count as evidence against H_0 in favor of H_a.

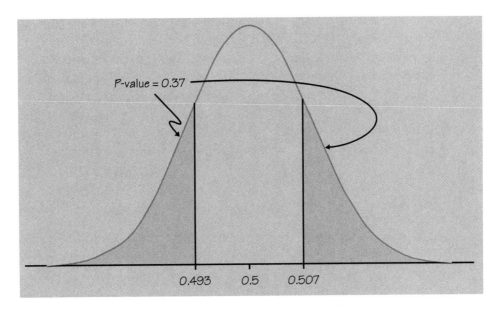

Figure 8-10 The *P*-value for Count Buffon's coin-tossing result $\hat{p} = 0.507$ is the probability that a fair coin would give a proportion of heads this far from 0.5 in either direction. That probability is the shaded area under the normal curve.

STATISTICAL SIGNIFICANCE

We sometimes take one final step to assess the evidence against H_0. We can compare the *P*-value with a fixed value that we regard as decisive. This amounts to announcing in advance how much evidence against H_0 we will insist on. The decisive value of *P* is called the **significance level.** We write it as α, the Greek letter alpha. If we choose $\alpha = 0.05$, we are requiring that the data give evidence against H_0 so strong that it would happen no more than 5% of the time (1 time in 20) when H_0 is true. If we choose $\alpha = 0.01$, we are insisting on stronger evidence against H_0, evidence so strong that it would appear only 1% of the time (1 time in 100) if H_0 is in fact true.

> **Statistical significance**
>
> If the *P*-value is as small or smaller than α, we say that the data are **statistically significant at level α.**

"Significant" in the statistical sense does not mean "important." It means simply "not likely to happen just by chance." The significance level α makes "not likely" more exact. Significance at level 0.01 is often expressed by the statement "The results were significant ($P < 0.01$)." Here P stands for the P-value. The P-value is more informative than a statement of significance, because it allows us to assess significance at any level we choose. For example, a result with $P = 0.03$ is significant at the $\alpha = 0.05$ level, but not significant at the $\alpha = 0.01$ level.

SECTION 3 EXERCISES

8.28   A social psychologist reports that "In our sample, ethnocentrism was significantly higher ($P < 0.05$) among church attenders than among nonattenders." Explain to someone who knows no statistics what this means.

8.29   The financial aid office of a university asks a sample of students about their employment and earnings. The report says that "For academic year earnings, a significant difference ($P = 0.038$) was found between the sexes, with men earning more on the average. No difference ($P = 0.476$) was found between the earnings of black and white students." Explain both of these conclusions, for the effects of sex and of race on mean earnings, in language understandable to someone who knows no statistics.

8.30   **Risky behavior.** How common is behavior that puts people at risk of AIDS? The National AIDS Behavioral Surveys interviewed a random sample of 2673 adult heterosexuals. Of these, 170 had more than one sexual partner in the past year. We might use a test to see if these data give good evidence that more than 5% of adult heterosexuals had more than one sexual partner.

(a)   Explain in words what the parameter p is in this setting.

(b)   What are the null and alternative hypotheses H_0 and H_a?

(c)   What is the numerical value of the sample proportion $\hat{p}$? Of what event is the P-value the probability?

(d)   The P-value is $P = 0.002$. Explain carefully why this is very strong evidence that H_0 is not true and that H_a is true.

8.31 **We want to be rich.** In a recent year, 73% of first-year college students responding to a national survey identified "being very well-off financially" as an important personal goal. A state university finds that 132 of an SRS of 200 of its first-year students say that this goal is important. We wonder if the proportion of first-year students at this university who think being very well-off is important differs from the national value, 73%.

(a) Explain in words what the parameter p is in this setting.

(b) What are the null and alternative hypotheses H_0 and H_a?

(c) What is the numerical value of the sample proportion $\hat{p}$? Of what event is the P-value the probability?

(d) The P-value is $P = 0.037$. Explain carefully why this is reasonably good evidence that H_0 is not true and that H_a is true.

8.32 Is the result of Exercise 8.30 statistically significant at the $\alpha = 0.01$ level? Is it significant at the $\alpha = 0.001$ level?

8.33 Is the result of Exercise 8.31 statistically significant at the 5% level? At the 1% level?

8.34 A sociologist asks a large sample of high school students which academic subject is their favorite. She suspects that a lower percentage of females than of males will say that mathematics is their favorite subject. State in words the sociologist's null hypothesis H_0 and alternative hypothesis H_a.

8.35 An educational researcher randomly divides sixth-grade students into two groups for gym class. He teaches both groups basketball skills with the same methods of instruction. He encourages Group A with compliments and other positive behavior but acts cool and neutral toward Group B. He hopes to show that Group A does better (on the average) than Group B on a test of basketball skills at the end of the instructional unit. State in words the researcher's null hypothesis H_0 and alternative hypothesis H_a.

8.36 A political scientist hypothesizes that among registered voters there is a negative correlation between age and the percent who actually vote. To test this, she draws a random sample from public records on registration and voting. State in words the null hypothesis H_0 and alternative hypothesis H_a.

8.37   Explain why a result that is statistically significant at the 1% level is always significant at the 5% level. Then explain why a result that is significant at the 5% level may or may not be significant at the 1% level.

8.38   **Finding a *P*-value by simulation.**   Is a new method of teaching reading to first graders (method B) more effective than the method now in use (method A)? You design a matched pairs experiment to answer this question. You form 20 pairs of first graders, with the two children in each pair carefully matched in IQ, socioeconomic status, and reading-readiness score. You assign at random one student from each pair to method A. The other student in the pair is taught by method B. At the end of first grade, all the children take a test of reading skill. Let p stand for the proportion of all possible matched pairs of children for which the child taught by method B will have the higher score. Your hypotheses are:

$$H_0: p = 0.5 \quad \text{(no difference in effectiveness)}$$

$$H_a: p > 0.5 \quad \text{(method B is more effective)}$$

The result of your experiment is that method B gave the higher score in 12 of the 20 pairs, or $\hat{p} = 12/20 = 0.6$.

(a)   If H_0 is true, the 20 pairs of students are just like 20 fair coins. That is, they are 20 independent trials with probability 0.5 that method B wins each trial. Explain how to use Table A to simulate these 20 trials if we assume for the sake of argument that H_0 is true.

(b)   Use Table A starting at line 105 to simulate 10 repetitions of the experiment. Estimate from your simulation the probability that method B will do better in 12 or more of the 20 pairs when H_0 is true. (Of course, 10 repetitions is not enough to estimate the probability reliably. Once you see the idea, more repetitions are easy.)

(c)   Explain why the probability you simulated in (b) is the *P*-value for your experiment. With enough patience, you could find all the *P*-values in this section by doing simulations similar to this one.

8.39 **Finding a *P*-value by simulation.** A classic experiment to detect extra-sensory perception (ESP) uses a shuffled deck of cards containing five suits (waves, stars, circles, squares, and crosses). As the experimenter turns over each card and concentrates on it, the subject guesses the suit of the card. A subject who lacks ESP has probability 1/5 of being right by luck on each guess. A subject who has ESP will be right more often. Julie is right in 5 of 10 tries. (Actual experiments naturally use much longer series of guesses so that weak ESP could be spotted. No one has ever been right half the time in a long experiment!)

(a) Give H_0 and H_a for a test to see if this result is significant evidence that Julie has ESP.

(b) Explain how to simulate the experiment if we assume for the sake of argument that H_0 is true.

(c) Simulate 20 repetitions of the experiment; begin at line 121 of Table A.

(d) The actual experimental result was 5 right in 10 tries. What is the event whose probability is the *P*-value for this experimental result? Give an estimate of the *P*-value based on your simulation. How convincing was Julie's performance?

8.40 **Predicting job performance.** A study on predicting job performance reports that:

An important predictor variable for later job performance was the score x on a screening test given to potential employees. The variable being predicted was the employee's score y on an evaluation made after a year on the job. In a sample of 70 employees, the correlation between x and y was r = 0.4, which is statistically significant at the 1% level.

(a) Explain to someone who knows no statistics what information "$r = 0.4$" carries about the connection between the screening test and the later evaluation score.

(b) The null hypothesis in the test is that there is *no association* between x and y when the population of all employees is considered. To what value of the correlation for the entire population does this null hypothesis correspond?

(c)   In order to use the screening test, the employer must show that screening scores are positively associated with on-the-job performance. What is the alternative hypothesis?

(d)   Explain to someone who knows no statistics why "statistically significant at the 1% level" means there is good reason to think that there is a positive association between the two scores.

▶ 4 SIGNIFICANCE TESTS FOR PROPORTIONS AND MEANS*

As is the case for confidence intervals, statisticians have developed tests of significance for use in many different circumstances. The recipes for carrying out these tests generally have two parts: a statistic that measures the effect being sought and a probability distribution for the statistic that gives the corresponding P-value. Tests are now usually implemented by computer routines that compute the test statistic and its P-value starting from the raw data. The computer will not explain to you what statistical significance means, however. Your understanding of the reasoning common to all tests is therefore more valuable than knowledge of the specific procedures discussed in this section.

TESTS FOR A POPULATION PROPORTION

Inspired by Senator Bean's use of polls and statistics, his colleague Senator Caucus commissions an SRS of 1200 registered voters in her state. The poll finds that 53% of the sample plan to vote for Caucus. We know how to use this sample result $\hat{p} = 0.53$ to give a confidence interval for the unknown proportion p of all voters who favor Caucus. But the senator just wants to know if she's ahead. Does the sample provide strong evidence that a majority of the population plan to vote for Caucus? We wish to test the null hypothesis

$$H_0 : p = 0.5 \quad \text{(the race is even)}$$

---

*This section is more technical than the rest of the book. It builds on material from Section 2.

against the one-sided alternative hypothesis

$$H_a : p > 0.5 \quad \text{(Caucus is ahead)}$$

As usual, H_a states that the effect we seek (a majority for Caucus) is present, and H_0 that the effect is absent.

> **EXAMPLE 11. A test for Senator Caucus.** The statistic is $\hat{p}$. We recorded the sampling distribution of $\hat{p}$ on page 472. Remember that to find the P-value, we use the sampling distribution *assuming for the sake of argument that the null hypothesis is true.* Because H_0 states that $p = 0.5$, the distribution of $\hat{p}$ is then approximately normal with mean 0.5 and standard deviation
>
> $$\sqrt{\frac{p(1-p)}{n}} = \sqrt{\frac{(0.5)(0.5)}{1200}} = 0.014$$
>
> The P-value is the probability that a statistic having this distribution will take a value at least as large as 0.53, the observed value of $\hat{p}$. Figure 8-11 illustrates this probability as an area under a normal curve.

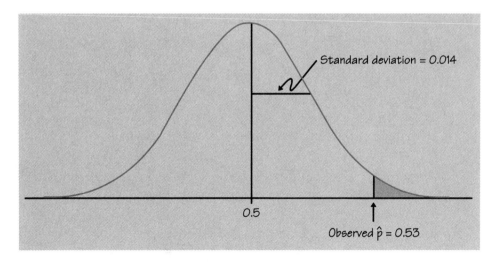

Figure 8-11 The P-value for Senator Caucus' poll. The P-value is the probability of observing an outcome $\hat{p} = 0.53$ or larger if in fact the null hypothesis that $p = 0.5$ is true.

To find this probability, convert $\hat{p} = 0.53$ to a standard score. It is usual in formal inference to call a standard score z.

$$\text{standard score } z = \frac{0.53 - 0.5}{0.014} = 2.1$$

Table B shows that a standard score of 2.1 corresponds to the 98.2 percentile. The P-value is therefore $P = 0.018$, for if 98.2% of the distribution lies below 2.1, the remaining 1.8% lies above. Senator Caucus is pleased. If the race were even, there is only probability 0.018 that the poll would show her as far ahead as it did. This is good evidence that she actually is ahead.

Example 11 obtained a test statistic (the standard score for $\hat{p}$) and its P-value (from Table B of normal percentiles). Here is the recipe for the test in general form, including both one-sided and two-sided alternatives.

Tests for a population proportion

Choose an SRS of size n from a population of individuals of which proportion p have some characteristic of interest. To test the hypothesis $H_0 : p = p_0$ for a specified value p_0, compute the standard score

$$z = \frac{\hat{p} - p_0}{\sqrt{\dfrac{p_0(1 - p_0)}{n}}}$$

The approximate P-value for a test of H_0 against

$H_a : p > p_0$ is the area under a normal curve z standard deviations or more to the right of the mean.

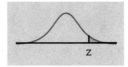

$H_a : p < p_0$ is the area under a normal curve z standard deviations or more to the left of the mean.

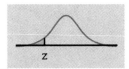

$H_a : p \neq p_0$ is the area under a normal curve z standard deviations or more away from the mean in either direction.

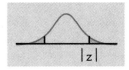

In all three cases, the direction of the alternative hypothesis H_a determines which deviations of z from zero count as evidence against H_0.

EXAMPLE 12. **Spinning pennies.**   We suspect that spinning a penny, unlike tossing it, does not give heads and tails equal probabilities. I spun a penny 200 times and got 83 heads. How significant is this evidence against equal probabilities?

Think of this as an SRS of size $n = 200$ from the population of all spins of the penny. If p is the probability of a head, we must test the hypotheses

$$H_0 : p = 0.5$$
$$H_a : p \neq 0.5$$

because we are seeking evidence of imbalance in either direction. The sample gave $\hat{p} = 83/200 = 0.415$, so the standard score is

$$z = \frac{0.415 - 0.5}{\sqrt{\dfrac{(0.5)(0.5)}{200}}} = -2.4$$

Figure 8-12 shows the two-sided P-value. The normal curve is labeled in z units, that is, the scale is standard deviations away from the mean. This is the sampling distribution of z when H_0 is true.

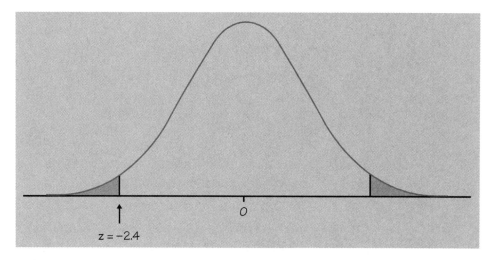

Figure 8-12 The *P*-value for spinning a penny. The normal curve pictured is the sampling distribution of the standard score *z*. The area is two-tailed because the alternative hypothesis is two-sided.

Table B tells us that the standard score -2.4 is the 0.82 percentile. Probability 0.0082 lies outside -2.4 and another 0.0082 outside 2.4. Adding these, $P = 0.0164$ is the probability that *z* takes a value farther from zero in either direction than that observed. We have good evidence (significant at the 5% level but not quite at the 1% level) that the probability of a head is less than 0.5.

TESTS FOR A POPULATION MEAN

The reasoning that led us to the test statistic *z* for a population proportion is remarkably general. It applies without change whenever a parameter is estimated by a statistic having a normal distribution with known standard deviation when the null hypothesis is true. When we wish to test a hypothesis about the mean μ of a population, the natural statistic is the sample mean $\overline{x}$. When the sample size *n* is large and the population has standard deviation σ, we know (see page 477) that the distribution of $\overline{x}$ is approximately normal with mean μ and standard deviation $\sigma/\sqrt{n}$. This leads to tests for hypotheses about μ.

Tests for a population mean

Choose an SRS of size n from a population of individuals having unknown mean μ and known standard deviation σ. To test the hypothesis $H_0 : \mu = \mu_0$ for a specified value μ_0, compute the standard score

$$z = \frac{\overline{x} - \mu_0}{\sigma/\sqrt{n}}$$

The approximate P-value for a test of H_0 against

$H_a : p > p_0$ is the area under a normal curve z standard deviations or more to the right of the mean.

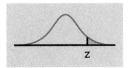

$H_a : p < p_0$ is the area under a normal curve z standard deviations or more to the left of the mean.

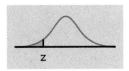

$H_a : p \neq p_0$ is the area under a normal curve z standard deviations or more away from the mean in either direction.

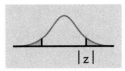

The process of finding the P-value from Table B is the same as for tests about proportions. The common notation z for a standard score emphasizes the similarity of these tests.

EXAMPLE 13. A camera company buys a machined metal part from a supplier. The length of a slot in the part is supposed to be 0.550 inch. No manufacturing process is perfectly precise, so the actual lengths vary according to a distribution with $\sigma = 0.001$ inch. This standard deviation measures the precision of the supplier's work and is known from long experience. The mean μ of the current shipment of 4000 parts may have moved away from the desired value $\mu_0 = 0.550$ inch because of improper adjustment, tool wear, or other reasons. The camera company measures a sample of $n = 80$ parts and finds that the mean slot length is $\bar{x} = 0.5501$. Is there reason to believe that the mean of the shipment is not correct?

We must test against a two-sided alternative, because slot lengths either too short or too long are undesirable. The hypotheses are

$$H_0 : \mu = 0.550$$

$$H_a : \mu \neq 0.550$$

The standard score is

$$z = \frac{0.5501 - 0.550}{0.001/\sqrt{80}} = 0.9$$

Figure 8-13 displays the *P*-value, using the z scale for a normal curve.

Table B shows that the probability to the right of 0.9 is 0.1841. The area to the left of -0.9 is the same, so $P = 0.3682$. About 37% of all samples would have an $\bar{x}$ at least this far from 0.550 even if μ were exactly equal to 0.550. There is no reason to think that μ is incorrect, and the lot should be accepted.

If the population standard deviation σ is unknown, we can estimate it by the sample standard deviation s and replace σ by s in the formula for z. When n is large, the test remains approximately correct with this substitution. Small samples (say n less than 30) are a more complex matter. Now the distribution of z depends on the distribution of the population, and replacing σ by s changes the distribution of z. Table B may not be accurate in this case.

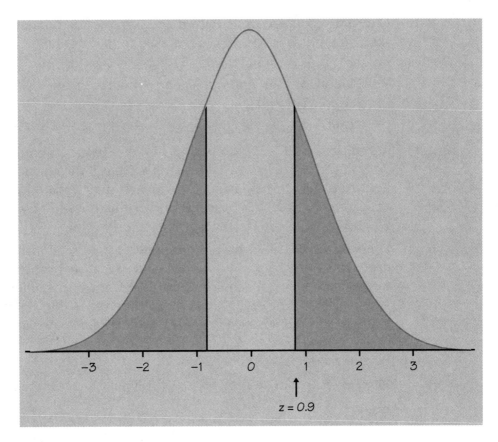

Figure 8-13 The *P*-value for Example 13. The observed standard score is $z = 0.9$, and the alternative hypothesis is two-sided.

SECTION 4 EXERCISES

8.41 **Is the coffee fresh?** Return to Example 9 on page 483. Verify the following facts that we used in that example.

(a) When the null hypothesis $H_0 : p = 0.5$ is true, the sampling distribution of $\hat{p}$ is approximately normal with mean 0.5 and standard deviation 0.0707.

(b) The *P*-value for the observed $\hat{p} = 0.72$ is about $P = 0.001$. If we had observed $\hat{p} = 0.56$, the *P*-value would be about $P = 0.2$.

8.42   **Count Buffon's coin.**   Return to Example 10 on page 487. Verify the following facts that we used in that example.

(a)   When the null hypothesis $H_0 : p = 0.5$ is true, the sampling distribution of $\hat{p}$ is approximately normal with mean 0.5 and standard deviation 0.008.

(b)   The P-value for the observed $\hat{p} = 0.507$ is about $P = 0.37$.

8.43   **A political poll.**   A pre-election poll of 1500 registered voters finds that 781 would vote for Senator Caucus if the election were held today. Is this convincing evidence that a majority of all voters would vote for Caucus? Answer this question by stating H_0 and H_a and using Table B to give the P–value.

8.44   **Dogwood trees.**   A wholesale supplier of tree seedlings advertises that 90% of its seedlings will survive a year when given proper care. A retail nursery buys 250 dogwood seedlings to grow for later sale. After a year, 207 survive. Is this significant evidence that less than 90% of the wholesaler's seedlings will survive? State H_0 and H_a and use Table B to give the P-value. Be sure to state your conclusion.

8.45   **Playing craps.**   The probability of rolling a 7 or 11 as the sum of the faces of two fair dice is 8/36, or 0.222. You decide to test this by watching a casino craps table. Of the first 300 rolls, 61 give 7 or 11. Is this good evidence against the hypothesis that 0.222 is the correct probability? State H_0 and H_a and use Table B to give the P-value. What do you conclude?

8.46   **Attitudes toward school.**   The Survey of Study Habits and Attitudes (SSHA) is a psychological test that measures the attitude toward school and study habits of students. Scores range from 0 to 200. The mean score for U.S. college students is about 115, and the standard deviation is about 30. A teacher suspects that older students have better attitudes toward school. She gives the SSHA to 25 students who are at least 30 years of age. Assume that scores in the population of older students are normally distributed with standard deviation $\sigma = 30$. The teacher wants to test the following hypotheses:

$$H_0 : \mu = 115$$
$$H_a : \mu > 115$$

(a)   What is the sampling distribution of the mean score $\bar{x}$ of a sample of 25 older students if the null hypothesis is true? Sketch the density curve of this distribution. (Hint: sketch a normal curve first, then mark the axis using what you know about locating μ and σ on a normal curve.)

(b)   Suppose that the sample data give $\bar{x} = 118.6$. Mark this point on the axis of your sketch. In fact, the result was $\bar{x} = 125.7$. Mark this point on your sketch. Using your sketch, explain in simple language why one result is good evidence that the mean score of all older students is greater than 115 and why the other outcome is not.

(c)   Shade the area under the curve that is the P-value for the sample result $\bar{x} = 118.6$.

(d)   Use Table B to find the P-values for both these observed values of $\bar{x}$. Explain how these probabilities agree with your conclusions in (b).

8.47   **Water quality.**   An environmentalist group collects a liter of water from each of 45 locations along a stream and measures the amount of dissolved oxygen in each specimen. The mean is 4.62 mg and the standard deviation is 0.92 mg. Is this strong evidence that the stream has a mean oxygen content of less than 5 mg per liter? State the hypotheses, use Table B to give the P-value, and give your conclusion.

8.48   **SAT scores.**   Scores on the SAT are approximately normally distributed with standard deviation $\sigma = 100$. When the population itself has a normal distribution, our recipe for tests about the mean μ is correct even for small samples (as long as σ is known). A random sample of 20 students from a large school system takes a training course before the SAT. Their SAT math scores are as follows:

452	438	577	421
498	450	396	743
514	520	483	328
508	398	429	450
449	547	788	593

Is there evidence that the training has raised the mean above the system mean of $\mu_0 = 483$?

8.49   **Frozen broccoli.**   Packages of frozen broccoli are supposed to weigh 12 ounces. Packages are automatically weighed at the end of the production line. For the last 50, $\bar{x} = 11.92$ ounces and $s = 0.11$ ounce. Is this convincing evidence that the mean weight being produced is not equal to the desired 12 ounces?

▶ 5   ISSUES: UNDERSTANDING STATISTICAL SIGNIFICANCE

Carrying out a test of significance is often quite simple, especially if you get the *P*-value effortlessly from a computer. Using tests wisely is not so simple. Some hesitation about the unthinking use of significance tests is a sign of statistical maturity.

Significance tests are widely used in reporting the results of research. Some products (such as pharmaceuticals) require significant evidence of effectiveness and safety. Courts inquire about statistical significance in hearing class action discrimination cases and in other legal proceedings. Marketers want to know whether a new ad campaign significantly outperforms the old one, and medical researchers want to know whether a new therapy performs significantly better. In all these uses, statistical significance is valued because it points to an effect that is unlikely to occur simply by chance. Here are some points to keep in mind when using or interpreting significance tests.

STATISTICAL SIGNIFICANCE AND PRACTICAL SIGNIFICANCE

When a null hypothesis ("no effect" or "no difference") can be rejected at the usual levels ($P < 0.05$ or $P < 0.01$), there is good evidence that an effect is present. But that effect may be very small. When large samples are available, even tiny deviations from the null hypothesis will be significant.

Reproduced by permission of Bob Schochet

"It was a numbers explosion."

EXAMPLE 14. We are testing the hypothesis of no correlation between two variables. With 1000 observations, an observed correlation of only $r = 0.08$ is significant evidence at the $\alpha = 0.01$ level that the correlation in the population is not zero but positive. The low significance level does not mean there is a strong association, only that there is strong evidence of some association. The true population correlation is probably quite close to the observed sample value, $r = 0.08$. We might well conclude that for practical purposes we can ignore the association between these variables, even though we are confident (at the 1% level) that the correlation is positive.

> **A wise saying**
>
> Statistical significance is not the same thing as practical significance.

The remedy for attaching too much importance to statistical significance is to pay attention to the actual data as well as to the *P*-value. Plot your data and examine them carefully. Are there outliers or other deviations from a consistent pattern? A few outlying observations can produce highly significant results if you blindly apply common tests of significance. Outliers can also destroy the significance of otherwise convincing data. The foolish user of statistics who feeds the data to a computer without graphing them first will often be embarrassed. Is the effect you are seeking visible in your plots? If not, ask yourself if the effect is large enough to be practically important. It is usually wise to give a confidence interval for the parameter in which you are interested. A confidence interval actually estimates the size of an effect, rather than simply asking if it is too large to occur often by chance alone. Confidence intervals are not used as often as they should be, while tests of significance are perhaps overused.

CHOOSING A LEVEL OF SIGNIFICANCE

The purpose of a test of significance is to give a clear statement of the strength of evidence provided by the sample against the null hypothesis. The *P*-value does this. Sometimes, however, you will make some decision or take some action if your evidence reaches a certain standard. A level of significance α sets such a standard. Perhaps you will publish a research finding if the effect is significant at the $\alpha = 0.01$ level. Perhaps your company will lose a lawsuit alleging racial discrimination if the percent of blacks hired is significantly below the percent of blacks in the pool of potential employees at the $\alpha = 0.05$ level. Courts have in fact tended to accept this standard in discrimination cases.[3]

Making a decision is different in spirit from testing significance, though the two are often mixed in practice. Choosing a level α in advance makes sense if you must make a decision, but not if you wish only to describe the strength of your evidence. If you do use a fixed α significance test to make a decision, choose α by asking how much evidence is required to reject H_0. This depends mainly on two circumstances:

▶ *How plausible is H_0?* If H_0 represents an assumption that the people you must convince have believed for years, strong evidence (small α) will be needed to persuade them.

▶ *What are the consequences of rejecting H_0?* If rejecting H_0 in favor of H_a means making an expensive changeover from one type of product packaging to another, you need strong evidence that the new packaging will boost sales.

Both the plausibility of H_0 and H_a and the consequences of any action inspired by rejection of H_0 are somewhat subjective. Different people may want to use different levels of significance. It is better to report the P-value, which allows each of us to decide individually if the evidence is sufficiently strong.

Users of statistics have often emphasized certain standard levels of significance, such as 10%, 5%, and 1%. The 5% level ($\alpha = 0.05$) is particularly common. *There is no sharp border between "significant" and "insignificant," only increasingly strong evidence as the P-value decreases.* There is no practical distinction between the P-values 0.049 and 0.051. It makes no sense to treat $\alpha = 0.05$ as a universal rule for what is significant.

DON'T IGNORE LACK OF SIGNIFICANCE

Researchers typically have in mind the research hypothesis that some effect exists. Following the peculiar logic of tests of significance, they set up as H_0 the null hypothesis that no such effect exists and try their best to get evidence against H_0. A perverse effect of the traditional emphasis on $\alpha = 0.05$ is that research in some fields has rarely been published unless significance at that level is attained. Studies of research journals show that in the social sciences almost all published studies achieved $P < 0.05$.[4]

Medical journals are somewhat more likely to report that a promising therapy *failed* to give significant benefits. This is often an important finding—it saves money and avoids use of an unproved treatment. If there is good reason to think that an effect is present, and a well-designed study fails to find it, that is news that ought to be published. Keeping failures quiet leads others to repeat them.

> *There's this desert prison, see, with an old prisoner, resigned to his life, and a young one just arrived. The young one talks constantly of escape, and, after a few months, he makes a break. He's gone a week, and then he's brought back by the guards. He's half dead, crazy with hunger and thirst. He describes how awful it was to the old prisoner. The endless stretches of sand, no oasis, no signs of life anywhere. The old prisoner listens for a while, then says, "Yep. I know. I tried to escape myself, twenty years ago." The young prisoner says, "You did? Why didn't you tell me, all these months I was planning my escape? Why didn't you let me know it was impossible?" And the old prisoner shrugs, and says, "So who publishes negative results?"*[5]

Even worse (for researchers if not for prisoners), repeated attempts can eventually give a success just by chance. That chance success is then published. The result is *publication bias*: published studies in medicine, for example, suggest that a higher percentage of new therapies are successful than is actually the case.

STATISTICAL INFERENCE IS NOT VALID FOR ALL SETS OF DATA

We learned long ago that badly designed surveys or experiments often produce invalid results. Formal statistical inference cannot correct basic flaws in the design.

> EXAMPLE 15. **Studying foreign languages.** There is no doubt a significant difference between the English vocabulary scores of high school seniors who have studied foreign languages and those who have not. As long as students choose whether or not to study foreign languages, however, statistical significance has little meaning. It does indicate that the difference in English scores is greater than

would often arise by chance alone. That leaves unsettled the issue of *what* other than chance caused the difference. No doubt the students who chose foreign language study were already ahead of the others in their English language skills. Confounding, not the actual effect of studying a new language, explains the significant outcome.

Both tests of significance and confidence intervals are based on the laws of probability. Randomization in sampling or experimentation assures that these laws apply. When these statistical strategies for collecting data cannot be used, statistical inference from the data obtained should be done only with caution. Many data in the social sciences by necessity are collected without randomization. It is universal practice to use tests of significance on such data. Significance does point to the presence of an effect greater than would be likely by chance, but that alone is little evidence against H_0 and in favor of the research hypothesis H_a. Do not allow the wonders of this chapter to obscure the common sense of Chapters 1 and 2.

BEWARE OF SEARCHING FOR SIGNIFICANCE

Statistical significance is a commodity much sought after. It ought to mean that you have found an effect that you were looking for. The reasoning behind statistical significance works well if you decide what effect you are seeking, design a study to search for it, and use a test of significance to weigh the evidence you get. In other settings, significance may have little meaning.

EXAMPLE 16. Predicting managerial ability. You want to learn what distinguishes managerial trainees who eventually become executives from those who, after expensive training, fail to advance and leave the company. You have abundant data on past trainees— data on their personalities and goals, their college preparation and performance, even their family backgrounds and their hobbies. Statistical software makes it easy to perform dozens of significance tests on these dozens of variables to see which ones best predict later success. Aha! You find that future executives are significantly more likely than washouts to have an urban or suburban upbringing and an undergraduate degree in a technical field.

Before basing future recruiting on these findings, pause for a moment of reflection. When you make dozens of tests at the 5% level, you expect a few of them to be significant by chance alone. After all, results significant at the 5% level do occur 5 times in 100 in the long run even when H_0 is true. Running one test and finding $P < 0.05$ is reasonably good evidence that you have found something. Running several dozen tests and reaching that level once or twice is not.

Similarly, searching the trainee data for the variable with the biggest difference between future washouts and future executives, then testing whether that difference is significant, is bad statistics. The P-value assumes you had that specific difference in mind before you looked at the data. It is very misleading when applied to the largest of many differences.

Searching data for suggestive patterns is certainly legitimate. But the reasoning of formal inference does not apply when your search for a striking effect in the data is successful. The remedy is clear. Once you have a hypothesis, design a study to search specifically for the effect you now think is there. If the result of this study is statistically significant, you have real evidence.

SECTION 5 EXERCISES

8.50 Which of the following questions does a test of significance help answer? Explain.

(a) Is the sample or experiment properly designed?

(b) Is the observed effect due to chance?

(c) Is the observed effect important?

8.51 **She has ESP!** A researcher looking for ESP tests 500 subjects. Four of these subjects do significantly better ($P < 0.01$) than random guessing.

(a) Is it proper for the researcher to conclude that these four people have ESP? Explain.

(b) Is it proper to choose these four people for more tests, conducted independently of the previous tests? Explain.

8.52 **Statistical versus practical significance.** Every user of statistics should understand the distinction between statistical significance and practical significance. A sufficiently large sample will declare very small effects statistically significant.

Suppose you are trying to decide whether a coin is fair when tossed. Let p be the probability of a head. If $p = 0.505$ rather than 0.500, this is of no practical significance to you. Test $H_0 : p = 0.5$ versus $H_a : p \neq 0.5$ in each of the following cases. Find the standard score z and use Table B to get approximate P-values.

(a) 1000 tosses give 505 heads ($\hat{p} = 0.505$).

(b) 10,000 tosses give 5050 heads ($\hat{p} = 0.505$).

(c) 100,000 tosses give 50,500 heads ($\hat{p} = 0.505$).

8.53 The previous exercise shows that simply reporting a P-value can be misleading. A confidence interval is more informative. Give a 95% confidence interval for p in each of the cases in the previous exercise. You see that for large n the confidence interval says, "Yes, p is larger than 1/2, but it is very little larger."

8.54 **Bias in academic hiring.** Researchers studying possible gender bias sent applications of equally qualified male and female Ph.D.s to academic department chairs. The chairs were asked which applicant they would hire. The number who chose the male was somewhat greater than the number who would hire the female. The authors wrote in the journal *Science* that "the results, although not statistically significant, showed definite trends that confirm our hypothesis that discrimination against women does exist at the time of the hiring decision."[6] This conclusion was strongly attacked in letters to the editor. One irate statistician wrote, "Are the standards of *Science* the standards of science?" Discuss the validity of the conclusion that the survey results confirm the existence of discrimination.

NOTES

1. Inductive inference from uncertain empirical data is a tricky business. It is not surprising that not all statisticians agree on the proper conceptual approach. In the interest of both simplicity and usefulness, I present the version of inference that dominates actual statistical practice. You can find brief comments on "inference as decision" in Section 5.4 of D. S. Moore, *The Basic Practice of Statistics*, Freeman, 1995. My reasons for avoiding the "Bayesian" approach in an introduction to statistics appear in D. S. Moore, "Bayes for beginners? Some reasons to hesitate," to appear shortly in S. Panchapakesan and N. Balakrishnan (eds.), *Advances in Statistical Decision Theory*, Birkhaüser, Boston. Email me at dsm@stat.purdue.edu to obtain a copy of this paper.

2. Data from J. H. Catania et al., "Prevalence of AIDS-related risk factors and condom use in the United States," *Science*, Volume 258 (1992), pp. 1101–1106.

3. For a discussion of statistical significance in the legal setting, see D. H. Kaye, "Is proof of statistical significance relevant?" *Washington Law Review*, Volume 61 (1986), pp. 1333–1365. Kaye argues that "Presenting the P-value without characterizing the evidence by a significance test is a step in the right direction. Interval estimation, in turn, is an improvement over P-values."

4. T. D. Sterling, W. L. Rosenbaum, and J. J. Weinkam, "Publication decisions revisited: the effect of the outcome of statistical tests on the decision to publish and vice versa," *American Statistician*, Volume 49 (1995), pp. 108–112.

5. From J. Hudson, *A Case of Need*, New American Library, 1968. Quoted in G. W. Walster and T. A. Cleary, "A proposal for a new editorial policy in the social sciences," *The American Statistician*, Volume 24 (1970), p. 16.

6. A. Y. Lewin and L. Duchan, "Women in academia," *Science*, Volume 173 (1971), pp. 892–895.

REVIEW EXERCISES

8.55 A roulette wheel has 18 red slots among its 38 slots. You observe many spins and record the number of times that red occurs. Now you want to use these data to test whether the probability p of a red has the value that is correct for a fair roulette wheel. State the hypotheses H_0 and H_a that you will test.

8.56 A randomized comparative experiment examined whether a calcium supplement in the diet will reduce the blood pressure of healthy men. The subjects received either a calcium supplement or a placebo for twelve weeks. The statistical analysis was quite complex, but one conclusion was that "The calcium group had lower seated systolic blood pressure ($P = .008$) compared with the placebo group." Explain this conclusion, especially the P-value, as if you were speaking to a doctor who knows no statistics.

8.57 When asked to explain the meaning of "statistically significant at the $\alpha = 0.05$ level," a student says, "This means there is only probability 0.05 that the null hypothesis is true." Is this an essentially correct explanation of statistical significance? Explain your answer.

8.58 Another student, when asked why statistical significance appears so often in research reports, says, "Because saying that results are significant tells us that they cannot easily be explained by chance variation alone." Do you think that this statement is essentially correct? Explain your answer.

8.59 **Making sense of a statistical report.** A study compares two groups of mothers with young children who were on welfare two years ago. One group attended a voluntary training program offered free of charge at a local vocational school and advertised in the local news media. The other group did not choose to attend the training program. The study finds a significant difference ($P < .01$) between the proportions of the mothers in the two groups who are still on welfare. The difference is not only significant but quite large. The report says that with 95% confidence the percent of the nonattending group still on welfare is 21% ± 4% higher than that of the group who attended the program. You are on the staff of a member of Congress who is interested in the plight of welfare mothers, and who asks you about the report.

(a) Explain in simple language what "a significant difference ($P < .01$)" means.

(b) Explain clearly and briefly what "95% confidence" means.

(c)   Is this study good evidence that requiring job training of all welfare mothers would greatly reduce the percent who remain on welfare for several years?

8.60   You are planning to test a new vaccine for a virus that now has no vaccine. Since the disease is usually not serious, you will expose 1000 volunteers to the virus. After some time, you will record whether or not each volunteer has been infected.

(a)   Explain how you would use these 1000 volunteers in a designed experiment to test the vaccine. Include all important details of designing the experiment (but don't actually do any random allocation).

(b)   You hope to show that the vaccine is more effective than a placebo. State H_0 and H_a.

(c)   The experiment gave a P-value of 0.25. Explain carefully what this means.

(d)   Your fellow researchers do not consider this evidence strong enough to recommend regular use of the vaccine. Do you agree?

8.61   Statisticians prefer large samples. Describe briefly the effect of increasing the size of a sample (or the number of subjects in an experiment) on each of the following.

(a)   The width of a confidence interval with confidence level held fixed.

(b)   The P-value of a test, when H_0 is false and all facts about the population remain unchanged as n increases.

The following exercise concerns the optional Sections 2 and 4.

8.62   **This wine smells.**   Sulfur compounds cause "off-odors" in wine, so winemakers want to know the odor threshold, the lowest concentration of a compound that the human nose can detect. The odor threshold for dimethyl sulfide (DMS) in trained wine tasters is about 25 micrograms per liter of wine (μg/l). The untrained noses of consumers may be less sensitive, however. Here are the DMS odor thresholds for 10 untrained students:

31 31 43 36 23 34 32 30 20 24

Assume that the standard deviation of the odor threshold for untrained noses is known to be $\sigma = 7\ \mu g/l$.

(a)   Make a stemplot to verify that the distribution is roughly symmetric with no outliers.

(b)   Give a 95% confidence interval for the mean DMS odor threshold among all students.

(c)   Are you convinced that the mean odor threshold for students is higher than the published threshold, 25 $\mu g/l$? Carry out a significance test to justify your answer.

WRITING PROJECTS

8.1   **Reporting a medical study.**   Look through recent issues of the *Journal of the American Medical Association* or the *New England Journal of Medicine* in your college library. Many of the major articles in these medical journals concern statistically designed studies and report the results of inference, especially *P*-values. Choose an article that describes a medical experiment on a topic that is understandable to those of us who lack medical training—aspirin to prevent heart attacks and calcium to lower blood pressure are two examples I have used in this book. Write a news article explaining the results of the study. Be sure to describe the design of the experiment and to interpret for your readers the statistical and medical significance of the results.

8.2   **How random are random digits?**   The table of random digits (Table A) was produced by a computer program that is supposed to give each digit probability 1/10 of being a zero. Let's see how well the randomization worked. The first 200 digits in Table A are an SRS of size 200 from the population of all digits produced by the program. What proportion of these first 200 digits are zeros? Based on this sample proportion $\hat{p}$, give a 95% confidence interval for the probability p that a digit generated by the program is a zero.

Now write an account of your study that is understandable to someone who knows no statistics. What are random digits?

Why aren't there exactly 20 zeros among the first 200 digits? What can we conclude from a 95% confidence interval? Did your confidence interval contain 0.1, which is supposed to be the true p? Did your study give good reason to doubt that the computer program gives each digit probability 1/10 of being a zero?

8.3   **Use and abuse of inference.** Few accounts of really complex statistical methods are readable without extensive training. One that is, and that is also an excellent essay on the abuse of statistical inference, is "The Real Error of Cyril Burt," a chapter in Stephen Jay Gould's *The Mismeasure of Man* (W. W. Norton, 1981). We met Cyril Burt under suspicious circumstances in Exercise 3.39 on page 191. Gould's long chapter shows Burt and others engaged in discovering dubious patterns by complex statistics. Read it, and write a brief explanation of why "factor analysis" failed to give a firm picture of the structure of mental ability.

TABLE A. Random digits

Line								
101	19223	95034	05756	28713	96409	12531	42544	82853
102	73676	47150	99400	01927	27754	42648	82425	36290
103	45467	71709	77558	00095	32863	29485	82226	90056
104	52711	38889	93074	60227	40011	85848	48767	52573
105	95592	94007	69971	91481	60779	53791	17297	59335
106	68417	35013	15529	72765	85089	57067	50211	47487
107	82739	57890	20807	47511	81676	55300	94383	14893
108	60940	72024	17868	24943	61790	90656	87964	18883
109	36009	19365	15412	39638	85453	46816	83485	41979
110	38448	48789	18338	24697	39364	42006	76688	08708
111	81486	69487	60513	09297	00412	71238	27649	39950
112	59636	88804	04634	71197	19352	73089	84898	45785
113	62568	70206	40325	03699	71080	22553	11486	11776
114	45149	32992	75730	66280	03819	56202	02938	70915
115	61041	77684	94322	24709	73698	14526	31893	32592
116	14459	26056	31424	80371	65103	62253	50490	61181
117	38167	98532	62183	70632	23417	26185	41448	75532
118	73190	32533	04470	29669	84407	90785	65956	86382
119	95857	07118	87664	92099	58806	66979	98624	84826
120	35476	55972	39421	65850	04266	35435	43742	11937
121	71487	09984	29077	14863	61683	47052	62224	51025
122	13873	81598	95052	90908	73592	75186	87136	95761
123	54580	81507	27102	56027	55892	33063	41842	81868
124	71035	09001	43367	49497	72719	96758	27611	91596
125	96746	12149	37823	71868	18442	35119	62103	39244

TABLE A. Random digits

Line								
126	96927	19931	36809	74192	77567	88741	48409	41903
127	43909	99477	25330	64359	40085	16925	85117	36071
128	15689	14227	06565	14374	13352	49367	81982	87209
129	36759	58984	68288	22913	18638	54303	00795	08727
130	69051	64817	87174	09517	84534	06489	87201	97245
131	05007	16632	81194	14873	04197	85576	45195	96565
132	68732	55259	84292	08796	43165	93739	31685	97150
133	45740	41807	65561	33302	07051	93623	18132	09547
134	27816	78416	18329	21337	35213	37741	04312	68508
135	66925	55658	39100	78458	11206	19876	87151	31260
136	08421	44753	77377	28744	75592	08563	79140	92454
137	53645	66812	61421	47836	12609	15373	98481	14592
138	66831	68908	40772	21558	47781	33586	79177	06928
139	55588	99404	70708	41098	43563	56934	48394	51719
140	12975	13258	13048	45144	72321	81940	00360	02428
141	96767	35964	23822	96012	94591	65194	50842	53372
142	72829	50232	97892	63408	77919	44575	24870	04178
143	88565	42628	17797	49376	61762	16953	88604	12724
144	62964	88145	83083	69453	46109	59505	69680	00900
145	19687	12633	57857	95806	09931	02150	43163	58636
146	37609	59057	66967	83401	60705	02384	90597	93600
147	54973	86278	88737	74351	47500	84552	19909	67181
148	00694	05977	19664	65441	20903	62371	22725	53340
149	71546	05233	53946	68743	72460	27601	45403	88692
150	07511	88915	41267	16853	84569	79367	32337	03316

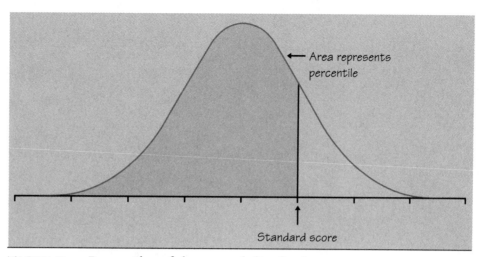

Area represents percentile

Standard score

TABLE B.   Percentiles of the normal distributions

Standard score	Percentile	Standard score	Percentile	Standard score	Percentile
−3.4	0.03	−1.1	13.57	1.2	88.49
−3.3	0.05	−1.0	15.87	1.3	90.32
−3.2	0.07	−0.9	18.41	1.4	91.92
−3.1	0.10	−0.8	21.19	1.5	93.32
−3.0	0.13	−0.7	24.20	1.6	94.52
−2.9	0.19	−0.6	27.42	1.7	95.54
−2.8	0.26	−0.5	30.85	1.8	96.41
−2.7	0.35	−0.4	34.46	1.9	97.13
−2.6	0.47	−0.3	38.21	2.0	97.73
−2.5	0.62	−0.2	42.07	2.1	98.21
−2.4	0.82	−0.1	46.02	2.2	98.61
−2.3	1.07	0.0	50.00	2.3	98.93
−2.2	1.39	0.1	53.98	2.4	99.18
−2.1	1.79	0.2	57.93	2.5	99.38
−2.0	2.27	0.3	61.79	2.6	99.53
−1.9	2.87	0.4	65.54	2.7	99.65
−1.8	3.59	0.5	69.15	2.8	99.74
−1.7	4.46	0.6	72.58	2.9	99.81
−1.6	5.48	0.7	75.80	3.0	99.87
−1.5	6.68	0.8	78.81	3.1	99.90
−1.4	8.08	0.9	81.59	3.2	99.93
−1.3	9.68	1.0	84.13	3.3	99.95
−1.2	11.51	1.1	86.43	3.4	99.97

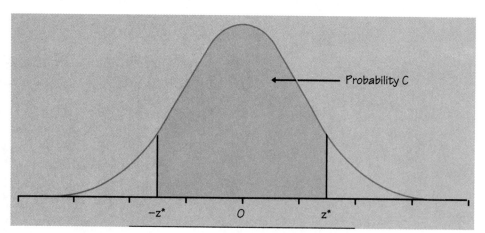

Probability C

−z*   O   z*

TABLE C.   Critical values of the
               normal distributions

C	z*	C	z*
0.50	0.67	0.80	1.28
0.55	0.76	0.85	1.44
0.60	0.84	0.90	1.64
0.65	0.93	0.95	1.96
0.70	1.04	0.99	2.58
0.75	1.15	0.999	3.29

DATA TABLE INDEX

SUBJECT INDEX